This book is to be returned on or before
the last date stamped below

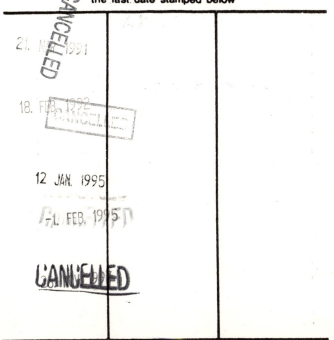

## PLYMOUTH POLYTECHNIC
## LEARNING RESOURCES CENTRE
Telephone: (0752) 221312 ext: 5413
(After 5p.m. (0752) 264661 weekdays only)

This book is subject to recall if required by another reader.
Books may be renewed by phone, please quote Telepen number.
CHARGES WILL BE MADE FOR OVERDUE BOOKS

# COMPOSITE STRUCTURES

## 4

### VOLUME 2

*Damage Assessment and Material Evaluation*

*Proceedings of the 4th International Conference on Composite Structures, held at Paisley College of Technology, Scotland, from 27th to 29th July 1987, co-sponsored by the Scottish Development Agency, the National Engineering Laboratory, the US Air Force European Office of Aerospace Research and Development and the US Army, Research, Development and Standardisation Group, UK.*

Volume 1: Analysis and Design Studies
Volume 2: Damage Assessment and Material Evaluation

Also published by Elsevier Applied Science Publishers:

COMPOSITE STRUCTURES

(*Proceedings of 1st International Conference, 1981*)

COMPOSITE STRUCTURES 2

(*Proceedings of 2nd International Conference, 1983*)

COMPOSITE STRUCTURES 3

(*Proceedings of 3rd International Conference, 1985*)

# COMPOSITE STRUCTURES

# 4

## VOLUME 2
### Damage Assessment and Material Evaluation

*Edited by*

## I. H. MARSHALL

*Department of Mechanical and Production Engineering,*
*Paisley College of Technology, Scotland, UK*

ELSEVIER APPLIED SCIENCE
LONDON and NEW YORK

ELSEVIER APPLIED SCIENCE PUBLISHERS LTD
Crown House, Linton Road, Barking, Essex IG11 8JU, England

*Sole Distributor in the USA and Canada*
ELSEVIER SCIENCE PUBLISHING CO., INC.
52 Vanderbilt Avenue, New York, NY 10017, USA

WITH 51 TABLES AND 211 ILLUSTRATIONS

© ELSEVIER APPLIED SCIENCE PUBLISHERS LTD 1987

© CROWN COPYRIGHT 1987—Chapters 16, 22 and 26

**British Library Cataloguing in Publication Data**

International Conference on Composite Structures
(4th: *Paisley College of Technology*)
Composite structures 4.
1. Composite materials 2. Composite construction
I. Title II. Marshall, I. H. III. Scottish
Development Agency
624.1'8 TA664

**Library of Congress Cataloging-in-Publication Data**

Composite structures 4.

Proceedings of the 4th International Conference
on Composite Structures, held at Paisley College of
Technology, Scotland, from July 27–29, 1987, and
co-sponsored by the Scottish Development Agency and
others.
Bibliography: p.
Includes index.
1. Composite construction—Congresses. 2. Composite
materials—Congresses. I. Marshall, I. H. (Ian H.)
II. Scottish Development Agency. III. International
Conference on Composite Structures (4th: 1987:
Paisley College of Technology, Scotland, UK)
IV. Title: Composite structures four.
TA664.C636 1987 620.1'18 87-13475

ISBN 1-85166-126-3 (Volume 1)
ISBN 1-85166-127-1 (Volume 2)
ISBN 1-85166-128-X (the set)

The selection and presentation of material and the opinions expressed are the sole
responsibility of the author(s) concerned.

**Special regulations for readers in the USA**

Printed in Great Britain by Galliard (Printers) Ltd, Great Yarmouth

# Preface

The papers contained herein were presented at the Fourth International Conference on Composite Structures (ICCS/4) held at Paisley College of Technology, Scotland in July 1987. The Conference was organised and sponsored by Paisley College of Technology. It was co-sponsored by the Scottish Development Agency, the National Engineering Laboratory, the US Air Force European Office of Aerospace Research and Development and the US Army Research, Development and Standardisation Group—UK. It forms a natural and ongoing progression from the highly successful First, Second and Third International Conferences on Composite Structures (ICCS/1, ICCS/2 and ICCS/3) held at Paisley in 1981, 1983 and 1985 respectively.

There is little doubt that composite materials are rightfully claiming a prominent role in structural engineering in the widest sense. Moreover, the range and variety of useful composites has expanded to a level inconceivable a decade ago. However, it is also true that this increasing utilisation has generated an enhanced awareness of the manifold factors which dictate the integrity of composite structures.

This is indeed a healthy attitude to a relatively new dimension in structural engineering which will have an increasingly dominant role as the century progresses. Both the diversity of application of composites in structural engineering and the endeavours which will ensure their fitness for purpose are reflected herein.

It is also inescapably true that traditional engineering design principles

v

and practices are largely inappropriate when contemplating the viability of composites as alternatives to traditional engineering materials. Consequently, engineers and scientists are being confronted with a new, and sometimes daunting, range of design concepts which they are, by and large, ill equipped to appreciate. Only by critically reassessing the education and training of today's engineers and scientists can proper advances in composite structures be sustained. Moreover, this also demands that the fruits of current research and development are available in a realistically digestible form. Without both of these ingredients there is little doubt that the projected future of composites in structural engineering will be impeded, or indeed compromised. To this end it is hoped that the present volume will provide both an overall appreciation of the current and future potential of composite structures and an awareness of the factors which dictate their safe usage.

Authors and delegates from in excess of twenty countries have combined to make the present conference a truly international forum of specialists in composite structures, reflecting a truly global appreciation of this expanding area of technology. Topics addressed range from the use of composites in wind turbine engineering to their usage in outer space, a truly expansive spectrum.

As always, an international conference can only take place and contribute to knowledge through the willing and enthusiastic efforts of a number of individuals. In particular, thanks are due to the following:

*The International Advisory Panel*

| | |
|---|---|
| E. Anderson | Battelle Laboratories (Switzerland) |
| J. Anderson | Paisley College of Technology (UK) |
| W. M. Banks | University of Strathclyde (UK) |
| A. M. Brandt | Polish Academy of Sciences (Poland) |
| A. R. Bunsell | Ecole des Mines de Paris (France) |
| W. S. Carswell | National Engineering Laboratory (UK) |
| T. Hayashi | Japan Plastic Inspection Association (Japan) |
| R. M. Jones | Virginia Polytechnic & State University (USA) |
| L. N. Phillips, OBE | Consultant, Farnborough (UK) |
| J. Rhodes | University of Strathclyde (UK) |
| S. W. Tsai | Air Force Materials Laboratory (USA) |
| J. A. Wylie | Paisley College of Technology (UK) |

*The Local Organising Committee*

S. K. Harvey
J. Kirk
G. Macaulay
J. S. Paul

*The Conference Secretary*

Mrs C. A. MacDonald

Grateful thanks are due to other individuals who contributed to the success of the event. A final thanks to Nan, Simon, Louise and Richard for their support during the conference.

I. H. MARSHALL

# Contents

## Damage Tolerance 2

(*Chairman:* C. T. SUN, *School of Aeronautics and Astronautics, West LaFayette, USA*)

## Metal Matrix Composites

(*Chairman:* A. M. BRANDT, *Institute of Fundamental Technological Research, Warsaw, Poland*)

## Material Characteristics

(*Chairman:* D. HULL, *University of Cambridge, UK*)

**Failure Analysis**

(*Joint Chairmen*: C. S. SMITH, *Admiralty Research Establishment,
Dunfermline, Scotland, UK* and J. F. WILLIAMS, *University of Melbourne,
Parkville, Australia*)

**Fabrication and Processing**

(*Chairman:* M. J. SEAMARK, *Balmoral Group Ltd, Aberdeen, Scotland, UK*)

## Fracture Mechanics

(*Chairman:* J. ANDERSON, *Paisley College of Technology, Scotland, UK*)

## Experimental Studies

(*Joint Chairmen:* R. G. WHITE, *Institute of Sound and Vibration Research,
Southampton, UK* and A. R. BUNSELL, *Ecole Nationale Supérieure des Mines
de Paris, Evry, Cedex, France*)

**Orthopaedic Applications**

(*Chairman:* W. S. CARSWELL, *National Engineering Laboratory,
East Kilbride, Scotland, UK*)

**Environmental Influences**

(*Chairman:* J. SUMMERSCALES, *Royal Naval Engineering College,
Plymouth, UK*)

## Fatigue and Creep

(*Chairman:* Y. RAJAPAKSE, *Office of Naval Research Mechanics Division,
Arlington, USA*)

## Theoretical Studies

(*Chairman:* W. M. BANKS, *University of Strathclyde,
Glasgow, Scotland, UK*)

**Plenary Paper**

# Recent Advances in Dynamics of Composite Structures

CHARLES W. BERT

*Department of Engineering, University of Oklahoma,
Norman, Oklahoma 73019, USA*

## ABSTRACT

*This chapter comprises a survey on the dynamics of composite structures, with emphasis on research completed since 1980. After a brief introduction to applications in which the subject is of importance, the following specific topics are addressed: (1) characterization of the dynamic stiffness and damping in composites reinforced with either continuous or short fibers, (2) vibratory response of composite beams, plates, and shells, (3) low-velocity transverse impact of composite plates, (4) dynamic instabilities including aeroelastic phenomena, and (5) nonlinear effects. Experimental, analytical, and numerical investigations are included and the chapter is concluded by some suggestions for future research directions.*

## 1. INTRODUCTION

The term, composite structures, as used here means any structural components, for either primary or secondary structure, which are constructed of fiber-reinforced composite materials. Such structures are being made in a greater diversity of sizes and shapes for a wide range of important applications. These include:

- Golf-club shafts and tennis rackets.
- Automotive components such as suspensions, driveshafts, and body panels.
- Pipelines, storage tanks, and pressure vessels.

- Blades for wind turbines, compressor blades, propellers, and rotorcraft.
- Military ordnance, such as gun barrels.
- Aircraft control surfaces, empennages, and wing and fuselage structures.
- Conventional surface ship structures, hydrofoils, ground-effect machine skirts, and possibly submersible hulls.
- Rocket-motor casings, nozzles, interstage structures, and re-entry vehicles.
- Large space structures, currently exemplified by the Space Shuttle payload-bay doors.

In many of the applications mentioned, dynamic loadings are very important. There is a wide range of dynamic loadings encountered, depending upon the application, including:

- Acoustic loading due to propulsion system sources.
- Random loadings such as atmospheric turbulence and highway/off-road terrain.
- Impact loadings:
  Low-velocity impact due to accidental tool drops.
  Aircraft landing impact.
  Vehicle crashes.
  Ballistic impact in military applications.
  Hypervelocity impact due to space debris or space weapons.
- Blast loadings due to:
  Conventional explosives.
  Sonic booms.
  Nuclear weapons.
- Thermal shock due to:
  Re-entry heating.
  Lasers.
  Electron beams.
- Aerodynamic loadings.
- Moving loadings, exemplified by loading of a missile during launch from a silo.

Due to the wide range of applications, loadings encountered, and structural component configurations (beams, rings, plates, and shells), it is not possible to discuss all aspects of the dynamics of composite structures within a single plenary lecture. Therefore, certain selected problem areas considered to be important and interesting by the present investigator have been singled out for exposition here:

- Characterization of dynamic stiffness and damping.
- Vibratory response of basic structural components.
- Low-velocity transverse impact.
- Dynamic instabilities including aeroelastic phenomena.
- Nonlinear effects.

In the current presentation, emphasis is placed on recent analytical, numerical, and experimental advances in the aforementioned problem areas. For historical and tutorial background, reference is made to various surveys of the field such as Refs 1–7. At this point, it is well to offer apologies for not being able to reference all of the interesting work which has taken place in the past few years.

## 2.  DYNAMIC STIFFNESS AND DAMPING

Those knowledgeable in the mechanics of composite materials are familiar with the complexities of the static stiffness of composite materials. These complexities include macroscopic anisotropic behavior (shear-normal coupling) at the scale of an individual layer and bending–stretching (actually bending/twisting–stretching/shearing) coupling at the scale of the entire laminate.[8] These same complexities carry over to dynamic behavior, where it is customary to characterize the behavior by a complex stiffness. The real component of the complex stiffness is called the 'storage' component, while the complex component is called the 'loss' or damping component. The micromechanics foundations of this behavior for continuous-fiber composites have been discussed by Hashin[9] and by Chang and Bert.[10] while the laminate ramifications of it were developed by Siu and Bert.[11] Various methods of experimentally characterizing the dynamic stiffness and damping behavior of composites were surveyed in Refs 12–16.

In automotive and other high-volume production applications, there is great interest in discontinuous-fiber composites, such as SMC (sheet molding compound). Such composites have been studied intensively by Gibson and his associates.[17–19] Due to the importance of tire damping on automotive fuel consumption as well as on tire dynamics, the damping of cord-rubber composites was investigated by Tabaddor and Clark.[20,21] There have been relatively few investigations of the damping characteristics of metal–matrix composites.[22]

For many years there has been sporadic interest in the possibility of using damping measurements to track material degradation. A recent study on this for composites was conducted by Mantena *et al.*[23]

## 3. VIBRATORY RESPONSE OF BASIC STRUCTURAL COMPONENTS

### 3.1. Straight and Curved Beams

There has been very little activity in this category since the 1980 survey,[5] to which one should refer for a survey of earlier work. In the category of anisotropic beam vibration, the works of Teh and Huang,[24,25] however, should be mentioned.

There have been a few recent investigations of the dynamics of laminated beams. For instance, Bratt[26] investigated the use of such beams as shock isolators, Mead and Markus[27,28] investigated coupled wave motion in laminated beams, while Miles and Reinhall[29] modeled the effects of both shear and thickness stretching deformation in the adhesive layer. Reddy and Mallik[30] analyzed vibration of a periodically supported, two-layer ring.

### 3.2. Plates

In view of the large amount of activity on this topic, attention is directed here to linear dynamic plate analyses published since the 1985 survey.[7] Probably the simplest category of composite plate is a rectangular one made of a material having rectangular orthotropy. Grace and Kennedy[31] analyzed fixed simply-supported and free-free plates by both beam and plate theory. Not surprisingly, they found that beam theory was not reliable for estimating higher vibration modes. Dickinson and his associates[32,33] very successfully used orthogonal polynomials in the Rayleigh–Ritz method to improve its accuracy. Nakahira *et al.*[34] demonstrated the use of Newmark's tabular numerical integration method. Grossi *et al.*[35] used the noninteger polynomial version of the Rayleigh method (the so-called Rayleigh–Schmidt method) and a semi-analytic method to analyze vibration of an orthotropic plate with one edge free. The results were in reasonably close agreement. It is noted that the noninteger-polynomial approach is increasing in popularity as evidenced by over 50 papers using it referenced in a recent survey.[36] Tomar and Gupta[37] investigated a varying thickness plate vibrating while subjected to a thermal gradient.

In the category of nonrectangular plates constructed of rectangularly orthotropic material, one should mention the work of Bianchi *et al.*[38] on annular circular plates and Narita[39,40] on elliptical plates. Cylindrical (or polar) orthotropy can be the result of filament winding around a mandrel. Plates constructed in this way, especially those of radially varying thickness, can be used as local reinforcements (nozzles) of pressure vessels or as turbine disks. Recent vibration analyses of such plates were made by Gorman[41,42] and Narita[43] for uniform-thickness circular planform, by Srinivasan and Thiruvenkatachari[44] for uniform-thickness annular sectorial plates, and by Gupta *et al.*[45-47] for varying-thickness annular plates.

Laminated composite plates are generally more complicated to analyze, yet they can be tailored to give greater resistance to stable static loading, to buckling, and to dynamic loading. There have been several recent interesting analyses of vibration of thin laminated plates, i.e. ones in which transverse shear deformation and rotatory inertia as well as thickness normal deformation, have been neglected. Dong and Huang[48] investigated the phenomenon of edge vibrations of semi-infinite laminates. It has been known for a long time that the presence of inplane compressive loads can significantly reduce the natural frequencies of flexural vibration. However, the presence of geometric imperfections can further affect the frequency versus load relationship. This was recently investigated in depth by Hui.[49] Reference 50 studied axisymmetric vibration of cylindrically orthotropic plates having a total thickness varying linearly with radius.

The concept of interlaminar hybridization means the use of different composite materials in different individual layers of the laminate. This concept allows an extra degree of freedom for designers seeking high performance at relatively lower cost. Recently, Iyengar and Umaretiya[51] analyzed the free vibration of such laminates with either rectangular or parallelogram planform.

Although there have been relatively few experimental investigations of laminated composite plates, the recent experimental work reported in Refs 52 and 53 should be mentioned. One of the causes for discrepancies between resonant frequencies predicted by thin plate theory and those measured in experiments is the presence of transverse shear deformation. Transverse shear deformation is generally more important for composite plates than for isotropic ones of the same geometry because composite materials have much lower ratios of transverse shear moduli to inplane elastic moduli. Recent composite plate vibration analyses incorporating transverse shear deformation include Lagrangian multiplier solutions by Ramkumar *et al.*,[54] Rayleigh–Ritz and finite-strip analyses by Craig and Dawe,[55,56]

Rayleigh–Ritz analysis by Kamal and Durvasula.[57] All of the shear-deformable analyses mentioned so far were based on consideration of the whole laminate as a Mindlin-type plate, in which the transverse shear deformation is assumed to be uniformly distributed through the entire thickness of the laminate and then 'corrected' by a Timoshenko-type shear correction factor. Several so-called higher order plate theories have been introduced in which the transverse shear strain is permitted to vary more realistically through the thickness, parabolically for instance. Analytical studies of plate vibration based on these theories have been conducted by Di Sciuva,[58] Kamal and Durvasula,[59] and by Stein and Jegley,[60] while Putcha and Reddy[61] made a FEA (finite element analysis) based on higher order theory. The whole subject of FEA for laminated plates was recently reviewed by Reddy[62] There have been a number of recent vibration analyses of sandwich plates, i.e. ones with thick, flexible cores and thin, stiff facings. However, most of these analyses have been limited to either isotropic or orthotropic (not generally anisotropic) materials; thus, they are not discussed here.

Most plate vibration analyses have been limited to elastic materials, i.e. material damping has been neglected. However, an important exception is the recent energy analysis due to Alam and Asnani.[63] Their analysis was based on the conventional Rayleigh–Ritz analysis and the use of the well-known elastic–viscoelastic correspondence principle. This is in contrast to the early work of Siu and Bert,[11] who used the generalized Hamilton's principle including directly the energy dissipated by damping. Unfortunately, it is not possible to compare the results, since Alam and Asnani were concerned with damped free vibration while Siu and Bert considered forced damped vibration.

### 3.3. Shells

Activity in the dynamics of composite shells has increased in the period 1980–87 compared to that of 1973–80, which was covered in the 1980 survey.[5] Thus, in the interests of brevity, discussion here is limited to cylindrically curved panels and complete cylindrical shells, just as it was in the 1980 survey. Further, to be consistent with the treatment of plates in Section 3.2, sandwich shells and composite shells with discrete stiffeners are not considered here.

Dynamic analyses of thin orthotropic cylindrical shells were limited to complicating features. Kozlov[64] analyzed the effect of concentrated attached masses, Chonan[65] considered the effect of moving loads on a liquid-filled shell, and Shirakawa[66] studied the response to rapid heating.

Several investigations[67-69] considered the free vibration of relatively thick, macroscopically homogeneous anisotropic shells, i.e. with transverse shear deformation included. References 67 and 69 were limited to orthotropic material and Ref. 67 to simply-supported ends. Reference 68 represents one of very few vibration investigations of anisotropic (monoclinic) material shells, i.e. shells in which all of the layers are oriented at the same acute angle to the cylinder axis (not alternating $+\theta$ and $-\theta$ as in filament-wound shells). The shell theory used was Mirsky's thick-shell theory modified to account for monoclinic material behavior. The method of solution used was one originally proposed by Flügge in 1934 and later implemented for vibration of thin, homogeneous, isotropic shells by Forsberg. The method is capable of handling arbitrary boundary conditions at each end of the shell. Experiments were conducted on a thick shell having free ends and made of glass–epoxy oriented at 30° to the axis. Reasonable agreement between theory and experiment was obtained for this case of highly exaggerated anisotropy. Marchuk and Shvets[70] analyzed free vibrations of a shell completely filled by a liquid, and Chonan[71] treated the response to a pressure pulse.

In the context of beam and plate theory, there are basically only four categories of theories:

(1) Thin, in which TSD (transverse shear deformation) is completely neglected.

(2) Bresse–Timoshenko type theory in which kinematically TSD is assumed to be uniformly distributed through the entire thickness of the laminate but 'corrected' in the transverse shear force by a shear correction factor.

(3) Reissner–Schmidt–Levinson type theory in which kinematically TSD varies parabolically through the entire thickness, so that no shear 'correction' is required.

(4) Three-dimensional theory of elasticity.

In shell theory, there are subcategories of theories which fall within categories (1) and (2). For thin shells, Bert and Kumar[72] identified five different theories which are listed here in order of ascending complexity and accuracy:

(1a) Donnell's very shallow shell theory.

(1b) Morley's moderately shallow shell theory.

(1c) Loo's theory.

(1d) Love's first-approximation theory.

(1e) Sanders–Koiter 'best' first-approximation theory.

However, Bert and Kumar showed that these five subcategories reduce to only three subcategories when TSD is incorporated. They are:

(2a) Donnell's and Morley's theories coinciding.
(2b) Loo's and Love's theories coinciding.
(2c) Sanders–Koiter theory, remaining a unique subcategory.

Further use of shell-theory tracers was employed by Soldatos.[73]

The work of Ref. 72 is also distinct in that it is believed to be the only existing vibration analysis of shells laminated of bimodular composite materials. These idealized materials have different elastic moduli in tension and in compression, and hence the adjective, bimodular. The most drastic example of such a material is aramid cord-reinforced rubber, such as used in automobile tires.[74,75] The mechanisms causing the bimodular action are straightening/untwisting of the cord by tie-bar action in tension and the tendency to buckle or collapse in compression due to the inability of the soft matrix to support the cords adequately.[76] However, even hard-matrix composites such as graphite–epoxy have been shown to be somewhat bimodular.[77]

Other recent analyses of laminated, thin cylindrical shells include the free vibration studies of complete cylinders in Refs 78–81 and of cylindrically curved panels in Refs 82 and 83. Reference 84 investigated the effect of axial compressive load on the natural frequencies, while Soldatos[85,86] and Hui and Du[87] considered noncircular (oval) cylinders.

Analyses of laminated shells with TSD considered as in category (2) above were conducted by Sharma and Darvizeh[88] and Greenberg and Stavsky.[89] A more refined theory, permitting the TSD distribution to be approximately parabolic, was introduced by Bhimaraddi,[90] while a more elaborate laminated theory, in which each individual layer is governed by three-dimensional elasticity theory, was introduced in Ref. 91.

## 4. LOW-VELOCITY TRANSVERSE IMPACT (LVTI)

Some persons feel that limited tolerance to LVTI damage is the 'Achilles heel' of laminated composite plates and panels as used in aircraft structures. Such impacts include technician's footfall or dropping of a small tool upon an upper wing panel or runway debris on lower wing skins or lower fuselage panels. Due to its technological importance, this topic is being researched very rapidly. For instance, Refs 7 and 92, covering 1978–85, contain numerous references on this topic. References 93–96, which were published after Refs 7 and 92 were written, report on four additional investigations in this rapidly expanding area. Due to the embryonic nature of the knowledge of the complicated phenomena involved in LVTI, the present investigator is hesitant to draw any general conclusions at this moment in time.

## 5. DYNAMIC INSTABILITY INCLUDING AEROELASTICITY

Many kinds of loadings induce dynamic instability in any structure. Composite structures are not immune to such instabilities, in fact due to their bending–stretching coupling, they may be less resistant to dynamic instability than homogeneous structures. Yet one can use aeroelastic tailoring, for instance, to make an aircraft wing less susceptible to flutter and/or divergence.

Some examples of dynamic instabilities and references to recent analyses in the context of composite structures include:

- Rapid axial compression; see Refs 97 and 98.
- Periodic external pressure; see Ref. 99.
- Periodic axial compression; see Refs 100–102 for plates and 103 and 104 for cylindrical shells.
- Rapid heating; see Refs 105 and 106.
- Panel flutter, see Refs 107 and 108.
- Wing flutter and divergence; see Refs 109 and 110.

## 6. NONLINEAR EFFECTS

It is well known that when plates or panels with inplane restraints at their edges undergo moderately large deflections, say one-half of the plate thickness, geometric nonlinearity occurs due to the development of membrane stresses in the plane of the plate. Composite plates and shells are not immune to geometric nonlinearity; in fact, there are indications in some recent research that modal coupling may play a larger role in composite nonlinear behavior than it does in the homogeneous isotropic case. Currently, there is considerable research activity on this topic; thus, for brevity, only a few illustrative examples, involving rectangular plates only, are mentioned here.

Chia considered many examples of geometric nonlinearity in laminated plates in his monograph.[111] More recently, he has considered the effects of nonuniform edge constraints,[112] an elastic foundation,[113] and, in conjunction with Sivakumaran, TSD and transverse normal stress.[114] Gray et al.[115] considered geometrically nonlinear random response.

## 7.  SUGGESTIONS FOR FUTURE RESEARCH

The present investigator has no magic crystal ball for foreseeing the future, yet in surveying practical needs in connection with increased use of composite structures in applications involving dynamic loading versus the technology base presently available, one can easily ascertain these needs:

1.  More realistic mathematical modeling of material behavior is highly desirable. This includes not only damping and stress–strain nonlinearity in shear but also material damage.
2.  Well-planned and meticulously conducted experiments and appropriately detailed data reduction and comparison with results of analysis are critically needed.
3.  A comprehensive and comparative assessment of the degree of detail needed in laminate analyses in order to predict failure (load, degree, and location) due to dynamic loadings.
4.  Much more extensive and intensive use of optimal design approaches to better tailor the design of composite structures. This probably will require much more attention to design details and thus optimal design guided by the results of numerical analyses. This is where the real 'payoff' in the use of composites is still yet to be realized.

## ACKNOWLEDGEMENTS

Helpful discussions with Professor Isaac Elishakoff of the Technion and Professor Victor Birman of the University of New Orleans are gratefully acknowledged. Also, the skilful typing of Rose Benda is sincerely appreciated.

## REFERENCES

1. BERT, B. W. and EGLE, D. M., Dynamics of composite, sandwich, and stiffened shell structures, *J. Spacecraft Rkts*, **6** (1969), 1345–1361.

2. BERT, C. W. and FRANCIS, P. H., Composite material mechanics: structural mechanics, *AIAA J.*, **12** (1974), 1173–1186.

3. BERT, C. W., Dynamics of composite and sandwich panels, parts I and II (corrected title), *Shock Vib. Digest*, **8** (10) (Oct. 1976), 37–48; **8** (11) (Nov. 1976), 15–24.

4. BERT, C. W., Recent research in composite and sandwich plate dynamics, *Shock Vib. Digest*, **11** (10) (Oct. 1979), 13–23.

5. BERT, C. W., Vibration of composite structures, *Recent Advances in Structural Dynamics* (Proc., Int. Conf., Univ. of Southampton, July 1980) (M. Petyt ed.), Vol. 2, pp. 693–712.

6. BERT, C. W., Research on dynamics of composite and sandwich plates, 1979–81, *Shock Vib. Digest*, **14** (10) (Oct. 1982), 17–34.

7. BERT, C. W., Research on dynamic behavior of composite and sandwich plates—IV, *Shock Vib. Digest*, **17** (11) (Nov. 1985), 3–15.

8. JONES, R. M., *Mechanics of Composite Materials*, New York, McGraw-Hill, 1975.

9. HASHIN, Z., Complex moduli of viscoelastic composites—II. Fiber reinforced materials, *Int. J. Solids Struct.*, **6** (1970), 797–807.

10. CHANG, S. and BERT, C. W., Analysis of damping for filamentary composite materials, *Composite Materials in Engineering Design* (Proc., 6th St. Louis Sympos., May 1972), American Society for Metals, Metals Park, OH, 1973, 51–62.

11. SIU, C. C. and BERT, C. W., Sinusoidal response of composite-material plates with material damping, *ASME J. Engng for Industry*, **96B** (1974), 603–610.

12. BERT, C. W. and CLARY, R. R., Evaluation of experimental methods for determining dynamic stiffness and damping of composite materials, *Composite Materials: Testing and Design ( 3rd Conference )*, ASTM STP, 546, 1974, 250–265.

13. GIBSON, R. F. and PLUNKETT, R., Dynamic stiffness and damping of fiber-reinforced composite materials, *Shock Vib. Digest*, **9** (2) (Feb. 1977), 9–17.

14. GIBSON, R. F. and WILSON, D. G., Dynamic mechanical properties of fiber-reinforced composite materials, *Shock Vib. Digest*, **11** (10) (Oct. 1979), 3–11.

15. BERT, C. W., Composite materials: a survey of the damping capacity of fiber reinforced composites, in: *Damping Applications for Vibration Control* (P. J. Torvik, ed.), New York, ASME, AMD Vol. 38, 1980, pp. 53–63.

16. GIBSON, R. F., Recent research on dynamic mechanical properties of fiber reinforced composite materials and structures, *Shock Vib. Digest*, **15** (2) (Feb. 1983), 3–15.

17. GIBSON, R. F. and YAU, A., Complex moduli of chopped fiber and continuous fiber composites: comparison of measurement with estimated bounds, *J. Comp. Mater.*, 14 (1980), 155–167.

18. GIBSON, R. F., YAU, A., MENDE, E. W., OSBORN, W. E. and REIGNER, D. A., The influence of environmental conditions on the vibration characteristics of chopped-fiber-reinforced composite materials, *J. Reinforced Plastics Comp.*, 1 (1982), 225–241.

19. SUAREZ, S. A., GIBSON, R. F., SUN, C. T. and CHATURVEDI, S. K., The influence of fiber length and fiber orientation on damping and stiffness of polymer composite materials, *Exp. Mech.*, **26** (1986), 175–184.

20. TABADDOR, F. and CLARK, S. K., Loss characteristics of cord rubber composites, *Proc. 4th Int. Conf. Vehicle Structural Mech.*, Detroit, SAE, 1981.
21. TABADDOR, F. and CLARK, S. K., Linear and nonlinear viscoelastic non-constitutive relations of cord reinforced elastomers, *Proc., Int. Conf. Constitutive Laws for Engng Mater.: Theory and Appl.*, Tucson, AZ, University of Arizona, 1983, pp. 143–150.
22. TIMMERMAN, N. S., *Damping Characteristics of Metal Matrix Composites*, Cambridge, MA, Bolt Beranek & Newman, Inc., AMMRC CTR-82-19, Apr. 1982.
23. MANTENA, R., GIBSON, R. F. and PLACE, T. A., Damping capacity measurements of degradation in advanced materials, *SAMPE Quarterly*, **17** (3) (Apr. 1986), 20–31.
24. TEH, K. K. and HUANG, C. C., The effects of fibre orientation on free vibrations of composite beams, *J. Sound Vib.*, **69** (1980), 327–337.
25. TEH, K. K. and HUANG, C. C., Wave propagation in generally orthotropic beams, *Fibre Sci. Technol.*, **14** (1981), 301–310.
26. BRATT, J. G., Light, reinforced beams used as shock isolators, *Trans. Canad. Soc. Mech. Engng*, **6** (1980–81), 41–46.
27. MEAD, D. J. and MARKUS, S., Coupled flexural-longitudinal wave motion in a periodic beam, *J. Sound Vib.*, **90** (1983), 1–24.
28. MEAD, D. J. and MARKUS, S., Coupled flexural, longitudinal and shear wave motion in two- and three-layered damped beams, *J. Sound Vib.*, **99** (1985), 501–519.
29. MILES, R. N. and REINHALL, P. G., An analytical model for the vibration of laminated beams including the effects of both shear and thickness deformation in the adhesive layer, *ASME J. Vib. Acoust., Stress, Reliab. in Design*, **108** (1986), 56–64.
30. REDDY, E. S. and MALLIK, A. K., Vibration of a two layered ring on periodic radial supports, *J. Sound Vib.*, **84** (1982), 417–430.
31. GRACE, N. F. and KENNEDY, J. B., Dynamic analysis of orthotropic plate structures, *J. Engng Mech.*, **111** (1985), 1027–1037.
32. KIM, C. S. and DICKINSON, S. M., Improved approximate expressions for the natural frequencies of isotropic and orthotropic rectangular plates, *J. Sound Vib.*, **103** (1985), 142–149.
33. DICKINSON, S. M. and DI BLASIO, A., On the use of orthogonal polynomials in the Rayleigh–Ritz method for the study of the flexural vibration and buckling of isotropic and orthotropic rectangular plates, *J. Sound Vib.*, **108** (1986), 51–62.
34. NAKAHIRA, N., Natsuaki, Y., Ozawa, K. and NARUOKA, M., Numerical vibration analysis of plate structures by Newmark's method, *J. Sound Vib.*, **99** (1985), 183–198.
35. GROSSI, R. O., Laura, P. A. A. and MUKHOPADHYAY, M., Fundamental frequency of vibration of orthotropic rectangular plates with three edges elastically restrained against rotation while the fourth is free, *J. Sound Vib.*, **103** (1985), 443–445.
36. BERT, C. W., Application of a version of the Rayleigh technique to problem of bars, beams, columns, membranes, and plates, *J. Sound Vib.* (in press).

37. TOMAR, J. S. and GUPTA, A. K., Effect of thermal gradient on frequencies of an orthotropic rectangular plate whose thickness varies in two directions *J. Sound Vib.*, **98** (1985), 257–262.

38. BIANCHI, A., AVALOS, D. R. and Laura, P. A. A., A note on transverse vibrations of annular, circular plates of rectangular orthotropy, *J. Sound Vib.*, **99** (1985), 140–143.

39. NARITA, Y., Natural frequencies of free, orthotropic elliptical plates, *J. Sound Vib.*, **100** (1985), 83–89.

40. NARITA, Y., Free vibration analysis of orthotropic elliptical plates resting on arbitrarily distributed point supports, *J. Sound Vib.*, **108** (1986), 1–10.

41. GORMAN, D. G., Thermal gradient effects upon the vibrations of certain composite circular plates, part I: plane orthotropic, *J. Sound Vib.*, **101** (1985), 325–336.

42. GORMAN, D. G., Thermal gradient effects upon the vibration of certain composite circular plates, part II: plane orthotropic with temperature dependent properties, *J. Sound Vib.*, **101** (1985), 337–345.

43. NARITA, Y., Natural frequencies of completely free annular and circular plates having polar orthotropy, *J. Sound Vib.*, **92** (1984), 33–38.

44. SRINIVASAN, R. S. and THIRUVENKATACHARI, V., Free vibration of annular sector plates by an integral equation technique, *J. Sound Vib.*, **89** (1983), 425–432.

45. GUPTA, U. S. and LAL, R., Axisymmetric vibrations of polar orthotropic Mindlin annular plates of variable thickness, *J. Sound Vib.*, **98** (1985), 565–573.

46. GUPTA, U. S., LAL, R. and VERMA, C. P., Effects of an elastic foundation on axisymmetric vibrations of polar orthotropic annular plates of variable thickness, *J. Sound Vib.*, **103** (1985), 159–169.

47. GUPTA, U. S., LAL, R. and VERMA, C. P., Buckling and vibrations of polar orthotropic annular plates of variable thickness, *J. Sound Vib.*, **104** (1986), 357–369.

48. DONG, S. B. and HUANG, K. H., Edge vibrations in laminated composite plates, *ASME J. appl. Mech.*, **52** (1985), 433–438.

49. HUI, D., Effects of geometric imperfections on frequency-load interaction of biaxially compressed antisymmetric angle ply rectangular plates, *ASME J. appl. Mech.*, **52** (1985), 155–162.

50. SANKARANARAYANAN, N., CHANDRASEKARAN, K. and RAMAIYAN, G., Axisymmetric vibrations of layered annular plates with linear variation in thickness, *J. Sound Vib.*, **99** (1985), 351–360.

51. IYENGAR, N. G. R. and UMARETIYA, J. R., Transverse vibrations of hybrid laminated plates, *J. Sound Vib.*, **104** (1986), 425–535.

52. GROSVELD, F. W. and METCALF, V. L., Modal response and noise transmission of composite panels, *Proc. 26th Structures, Struct. Dynamics and Mater. Conf.*, Orlando, FL, 1985, pt. 2, pp. 617–627.

53. YEN, S. C. and CUNNINGHAM, F. M., Vibration characteristics of graphite–epoxy composite plates, *Proc. Soc. Exp. Mech. Spring Conf.*, Las Vegas, NV, 1985, pp. 60–67.

54. RAMKUMAR, R. L., CHEN, P. C. and SANDERS, W. J., Free vibration solution for clamped orthotropic plates using Lagrangian multiplier technique, *AIAA J.*, **25** (1987), 146–151.

55. CRAIG, T. J. and DAWE, D. J., Flexural vibration of symmetrically laminated composite rectangular plates including transverse shear effects, *Int. J. Solids Struct.*, **22** (1986), 155–169.

56. DAWE, D. J. and CRAIG, T. J., The vibration and stability of symmetrically-laminated composite rectangular plates subjected to inplane stress, *Comp. Struct.*, **5** (1986), 281–307.

57. KAMAL, K. and DURVASULA, S., Some studies of free vibration of composite laminates, *Comp. Struct.*, **5** (1986), 177–202.

58. DI SCIUVA, M., Bending, vibration and buckling of simply supported thick multilayered orthotropic plates: an evaluation of a new displacement model, *J. Sound Vib.*, **105** (1986), 425–442.

59. KAMAL, K. and DURVASULA, S., Macromechanical behaviour of composite laminates, *Comp. Struct.* **5** (1986), 309–318.

60. STEIN, M. and JEGLEY, D. C., Effects of transverse shearing on cylindrical bending, vibration, and buckling of laminated plates, *AIAA J.*, **25** (1987), 123–129.

61. PUTCHA, N. S. and REDDY, J. N., Stability and natural vibration analysis of laminated plates by using a mixed element based on a refined plate theory, *J. Sound Vib.*, **104** (1986), 285–300.

62. REDDY, J. N., A review of the literature on finite-element modeling of laminated composite plates, *Shock Vib. Digest*, **17** (5) (Apr. 1985), 3–8.

63. ALAM, N. and ASNANI, N. T., Vibration and damping analysis of fibre reinforced composite material plates, *J. Comp. Mater.*, **20** (1986), 2–18.

64. KOZLOV, S. V., Determination of the natural frequencies and mode configurations for small vibrations of an orthotropic cylindrical shell with attached masses, *Sov. appl. Mech.*, **17** (1981), 138–142.

65. CHONAN, S., Moving load on initially stressed orthotropic cylindrical shells filled with liquid, *ASME J. appl. Mech.*, **52** (1985), 976–977.

66. SHIRAKAWA, K., Dynamic response of an orthotropic cylindrical shell to rapid heating, *J. Sound Vib.*, **83** (1982), 27–35.

67. SHUL'GA, N. A., Eigenfrequencies of axisymmetric vibrations of a hollow cylinder made out of a composite material, *Mech. Comp. Mater.*, **16** (1980), 349–352.

68. VANDERPOOL, M. E. and BERT, C. W., Vibration of a materially monoclinic, thick-wall circular cylindrical shell, *AIAA J.*, **19** (1981), 634–641.

69. SHARMA, C. B. and DIMAKAKOS, K., Natural frequencies of orthotropic cylindrical shells with shear deformation and rotary inertia, American Society of Mechanical Engineers, Paper 85-DET-174, Sept. 1985.

70. MARCHUK, R. A. and SHVETS, R. N., Small oscillations of an orthotropic cylindrical shell containing a liquid covered by rigid endplates, *J. appl. Math. Mech.*, **46** (1982), 237–243.

71. CHONAN, S., Response of a pre-stressed, orthotropic thick cylindrical shell subjected to a pressure pulse, *J. Sound Vib.*, **93** (1984), 31–38.

72. BERT, C. W. and KUMAR, M., Vibration of cylindrical shells of bimodulus composite materials, *J. Sound Vib.*, **81** (1982), 107–121.

73. SOLDATOS, K. P., A comparison of some shell theories used for the dynamic analysis of cross-ply laminated circular cylindrical panels, *J. Sound Vib.*, **97** (1984), 305–319.

74. BERT, C. W., Models for fibrous composites with different properties in tension and compression, *ASME J. Mater. Technol.*, **99** (1977), 344–349.
75. KUMAR, M. and BERT, C. W., Experimental characterization of mechanical behavior of cord-rubber composites, *Tire Sci. Technol.*, **10** (1982), 37–54.
76. BERT, C. W., Micromechanics of the different elastic behavior of filamentary composites in tension and compression, in: *Mechanics of Bimodulus Materials* (C. W. Bert ed)., New York, ASME, AMD Vol. 33, 1979, pp. 17–28.
77. SCHMUESER, D. W., Nonlinear stress–strain and strength response of axisymmetric bimodulus composite material shells, *AIAA J.*, **21** (1983), 1742–1747.
78. GREENBERG, J. B. and STAVSKY, Y., Vibrations of laminated filament-wound cylindrical shells, *AIAA J.*, **9** (1981), 1055–1062.
79. GULGAZARYAN, G. R. and LIDSKII, V. B., Density of free vibration frequencies for a thin anisotropic shell made of anisotropic layers, *Mechanics of Solids*, **17** (1982), 159–162.
80. SOLDATOS, K. P., On the buckling and vibration of antisymmetric angle-ply laminated circular cylindrical shells, *Int. J. Engng Sci.*, **21** (1983), 217–222.
81. DARVIZEH, M. and SHARMA, C. B., Natural frequencies of laminated orthotropic thin circular cylinders, *Thin-Walled Struc.*, **2** (1984), 207–217.
82. SOLDATOS, K. P. and TZIVENIDIS, G. J., Buckling and vibration of cross-ply laminated circular cylindrical panels, *Zeits. angew. Math. Phys.*, **33** (1982), 230–240.
83. SOLDATOS, K. P., Free vibrations of antisymmetric angle-ply laminated circular cylindrical panels, *Quart. J. Mech. appl. Math.*, **26** (1983), 207–221.
84. GREENBERG, J. B. and STAVSKY, Y., Buckling and vibration of orthotropic composite cylindrical shells, *Acta Mech.*, **36** (1980), 15–29.
85. SOLDATOS, K. P. and TZIVANIDIS, G. J., Buckling and vibration of cross-ply laminated non-circular cylindrical shells, *J. Sound Vib.*, **82** (1982), 425–434.
86. SOLDATOS, K. P., Equivalence of some methods used for the dynamic analysis of cross-ply laminated oval cylindrical shells, *J. Sound Vib.*, **91** (1983), 461–465.
87. HUI, D. and DU, I. H. Y., Effects of axial imperfections on vibrations of anti-symmetric cross-ply, oval cylindrical shells, *ASME J. appl. Mech.*, **53** (1986), 675–680.
88. SHARMA, C. B. and DARVIZEH, M., Vibration characteristics of laminated composite shells with transverse shear and rotary inertia, American Society of Mechanical Engineers, Paper 85-DET-173, Sept. 1985.
89. GREENBERG, J. B. and STAVSKY, Y., Vibrations of axially compressed laminated orthotropic cylindrical shells, including transverse shear deformation, *Acta Mech.*, **37** (1980), 13–28.
90. BHIMARADDI, A., Dynamic response of orthotropic, homogeneous, and laminated cylindrical shells, *AIAA J.*, **23** (1985), 1834–1837.
91. BAKULIN, V. N. and POTOPAKHIN, V. A., Use of the equations of the three-dimensional elasticity theory to solve multilayer shell dynamics problems, *Sov. Aeron.*, **28** (1985), 6–11.
92. BERT, C. W., Dynamic behavior of composites: an overview, *Proc. Spring Conf. on Exp. Mech.*, New Orleans, LA, 1986, pp. 747–751.
93. BISHOP, S. M., The mechanical performance and impact behaviour of carbon-fibre reinforced PEEK, *Comp. Struct.*, **3** (1985), 295–318.

94. POTTER, R. T., The interaction of impact damage and tapered-thickness sections in CFRP, *Comp. Struct.*, **3** (1985), 319–339.

95. CHEN, J. K. and SUN, C. T., Dynamic large deflection response of composite laminates subjected to impact, *Comp. Struct.*, **4** (1985), 59–73.

96. CHEN, P. C. and RAMKUMAR, R. L., Static and dynamic analysis of clamped orthotropic plates, *27th Struct., Struc. Dynamics & Matls. Conf.*, pt. 2, San Antonio, 1986, New York, AIAA.

97. BABICH, D. V. and Khoroshyn, L. P., Dynamic loss of stability of an instantaneously compressed layered cylindrical shell, *Sov. appl. Mech.*, **16** (1980), 586–590.

98. BIRMAN, V., Dynamic buckling of antisymmetrically laminated imperfect rectangular plates, *Proc. Fourth Int. Conf. Composite Structures*, Paisley College of Tech., Paisley, Scotland, 1987 (in press).

99. STUART, R. J., DHARMARAJAN, S. and PENZES, L. E. *Fibrous Composites in Structural Design* (Proc., 4th Conf.), San Diego, CA, 1978, New York, Plenum Press, 1980, pp. 329–340.

100. BIRMAN, V., Dynamic stability of unsymmetrically laminated rectangular plates, *Mech. Res. Communications*, **12** (1985), 81–86.

101. SRINIVASAN, R. S. and CHELLAPANDI, P., Dynamic stability of rectangular laminated composite plates, *Computer & Struct.*, **24** (1986), 233–238.

102. BERT, C. W. and BIRMAN, V., Dynamic stability of shear deformable antisymmetric angle-ply plates, *Int. J. Solids Struct.*, (in press).

103. GALAKA, P. I., KOVAL'CHUK, P. S., MENDELUTSA, V. M. and NOSACHENKO, A. M., Dynamic instability of glass–plastic shells carrying concentrated masses, *Sov. appl. Mech.*, **16** (1980), 686–690.

104. BERT, C. W. and BIRMAN, V., Dynamic stability of thick, orthotropic, circular cylindrical shells, in: *Refined Dynamical Theories of Beams, Plates and Shells and Their Applications*, Euromech Colloquium 219, Kassel, Federal Republic of Germany, Sept. 1986.

105. RAY, H., Dynamic instability of suddenly heated angle-ply laminated composite cylindrical shells, *Computers & Struct.*, **16** (1983), 119–124.

106. RAY, H. and BERT, C. W., Dynamic instability of suddenly heated, thick composite shells, *Int. J. Engng Sci.*, **22** (1984), 1259–1268.

107. KAMAL, K. and DURVASULA, S., Flutter of laminated composite skew panels in supersonic flow, *Developments in Mechanics*, Vol. 13 (Proc., 19th Midwestern Mech. Conf.), Columbus, OH, 1985, pp. 440–441.

108. LIBRESCU, L. and BEINER, L., Weight minimization of orthotropic flat panels subjected to a flutter speed constraint, *AIAA J.*, **24** (1986), 991–997.

109. LOTTATI, I., Flutter and divergence aeroelastic characteristics for composite forward swept cantilevered wing, *J. Aircraft*, **22** (1985), 1001–1007.

110. SHIRK, M. H., HERTZ, T. J. and WEISSHAAR, T. A., Aeroelastic tailoring—theory, practice, and promise, *J. Aircraft*, **23** (1986), 6–18.

111. CHIA, C. Y., *Nonlinear Analysis of Plates*, New York, McGraw-Hill, 1980.

112. CHIA, C. Y., Non-linear vibration of anisotropic rectangular plates with non-uniform edge constraints, *J. Sound Vib.*, **101** (1985), 539–550.

113. CHIA, C. Y., Nonlinear oscillation of unsymmetric angle-ply plate on elastic foundation having nonuniform edge supports, *Comp. Struct.*, **4** (1985), 161–178.

114. SIVAKUMARAN, K. S. and CHIA, C. Y., Large-amplitude oscillations of unsymmetrically laminated anisotropic rectangular plates including shear, rotatory inertia, and transverse normal stress, *ASME J. appl. Mech.*, **52** (1985), 536–542.

115. GRAY, C. E. JR., DECHA-UMPHAI, K. and MEI, C., Large deflection, large amplitude vibrations and random response of symmetrically laminated plates, *J. Aircraft*, **22** (1985), 929–930.

# 1

# Optimal Use of Adhesive Layers in Reducing Impact Damage in Composite Laminates

S. RECHAK and C. T. SUN

*Composite Materials Laboratory,*
*School of Aeronautics and Astronautics, Purdue University,*
*West Lafayette, Indiana 47907, USA*

## ABSTRACT

*Impact experiments on a group of cross-plied graphite/epoxy laminates with and without adhesive layers were conducted. Impact-damage modes were examined by sectioning the impacted specimen and examining the sections with an optimal microscope. The effectiveness of using adhesive layers to suppress impact-induced delamination was investigated. It was found that the presence of adhesive layers could toughen the interface between two laminae and also reduce matrix cracking. A guide to the optical use of adhesive layers for improving impact resistance properties was proposed.*

## INTRODUCTION

Composite laminates are susceptible to impact damage. Under low velocity impacts, the major forms of damage consist of matrix cracking and delamination. Delamination is the more critical, as it may cause significant reduction in compressive strength. Thus, to improve the impact resistance property of a composite laminate, the interfaces of the laminate must be toughened.

One method being considered for interface toughening is the use of adhesive layers. By placing tough adhesive films along the interfaces of the laminae, Chan et al.[1] demonstrated that edge delamination in coupon laminate specimens subjected to in-plane tension could be suppressed. Sun and Rechak[2] employed a similar method to delay the initiation of and reduce the extent of delamination induced by impact loading. In Ref. 2, the mechanisms provided by the presence of interlaminar adhesive layers in reducing impact damage was studied. It was found that adhesive layers included in a composite laminate could increase the contact area and, as a result, reduce the transverse shear concentration effect. This, together with the toughened interfacial property, resulted in less delamination.

However, it is not practical to add an adhesive layer along every interface of the laminate. This may cause the laminate a great weight penalty. Moreover, too many adhesive layers may result in poorer compressive properties. Hence, in order to avoid incurring other adverse side effects, the use of adhesive layers must be kept minimal. Thus, one needs to know the optimal placement for adhesive layers to obtain the greatest benefit in improving impact resistance.

## EXPERIMENTAL PROCEDURES

Hercules AS4/3501-6 graphite/epoxy prepreg tape was used to lay up $30 \times 30$ cm ($12 \times 12$ in) panels for impact experiments. No bleeder cloth was used during curing. The adhesive film used was adhesive FM 1000 manufactured by American Cyanamid. The film has a thickness of 0·013 cm (0·005 in) and can be co-cured with the composite.

Cured panels were cut into $2.5 \times 30$ cm ($1 \times 12$ in) beam-like specimens for the impact experiments.

The following laminates were used in the experiment:

Laminate 1: $[0_5/90_5/0_5]$
Laminate 2: $[90/0_2/90_2/0_2]_s$
Laminate 3: $[0_5/A/90_5/A/0_5]$
Laminate 4: $[0/A/0_4/90_5/0_4/A/0]$
Laminate 5: $[0_5/90_2/A]_s$
Laminate 6: $[90/0_2/90_2/A/0_2]_s$

Laminates 1 and 2 are the baseline laminates, and laminates 3–6 contain adhesive layers. The location of the adhesive layer is indicated by A.

Impact experiments were conducted using an air gun. The beam specimen was clamped at both ends, having a 15 cm (6 in) gage length. The impactor was a steel ball 1·27 cm (0·5 in) in diameter. The ball traveled a distance of 107 cm (42 in) in the barrel, and another 20 cm (8 in) before reaching the mounted beam specimen. Two photodiodes located 25·4 cm

(10 in) apart were connected to an electrical countertimer to measure the time the impactor traveled between the two photodiodes. Impacted specimens were sectioned transversely and longitudinally to observe the impact-inflicted damage with an optical microscope. In addition, maximum contact area after an impact was measured. A very thin layer of white paint was applied to the surface of the composite specimen at the impact site to highlight the impression left by the contact.

## EXPERIMENTAL RESULTS

Extensive experimental results on laminates 1 and 3 were reported in Ref. 2. The damage modes and the corresponding initiation impact velocities are summarized in Table 1. From the results of Ref. 2, it was found that the adhesive layer has no effect on the occurrence of matrix cracks due to bending stresses; it can reduce matrix cracking in the upper lamina near the impact site; and it can effectively suppress delamination.

The major forms of impact damage in laminate 1 due to low velocity impact are shown in Fig. 1. In the top and middle laminae, transverse shear cracks (matrix cracks that are inclined at approximately $45°$ to the loading) prevail, and in the bottom lamina only bending cracks are present. Delamination cracks can be induced by transverse shear and bending cracks along the two 0/90 interfaces.

TABLE 1
*Damage modes and corresponding initiation impact velocities for laminates 1 and 3*

| Damage mode | Initiation $[0_5/90_5/0_5]$ | Velocity (m/s) $[0_5/A/90_5/A/0_5]$ |
|---|---|---|
| Bending crack | 10·40 | 12·60 |
| Transverse shear crack in upper lamina | 14·90 | 28·40 |
| Transverse shear crack in middle lamina | 10·40 | 11·80 |
| Fibers breakage in the middle lamina due to bending crack | 42·0 | 28·40 |
| Delamination due to middle lamina transverse shear crack | 10·40 | 26·50 (some specimens) |
| Delamination due to upper lamina transverse shear crack | 20·50 | 41·50 |
| Delamination due to bending crack | 10·40 | none |
| Delamination in the upper interface within the impact region | 34·0 | none |

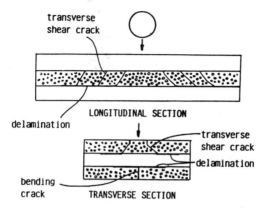

FIG. 1. Impact-induced matrix cracks and delamination in $[0_5/90_5/0_5]$ laminate.

The contact force between the impactor and the laminate is distributed in a small contact area. The size of the contact area affects the intensity of contact stress and, consequently, stress concentration in the laminate near the impact site.

The maximum contact area for a given impact velocity was determined. Four types of laminate specimens were considered: $[0_5/90_5/0_5]$, $[0_5/A/90_5/A/0_5]$, $[0_3/A/0_2/90_5/0_2/A/0_5]$ and $[0/A/0_4/90_5/0_4/A/0]$. Experimental results are shown in Fig. 2, from which a definite conclusion is reached: adhesive layers increase the contact area appreciably. In Fig. 2, the contact length is equal to the diameter of the contact area. The closer the adhesive layer is placed relative to the impact site, the larger the contact area is.

**Damage Modes in Laminate 4**

Figure 3 shows the transverse and longitudinal sections of laminate 4 subjected to impacts at 20 and 15 m/s, respectively. With the exception of some features, this laminate experienced similar damage modes as did laminate 1.

It was noted that the adhesive layer did not arrest the bending crack. The bending crack was able to cut through the adhesive layer, reach the 0/90 interface and branch out into delamination cracks. Possibly due to the large contact area that reduces the transverse stress concentration effect, no cracks occurred in the single ply above the adhesive layer within the 35 m/s impact velocity range. However, transverse shear cracking was not prevented beneath the adhesive layer.

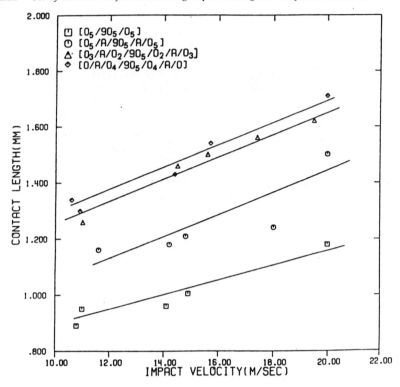

Fig. 2.   Size of contact area versus impact velocity for cross-plied laminates with and without adhesive layers.

From the longitudinal section of the laminate impacted at 15 m/s, as shown in Fig. 3, it is evident that adhesive layers do not seem to affect matrix cracking in the middle lamina.

**Damage Modes in Laminate 5**

In this laminate, two adhesive films were placed in the middle plane of the laminate separating the four 90° plies into two regions. The objective was to investigate whether or not the adhesive layer could stop transverse shear cracks in the middle lamina.

Figure 4 shows the longitudinal sections of the laminate impacted at 10·7 and 14·75 m/s, respectively. At the lower impact velocity, transverse shear cracks occurred beneath the adhesive layer but were absent in the region above the adhesive layer. At the higher impact velocity, some of the cracks

20.00 M/SEC

**15 M/SEC**

FIG. 3.    Transverse (top) and longitudinal (bottom) sections of laminate 4 subjected to impacts at 20 and 15 m/s, respectively.

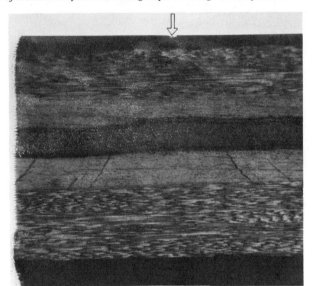

**10.7 M/SEC**

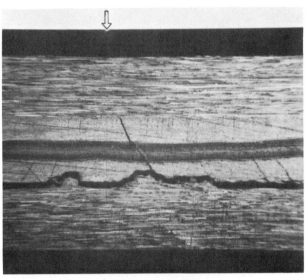

**14.75 M/SEC**

FIG. 4.    Longitudinal sections of laminate 5 subjected to impacts at 10·7 and 14·75 m/s, respectively.

near the impact point in the lower region broke through the adhesive layer and extended into the upper region. However, at greater distances from the impact point, transverse shear cracks in the lower region were stopped by the adhesive layer as they lacked enough fracture energy to drive through the adhesive layer.

## Damage Modes in Laminates 2 and 6

From the results of the previous laminates, we note that one of the major mechanisms causing delamination is the bending crack in the bottom lamina. It is conceivable that if the bending crack size is kept small, the driving force of the crack may become small and may cause less delamination. To test the validity of this concept, only a single 0° ply was laid up at the top and bottom.

The previous test results also indicate that adhesive layers have the effect of minimizing transverse shear cracks above them. In view of this, adhesive layers were embedded along the third 0/90 interfaces from the top and bottom in laminate 6.

Table 2 summarizes the impact velocities of transverse crack initiation in different laminae of the two laminates. The effect of adhesive layers shows in delaying matrix cracking in the laminae above the upper adhesive layer. The adhesive layers do not seem to have any effect on the bending crack and matrix cracks in other laminae beneath the upper adhesive layer. This is seen clearly in Figs 5 and 6, showing matrix cracks in the two laminates impacted at about 22 m/s. A sketch of the extent of delamination in laminate 2 is shown in Fig. 7.

Figures 8 and 9 show the results for an impact velocity of about 30 m/s. At this velocity, transverse shear cracks start to appear in the second and

TABLE 2
*Matrix cracks and the corresponding initiation impact velocities for laminates 2 and 6*

| Damage mode | Cross-section | Initiation velocity (m/s) | |
|---|---|---|---|
| | | Laminate 2 | Laminate 6 |
| Transverse shear crack in lamina 2 | Transverse | 22·0 | — |
| Transverse shear crack in lamina 3 | Longitudinal | 22·0 | 30·0 |
| Transverse shear crack in lamina 4 | Transverse | 18·0 | 22·0 |
| Transverse shear crack in lamina 5 | Longitudinal | 22·0 | 22·0 |
| Transverse shear crack in lamina 6 | Transverse | 18·0 | 18·0 |
| Bending crack | Longitudinal | 18·0 | 18·0 |

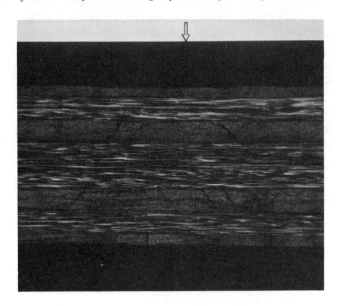

FIG. 5.  Longitudinal (top) and transverse (bottom) sections of laminate 2 impacted at 22·6 m/s.

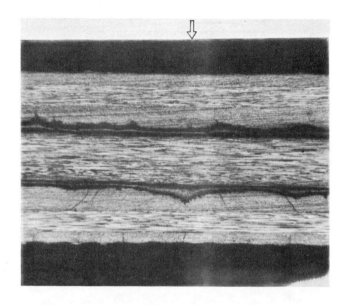

FIG. 6.    Longitudinal (top) and transverse (bottom) sections of laminate 6 impacted at 22 m/s.

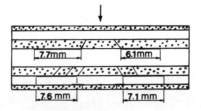

**a-LONGITUDINAL SECTION**

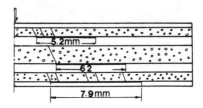

**b-TRANSVERSE SECTION**

FIG. 7. Matrix cracks and extent of delamination in laminate 2 impacted at 22 m/s.

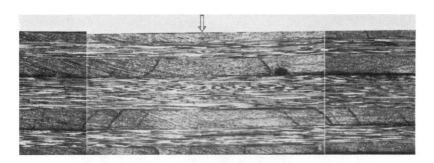

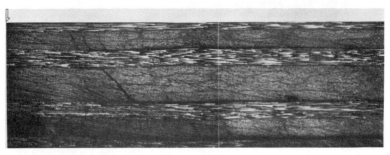

FIG. 8. Longitudinal (top) and transverse (bottom) sections of laminate 2 impacted at 30·5 m/s.

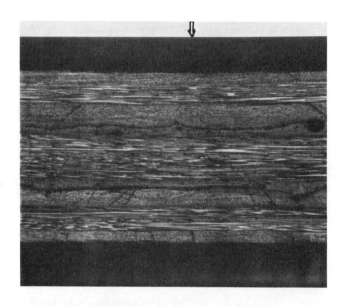

FIG. 9. Longitudinal (top) and transverse (bottom) sections of laminate 6 impacted at 29·75 m/s.

third laminae of laminate 6. However, no delamination is visible at the interfaces above the upper adhesive layer. It is interesting to note that transverse shear cracks in laminae above the upper adhesive layer occur at a larger distance from the impact center in laminate 6 than that in laminate 2. This could result from less severe transverse shear stress concentration in laminate 6.

## DISCUSSION AND CONCLUSION

The following are the major conclusions of this study.

1. Adhesive layers placed along the interface of two laminae of different fiber orientations can effectively suppress delamination in laminates subjected to impact loading. However, when embedded inside a lamina, adhesive layers cannot arrest matrix cracks.
2. When placed near the impacted surface of the laminate, the adhesive layer can increase the contact area between the projectile and the target. As a result, the transverse shear stress concentration, and consequently matrix cracking, is reduced in the laminae above the adhesive layer. However, this adhesive layer does not affect the matrix cracking in the laminae beneath it.
3. Bending crack initiation in the surface lamina opposite to the impact site cannot be delayed by using adhesive layers.
4. For low velocity impact, delamination seems to result from branching of matrix cracks (transverse shear cracks and bending cracks). Thus, suppression of matrix cracks can also lead to suppression of delamination.

From the above conclusions, a general guide to the optimal use of adhesive layers can be established. In general, an adhesive layer can be used to directly toughen an interface or to reduce matrix cracking. The following procedures are recommended.

1. Place the adhesive layer along the interface located below the impacted face at a distance equal to the size of the contact area. Above this adhesive layer, matrix cracking can be reduced and thus delamination. If the adhesive layer is placed too far down, its effectiveness in increasing the contact area, and thus reducing transverse shear stress concentration effect, may be diminished.
2. To minimize delamination induced by bending cracks in the surface laminae, these surface laminae should consist of a single ply. The

thinness of the lamina can reduce the fracture force of the bending crack and consequently the delamination. Alternatively, an adhesive layer can be placed along the interface below this lamina to arrest delamination cracks.

## ACKNOWLEDGEMENT

This research was supported by ONR Grant No. N00014-84-K-0554 to Purdue University. Dr Y. Rajapakse was Technical Monitor.

## REFERENCES

1. CHAN, W. S., ROGERS, C. and AKER, S., Improvement of edge delamination strength of composite laminates using adhesive layers, in: *Composite Materials Testing and Design, 7th Conference* (Whitney, J. M. ed.), ASTM STP 893, American Society for Testing and Materials, 1976, pp. 266–285.
2. SUN, C. T. and RECHAK, S., Effect of adhesive layers on impact damage in composite laminates, in: *Composite Materials Testing and Design, 8th Conference* (Whitcomb, J. D. ed.), American Society for Testing and Materials, to be published.

# 2

# The Crush Performance of Composite Structures

J. N. PRICE and D. HULL

*Department of Materials Science and Metallurgy, University of Cambridge, Pembroke Street, Cambridge CB2 3QZ, UK*

## ABSTRACT

*Compression tests have been carried out on a range of composite structures made from a csm/polyester laminate. Geometries tested included truncated cones, slotted and sectioned round tubes, square tubes and flat plates. For cones and tubular specimens failure is by progressive crush absorbing relatively large amounts of energy. A trigger is necessary to initiate crush in tubular specimens, however cone failure is self triggering with crush commencing at the small end. Plate specimens failed by catastrophic splitting and buckling with minimal energy absorption. The inability of plate specimens to support stable crush is thought to be due to the lack of transverse restraining forces in the crush zone.*

## INTRODUCTION

The impact response of structural composites has been examined in detail over the last few years.[1-5] This interest is fostered by the need for components which perform a structural duty under normal conditions but when overloaded collapse in a controlled manner absorbing kinetic energy in a well defined way, the prime example being the automobile.

Most work in this area has considered the crush performance of axisymmetric cylinders[1,2] and many composite materials have been examined. These tests have demonstrated that tubular specimens can fail by progressive crush under axial impact if a suitable trigger is used (e.g. a chamfer at one end of the specimen). This type of failure takes place over a wide range of impact speeds and the energy absorbed is higher than that of mild steel when compared on a weight basis.[1] The crush zones produced have been studied and it is noteworthy that the damage region is well defined, whilst away from the crush zone the material retains its integrity.

If composite materials are to be used in energy absorbing members, for example in a vehicle's 'crumple' zone, then structures other than axisymmetric cylinders will be required. The crush performance of composite structures has been shown to be geometry dependent.[1-4] Whilst there is an infinite range of specimen configurations that can be considered, the objective of this work was to examine the crush behaviour of structures which represent a simple step from the axisymmetric tube. Four geometries have been examined; truncated cones, slotted and sectional axisymmetric tubes, square section tubes and flat plates. Square tubes with various corner radii were tested.

Thornton[5] has published results of crush tests on square tubes and sectioned tubes made from an aligned glass epoxy material with various ply orientations. The tests reported in this chapter were made on specimens fabricated from a random chopped strand mat (csm)/polyester laminate. This material is representative of those which would be used in a high production application and its random orientation alleviates difficulties associated with changes in fibre orientation which arise when conical specimens are made from aligned plies.


## EXPERIMENTAL METHODS

All specimens were made by hand lay-up of $450 \text{ g/m}^2$ powder bound csm (Pilkington Brothers Limited) with Crystic 272 polyester resin (Scott Bader). Truncated cones were fabricated using hardwood mandrels to give cone angles $\Theta$ of 11, 22 and 45°. The range of cone geometries is shown in Fig. 1a. Some cones were tested square ended, i.e. without a trigger, and others were triggered with internal or external chamfers at their small ends. Tubes were laid up on 76·2 mm diameter steel mandrels. External surfaces of cones and tubes were lathe cut to give constant wall thicknesses. The tube sections tested are shown in Fig. 1b and include complete, slotted and sectioned tubes. All tube specimens had external chamfers at one end to trigger failure. Square tubes were laid up on polystyrene mandrels 300 mm high and 75 mm wide. Three corner radii were used as shown in Fig. 1c. It was not possible to machine the external surfaces of the square tubes. Square tubes were triggered using external chamfers.

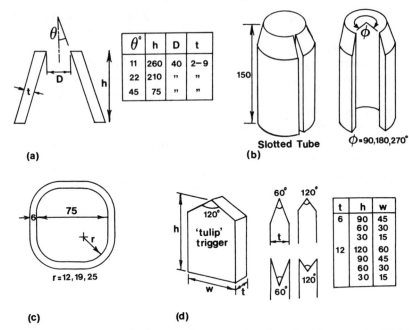

FIG. 1.    Specimen geometries; (a) cones, (b) slotted and section tubes, (c) square tubes, (d) flat plates. All dimensions in mm.

Flat laminates were laid up between steel plates separated by spacers to give thicknesses of 6 and 12 mm. The range of flat plate geometries is shown in Fig. 1d. Five trigger types were used (Fig. 1d), two double external bevels (60° and 120° included angle), two internal bevels (60° and 120°) and a so-called 'tulip' configuration.[5] All plate specimens were cut using a band saw and cut surfaces were finished by milling and grinding with 240 grit SiC paper.

All specimens were postcured for 4 h at 50°C and the nominal fibre volume fraction was 0·2. The specimens were compression tested at a displacement rate of 3 mm/min ($5 \times 10^{-5}$ m/s) using a MAND 250 kN servo hydraulic machine and a Schenk 50 kN screw driven machine.

## RESULTS

**Axisymmetric Tubes**

A typical load displacement curve for a circular tube with internal diameter 76·2 mm, wall thickness $t = 3$ mm and triggered by an external

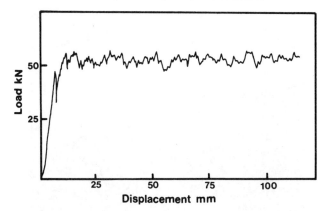

FIG. 2. Load–displacement curve for 76·2 mm diameter tube with wall thickness of 3 mm.

chamfer is shown in Fig. 2. The initial load rise is associated with elastic deformation of the tube and local crushing of the trigger region. During progressive crushing the load fluctuates about an average value and the crush zone moves down the specimen at the same speed as the cross head. Once progressive crushing is developed the crush load and crush zone morphology are independent of trigger geometry.[1]

A section through the crush zone of an axisymmetric tube is shown in Fig. 3. There are three main features; internal and external fronds of highly damaged debris (labelled A and B); a central wedge shaped bundle of fragmented debris (C); and a short intrawall crack lying below the debris wedge (D). This crush zone morphology has |been described in detail elsewhere.[1−4]

## Cones

The variation of crush load with displacement for a cone with $\Theta = 11°$ and $t = 4·9$ mm, triggered by an external chamfer is shown in Fig. 4. As for axisymmetric tubes, the initial load rise corresponds to elastic deformation and crushing of the trigger region. Once a crush zone has developed the average crush load rises linearly with displacement, however the load still fluctuates resulting in a serrated trace. An increase in load with displacement is expected since the cross-sectional area of the cone at the crush zone is increasing. If the materials crush stress is constant then the increase in load can be predicted and this is shown in Fig. 4 by a broken line. This line does not coincide with the mean crush load, hence the crush stress is not constant and depends on geometrical parameters.

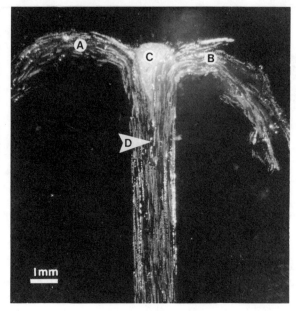

Fig. 3.   Section through crush zone of tube showing debris fronds (A and B), debris wedge (C) and intrawall crack (D).

Cones with internal chamfers and square ended cones also failed by progressive crush although there were variations in the initial load displacement traces. For square ended cones the load rises sharply to a peak before crushing commences. Once the crush zone has become established crushing proceeds in a manner independent of the shape of the initial chamfer and the load displacement curves are very similar.

As the cone angle is increased the load displacement traces show fewer serrations. Cones with $\Theta = 45°$ suffered catastrophic shell failure after only limited progressive crushing.

The energy absorbed by crushing is usually quantified by the energy absorbed per unit mass of crushed material.[1-4] If failure takes place at a mean stress $\bar{\sigma}$ and the density of the material is $\rho$, the specific energy absorption, $S_s$, for circular cross-section structures is given by:[2]

$$S_s = \frac{\bar{P}}{\rho[\pi(D_2^2 - D_1^2)/4]} = \frac{\bar{\sigma}}{\rho} \tag{1}$$

where $\bar{P}$ is the average crush load and $D_1$ and $D_2$ are the internal and external diameters of the tube.

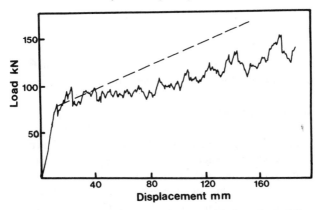

FIG. 4. Load–displacement curve for cone with $\Theta = 11°$, $t = 4·9°$.

For conical specimens the average load and specimen diameter change with crush distance. To calculate $S_s$ for cones $\bar{\sigma}$ is defined by

$$\bar{\sigma} = \frac{P}{A} \tag{2}$$

where $P$ is the applied load and $A$ is the cross-sectional area of the wall normal to the crush direction. The variations of $S_s$ with diameter for cones with $t \approx 8\,\text{mm}$ and $\Theta = 11, 22$ and $45°$ are shown in Fig. 5. It is apparent that the energy absorbed depends on both cone angle and diameter. Cones

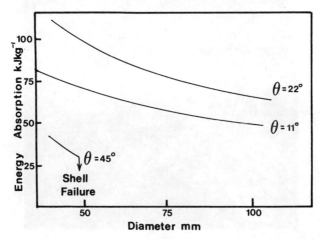

FIG. 5. Variation of specific energy absorption with diameter for 8 mm thick cones.

with $\Theta = 11°$, and those with $\Theta = 22°$ triggered by internal chamfers had crush zone morphologies similar to those of axisymmetric tubes (Fig. 2), however the internal debris fronds were wider than the external fronds. In cones with $\Theta = 22°$ triggered by an external chamfer and all $\Theta = 45°$ cones, all the debris passed to the inside of the specimen as can be seen from Fig. 6.

### Slotted and Sectioned Tubes

Slotted tubes failed in a similar way to complete tubes although there was some evidence of compressive buckling close to the slot. The normal frond/wedge/frond crush zone was less well defined close to the slot but was clearly visible around the rest of the crush section.

The load–displacement traces produced by slotted specimens were similar to those of complete tubes (Fig. 2). However in some cases the average crush load was slightly lower than that of the comparable complete tube.

The sectioned tubes all failed by progressive crush. Buckling at the cut edges was evident in all cases and significant bending of $\frac{1}{4}$ section specimens was observed. Failure produced well defined frond/wedge/frond fracture zones away from the cut edges but close to these edges all the debris tended to pass to the inside or outside of the specimen.

Typical load–displacement traces for tube sections are shown in Fig. 7. For the $\frac{3}{4}, \frac{1}{2}$ and $\frac{1}{4}$ section specimens the average crush loads are slightly less

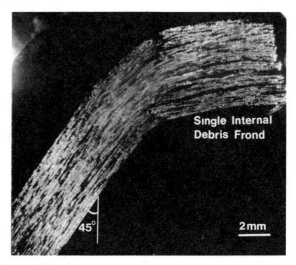

Fɪɢ. 6.   Polished section through crush zone of cone with $\Theta = 45°$.

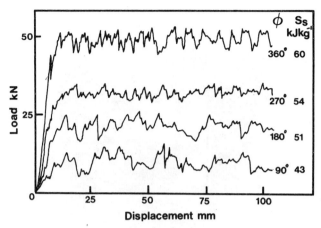

FIG. 7. Load–displacement traces for tube sections and complete tube.

than the relevant fractions of $\bar{P}$ for the complete tube. The specific energy absorbed for progressive crush is proportional to the average crush stress (eqn. (1)), thus the specific energies absorbed by slotted tubes and tube sections are lower than those for complete tubes. Values of $S_s$ for tube sections are given in Fig. 7.

**Flat Plates**

None of the specimens tested failed by progressive crushing comparable with that observed in tubes. Failure was normally catastrophic involving buckling of the test piece, an example is shown in Fig. 8. Failure has occurred by a combination of delamination (labelled A) and compressive shear (B).

The load displacement traces were not systematic and it was not possible to deduce any values of average crush load. In all cases the load rises to a maximum value and then drops to an insignificant level as catastrophic failure occurs.

**Square Tubes**

Square tubes failed by progressive crushing from the triggered end. A load displacement trace for a specimen with the smallest corner radius is shown in Fig. 9. As for round tubes the load rises, and then fluctuates about an average value. The degree of load fluctuation increased with decreasing corner radius. The average crush loads for square tubes were about

FIG. 8.   Plate specimen after test showing delamination (A) and compressive shear failure (B).

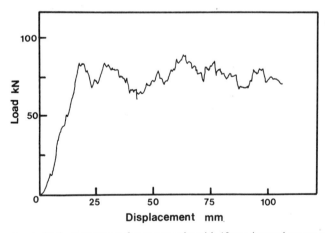

FIG. 9.   Load–displacement trace for square tube with 12 mm internal corner radius.

10–20% lower than those of round tubes with the same cross-sectional area, and hence the specific energy absorption values showed a corresponding reduction. The failure mode of the square tubes varied around their cross-sections. The corner regions produced well formed frond/wedge/frond crush zones, however, the sides of the tubes showed some buckling during testing and the crush zone was larger in these regions extending up to 20 mm from the crush platen.

## DISCUSSION

There are two important features of the results of tests on cones which relate to the use of composite materials in energy absorbing structures.

(1)  Cones with a wide range of geometries fail by progressive crushing, absorbing large amounts of energy comparable with values for axisymmetric cylinders made of the same material.

(ii)  Progressive crushing occurs without chamfers because fracture initiates at the narrow end of the cone in the region of highest stress, i.e. cones are self-triggering.

Comparison of energy absorption levels of different geometry cones and tubes can be made on the basis of a variety of parameters, for example wall thickness $(t)$, section diameter $(D)$ or the ratio of $t/D$. Previous workers have attempted to deduce whether there is a simple geometric factor that controls energy absorption, and some results suggest that the ratio $t/D$ is the controlling variable.[2] Results for $\Theta = 11°$ cones and axisymmetric tubes are shown in Fig. 10 compared on the basis of $t/D$. Each cone test covers a range of $t/D$ values since the section diameter increases as crush progresses, and by testing cones with different wall thicknesses a given $t/D$ value can be achieved at various diameters. These results show that energy absorption is not a function of $t/D$ alone and cone and tube results only in agreement for specimens which have the same $t$ and $D$.

The specific energy absorption of thick walled ($> 8.0$ mm) cones with $\Theta = 22°$ exceeds values for axisymmetric tubes. It is not clear why this is so, however there are variations between the crush zone morphologies of tubes and cones in particular the degree of frond deformation. Friction between the crush platen and debris is thought to absorb considerable amounts of energy.[3,4] Due to the increase in diameter with displacement the failure zone moves relative to the platen as crush proceeds, thus friction effects may

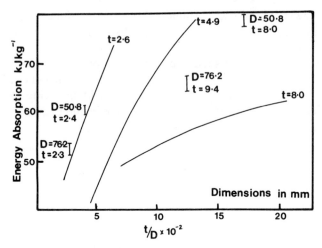

FIG. 10. Variation of specific energy absorption with $t/D$ for cones ($\Theta = 11°$) and axisymmetric tubes.

be greater for cones than for tubes. A complete investigation of these processes is not possible at present, a more detailed discussion has been published elsewhere.[6]

The most striking result of the tests on sectioned tubes, square tubes and flat plates is the fact that incomplete tubes, such as $\frac{1}{4}$ sections, and the sides of square tubes failed by progressive crush whilst all the plate specimens tested failed in a catastrophic manner. What are the vital differences between small sections of tube walls and flat plate specimens? Two criteria must be met if progressive crushing is to be developed. Firstly the structure must be stiff enough to resist buckling and catastrophic failure. This will depend on specimen height, wall thickness and cross-sectional geometry. It is apparent that the plate specimens were most likely to undergo compressive buckling especially since they were tested without any base support. The sides of square tube specimens did buckle to some extent but this buckling was limited by the rigidity of the corner regions and did not prevent stable crushing of these flat regions.

Short plate specimens did not fail by crushing even though buckling did not occur, highlighting the second criteria that must be met to produce stable crush. For a localised crush zone to exist, the central intrawall crack must extend only a short way ahead of the crush front. A schematic section through a crush zone is shown in Fig. 11. If the central crack extends too far the debris fronds will assume a greater radius of curvature with reduces the

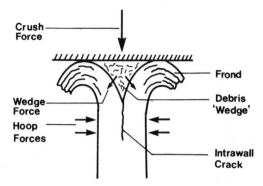

FIG. 11.   Schematic diagram of frond/wedge/frond crush zone.

amount of microfracture and energy absorption.[3,4] In a flat plate specimen such a cracking leads to catastrophic collapse.

The extent of intrawall fracture will depend on the transverse tensile strength of the laminate and the magnitude of forces tending to split the laminate open and forces opposing this splitting. Keal[3] suggested that splitting forces are caused by the debris wedge forcing the fronds apart, whilst restraining forces are generated by circumferential hoop deformation in the outer layers of a tube and compressive hoop forces on the inner face (see Fig. 11). Obviously the magnitude of these forces will depend on the size of the bundle wedge, tube dimensions and elastic properties. It is interesting to note that for certain geometries of epoxy glasscloth tubes there is no evidence of intrawall cracking below the crush zone.[5] In plate specimens there will be no hoop restraints and only the transverse tensile strength can oppose splitting. Thornton carried out compression tests on small plate specimens made from oriented glass epoxy,[5] and found that for this material stable crushing does occur. Presumably in this case the relative magnitudes of transverse tensile strength, frond stiffness and the crack opening forces lead to stable crushing. A more detailed consideration of the stresses in the crush region is necessary in order to investigate the conditions for stable crushing further.

## CONCLUSIONS

1.   For a wide range of geometries truncated conical shells made from csm/polyester laminate fail by progressive crush from the small end. No trigger is necessary to initiate crush.

2. The energy absorbed during crushing of cones varies in a complex way with cone angle, diameter and wall thickness and in some cases exceeds values recorded for axisymmetric tubes.
3. Slotted and sectioned tubes, and square tubes can fail by progressive crush, however the energy absorption levels are reduced compared with complete tubes.
4. None of a wide range of plate specimens fail by crushing when compressed. This is due to their inability to generate the stresses necessary to support a localised crush zone. Consideration of the stresses in the crush region is necessary in order to investigate the conditions for stable crushing further.

## ACKNOWLEDGEMENTS

The authors are grateful to the Polymer Engineering Directorate of the Science and Engineering Research Council, British Petroleum plc, Pilkington plc and the Ford Motor Company for financial support.

## REFERENCES

1. HULL, D., Axial crushing of fibre reinforced composite tubes, in: *Crashworthiness* (Jones, N. and Wierzbicki, T. eds), London, Butterworths, 1983, pp. 118–135.
2. HULL, D., Energy absorption of composite materials under crash conditions, in: *Progress in Science and Engineering of Composites, Proceedings of ICCM IV* (Hayashi, T., Kawata, K. and Umekawa, S., eds), Tokyo, Pergamon Press, 1982, pp. 861–870.
3. KEAL, R., Post failure energy absorbing mechanisms of axially crushed GRP tubes, Ph.D. thesis, University of Liverpool, 1984.
4. FAIRFULL, A. H., Scaling effects in the energy absorption of axially crushed composite tubes, Ph.D. thesis, University of Liverpool, 1986.
5. THORNTON, P. H., The crush behaviour of glass fibre reinforced plastic sections, *Comp. Sci. Technol.*, **27** (1986), 199–223.
6. PRICE, J. N. and HULL, D., Axial crushing of glass fibre-polyester composite cones, *Comp. Sci. Technol.*, **28** (1987), 211–230.

# 3

# An Experimental Investigation into Damage Propagation and its Effects Upon Dynamic Properties in CFRP Composite Material

R. C. DREW and R. G. WHITE

*Structures Group, Institute of Sound and Vibration Research, University of Southampton, Southampton SO9 5NH, UK*

## ABSTRACT

*With present concern being directed towards fatigue performance of CFRP materials, it is important to understand the causes of damage initiation and propagation, damage mechanisms and the effect of flaws upon material properties. Such information is needed in order to identify what actually constitutes 'failure' in CFRP materials, so that this may be adequately defined and reliable design criteria established. This chapter first describes a finite element (FE) analysis to predict damage-causing stresses and a suitable specimen configuration, and, thereafter, investigates the effect upon material properties of small specimens as damage progresses. The principal type of damage found was delamination, which caused an increase in damping and a reduction in static stiffness and natural frequency. Scanning electron microscopy, ultrasonic scanning and thermography techniques were used to investigate damage type and mechanism.*

## 1. INTRODUCTION

Present concern is directed towards the dynamic performance of CFRP materials when subjected to a combination of dynamic excitation in flexure and static in-plane stress. However, before such behaviour can be investigated it is first necessary to understand the simple case, where such material is subjected to fatigue in flexure with no in-plane stresses. In particular, in an effort to define 'failure', it was felt important to investigate:

(1) the levels of strain at which damage occurred;
(2) the number of fatigue cycles needed to induce damage at a given level of strain;
(3) the rate of damage propagation at a given level of strain;
(4) the type of damage and damage mechanisms that occurred;
(5) the effect of damage upon natural frequency, damping and static/ dynamic stiffness of the specimen.

The initial problem was to find a way in which damage through dynamic flexure could be induced in a small specimen whilst not also inducing undesirable edge delamination. It was felt that this could be accomplished by varying geometry of the clamping edge such that stresses were concentrated at the centre of the specimen and reduced at the edge. The first stage therefore involved a theoretical analysis to try to establish the best method of fatigue testing.

## 2. THEORETICAL ANALYSIS

Since it was necessary to undertake a 3-D analysis, to take the different laminar orientations into account, and to find stress variations over an entire 'cantilever' specimen which had a complicated clamping boundary shape, the finite element method was adopted. Using the ANSYS FE program, various boundary geometries were tried, in order to assess that which would be most likely to cause damage initiation in the middle of the specimen without simultaneously causing edge delamination. The specimen dimensions were: length 90 mm, width 70 mm and thickness 1 mm. The model has eight quadrilateral elements in its thickness with appropriate material properties appertaining to a $[0°, +45°, -45°, 0°]_s$ lay-up for XAS/914 CFRP. From scanning electron microscopy (see later) it was observed that thin resin-rich layers existed between laminae, having a thickness approximately one-tenth that of a lamina, but the consequent size of the FE model precluded the possibility of including this feature. Thus each layer thickness was modelled as being 0·125 mm. Stresses were calculated on the basis of a downward static deflection of 5·0 mm.

### 2.1. Results from the FE Analysis

Typical results are given in Figs 1, 2 and 3. Results were plotted as stress contours to show distribution across the surface, and as 3-D line plots to indicate stress distributions across the width and through-the-thickness for

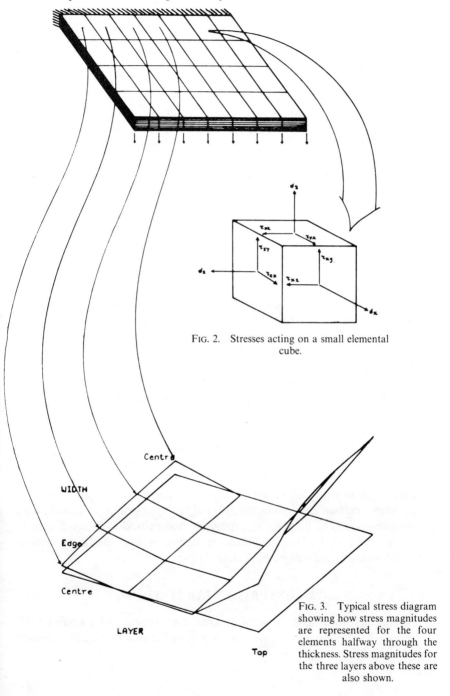

FIG. 1.   Finite element model of straight-edge
clamped cantilever showing element layers.

FIG. 2.   Stresses acting on a small elemental
cube.

Centre

WIDTH

Edge

Centre

LAYER

Top

FIG. 3.   Typical stress diagram
showing how stress magnitudes
are represented for the four
elements halfway through the
thickness. Stress magnitudes for
the three layers above these are
also shown.

the four element stacks nearest the clamped edge (see Fig. 1). The stress values of greatest interest were the direct $Y$ stress and the $XY$ shear stress. At the edge of the two specimens the $Y$ stresses, whilst being of similar magnitudes, were highest at the core for the straight-edge clamped case. However, the reverse was true for the half-sine clamped case. Also, $Y$ stresses became negative at the top and central layers. The $XY$ shear stresses were, surprisingly, around ten times less for the straight-edge clamped condition, although, through-the-thickness, there was an increase towards the core, reaching a maximum in the second layer up. This variation was not apparent in the half-sine clamped case.

At the middle portion of both specimens $Y$ stresses were of similar magnitudes. However, $XY$ stresses were around three times greater for the half-sine clamped case, being maximum nearest the core. This increase did not occur for the straight-edge clamped condition.

Without knowledge of the relative effect of the various stresses, it was difficult to predict whether or not the half-sine clamping would be a certain success and that it would be possible to induce delamination only at the middle of the specimen. High central $XY$ shear stresses were predicted for the half-sine clamped specimen, although they were lower at the edge than for the straight-edge clamped case. Edge variations showed that, for the straight-edge condition, both $Y$ and $XY$ stresses reached maximum values near the core. Since $Y$ stresses were of around a magnitude greater than the $XY$ stresses, it was concluded that they may be more important, causing an edge 'peeling' effect, for it was notable that in the half-sine clamped condition $Y$ stresses were low near the core and shear stresses were high. It is interesting that $Y$ stresses became negative at the top and central layers, tending to further increase any 'peeling' effect at the centre.

In order to verify the findings, it was decided to run an experimental trial. The results of this showed quite clearly that delamination could be induced at the centre, whilst edge delamination was prevented. It therefore appears that $Y$ stresses are significant in causing a 'peeling' effect, although it is perhaps a certain level of combination of both stresses that results in edge delamination. The half-sine clamping method proved to be an extremely effective way of obtaining delamination damage in small specimens without simultaneously inducing edge delamination.

## 3.  EXPERIMENTAL METHODS

Having first measured the static stiffness and damping of a small CFRP specimen, with dimensions given as above, and also ultrasonically scanned it to identify any initial flaws, the specimen was mounted, one end being held in a half-sine clamp and the other being lightly pinched between two rollers. The rollers were connected to a shaft, which in turn was connected to a $\frac{1}{2}$-hp motor and cam. As the rollers are only able to move vertically, the specimen could thus undergo bending fatigue with no in-plane stress, since it was able to slide between the rollers. Different displacements, and thus strain amplitudes, could be selected by varying the throw of the cam. Peak strain level was recorded by means of strain gauges attached to the specimen.

After subjecting the specimen to a given number of fatigue cycles, the strain gauges were removed, and it was re-scanned. Static stiffness, damping and damage area, if present, were all measured. Static stiffness was simply obtained by clamping the specimen in another half-sine clamp, hanging a series of weights at the free end, and measuring tip displacement with a dial gauge. Graphs of deflection versus load were then drawn, the relationship being found to be progressively more curvilinear, as shown in Fig. 4. Since changes in deflection with damage were more noticeable for deflection at high loads, the gradient, i.e. stiffness, was measured at the high-load end of the curve.

Damping measurements were accomplished by again attaching the specimen to a half-sine clamp and attaching a magnet to the free end. Below was positioned a coil, and thus the specimen/magnet system could be excited at its fundamental natural frequency. This frequency was recorded, and the current then switched off to allow the specimen vibration to decay. Since there were no connecting wires or drive shafts attached to the specimen, which would cause additional damping, it was hoped that good measurements would be obtained. The decaying signal was acquired and damping was calculated by the logarithmic decrement method. Readings were obtained from successive peaks and averaged to obtain the most representative value. Throughout the experiment, readings were also taken from a control specimen.

In order to investigate damage mechanisms, some scanning electron microscopy was carried out. When necessary, specimens were prepared by grinding on a series of fine pastes before being ultrasonically cleaned. Specimens were then treated *in vacuo* with a gold/platinum coating to ensure good electron conduction.

### 3.1. Experimental Results

In this fatigue programme, a total of 14 specimens were tested. These, along with their associated strain levels, are given in Table 1.

TABLE 1

| Specimen | Peak sinusoidal strain level (μ-strain) | Specimen | Peak sinusoidal strain level (μ-strain) |
|---|---|---|---|
| 1A | 4 000 | 1C[a] | 7 000 |
| 2A | 4 000 | 1D | 7 800 |
| 1B[a] | 6 000 | 2D | 7 800 |
| 2B | 6 000 | 1E[a] | 8 000 |
| 3B | 6 000 | 1F | 8 400 |
| 4B | 6 000 | 2F | 8 400 |
| 5B | 6 000 | 1G[a] | 9 000 |

[a] Here the specimen was cut from a plate made using the autoclaving method; otherwise the specimen was cut from a plate made using vacuum bags and hot-platten pressing.

### 3.1.1. Effect of fatigue upon damage area

Figure 5 shows a graph of damage area versus number of fatigue cycles for the 14 test specimens. From these results, the following observations may be made.

(a) Once damage begins, it progresses at a uniform rate, regardless of the level of strain, although this rate appears to be dependent upon the manufacturing process of the material, since it was greater for the autoclaved specimens than for those made by pressing. However, it is

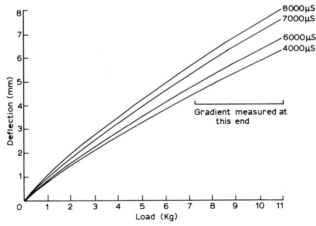

FIG. 4.    Typical deflections versus load diagram for four CFRP cantilever specimens fatigued at different strain levels.

important to recognise that fatigue cycling was carried out at constant displacement rather than at constant strain. Once damage initiates, the internal stress state changes. Since this is impossible to measure, strain was only measured initially on the surface near the clamp tip and was assumed constant during the experiment.

(b) The lower the strain level, the later the onset of damage. At the high strain levels damage was clearly visible after 100 cycles, but at the lower levels damage was not visible until after 10 000 cycles. In specimens 1A and 2B it was not even visible after 100 000 cycles. This leads to the possibility of the existence of a fatigue stress limit for CFRP materials.

(c) Typically, the damage shape was elliptical, the shape being preserved as damage advanced. However, at higher strain levels, fibre tearing was visible at the surface, leading to the development of a crescent-shaped damage zone. Damage tended to spread outwards around the clamping edge rather than forwards. This difference did not, however, lead to a reduction in overall damage area, which still increased at about the same rate as in samples with elliptically-shaped damage areas.

### 3.1.2. *Effect of fatigue upon static stiffness*
Static stiffness was chosen as a simple material parameter which might be used to give a reliable indication of damage within a CFRP material.

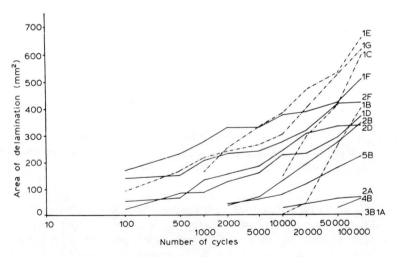

FIG. 5. Graph showing delamination area versus number of cycles for 14 CFRP cantilever specimens.

Measurements taken for each specimen at each of the ten fatigue cycle increments showed noticeable reductions in static stiffness with increasing strain levels (see Fig. 6). Such changes are obviously linked to the size of the damage area within the specimen. In order to assess the relationship between static stiffness and damage area, correlation coefficients were calculated separately for autoclaved and pressed specimens, using the following relationship:

$$r = \frac{n.\Sigma xy - \Sigma x.\Sigma y}{\sqrt{\{n.\Sigma x^2 - (\Sigma x)^2\}\{n.\Sigma y^2 - (\Sigma y)^2\}}}$$

where $x$ = static stiffness and $y$ = damage area. For pressed specimens, this value was $-0.85$, indicating a strong inverse correlation. For the autoclaved specimens, the value was $-0.81$.

An interesting observation is the apparent variation between specimens made using the autoclaving method and those made by hot-platten pressing. With the former, static stiffness values fell dramatically after between 1000 and 10000 cycles, whereas no such sudden change was exhibited by the latter. This may have important bearings upon the influence of different fabrication processes upon fatigue performance.

### 3.1.3. Effect of fatigue upon damping

Damping results are shown in Fig. 7. Again, an interesting discrepancy was shown between autoclaved and pressed specimens. In the former, there seemed to be a sudden increase in damping with increase in the number of fatigue cycles, followed by a fall, whereas in the latter, a gradual overall increase was observed.

Whilst damping values were relatively high, it should be remembered that they were not material values. The small specimen size necessitated clamping, and some frictional damping inevitably occurred at the clamp. A free–free test, only suitable for beam-type specimens, yields much more accurate results and will be adopted for use with 'coupon'-type specimens in subsequent work.

### 3.1.4. Effect of fatigue upon specimen fundamental natural frequency

Although changes were only of a few Hertz (see Fig. 8), they are clearly visible, being more significant for the autoclaved specimens. Changes were generally greater for specimens cycled at higher strain levels, indicating a positive relationship with stiffness, as would be expected.

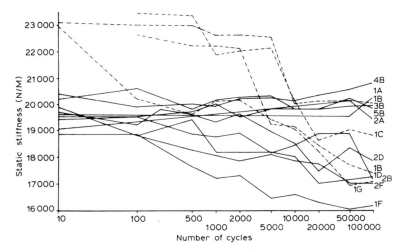

Fig. 6. Graph of static stiffness versus number of fatigue cycles for 14 CFRP cantilever specimens.

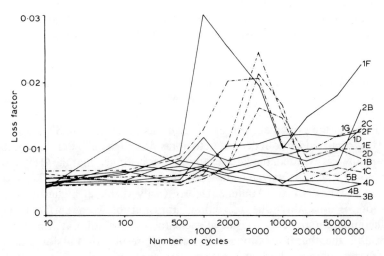

Fig. 7. Graph of loss factor versus number of fatigue cycles for 14 CFRP cantilever specimens.

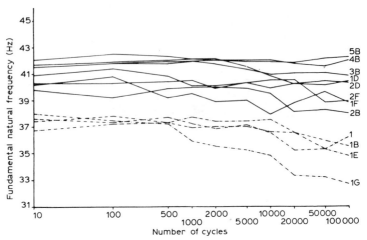

FIG. 8.    Graph showing natural frequency versus number of cycles for 14 CFRP cantilever specimens.

## 3.2. Examination of Damage Mechanisms

### 3.2.1. Thermography

Results from thermography tests showed that as fatigue cycling proceeded a 'hot spot' developed after about 4 s and reached a maximum of $21{\cdot}1\,°C$ at around 40 s. The peak level of strain was 8000 $\mu$Strain. This would indicate a high stress concentration in the material near the clamp tip, heat dissipation probably being especially due to high shear stresses.

### 3.2.2. Scanning electron microscopy

In order to examine damage mechanisms, scanning electron microscopy (SEM) specimens were prepared and photographed (see Fig. 9). Explanations are given on the individual photographs. These photographs clearly show that the principal form of damage was delamination. Separation of layers occurred without detriment to fibres, although on one crack face they were always exposed. In all the examinations it seemed that the O layer was exposed, the −45° layer being still covered by the resin-rich layer that always exists between laminae. Debris, easily visible at higher magnification, is probably due to sliding abrasion between surfaces. That delamination arose as a result of shear forces is corroborated by the FE results, the thermography results and the results of Purslow,[1] whose electron micrograph of a typical interlaminar shear failure shows marked similarity to those shown.

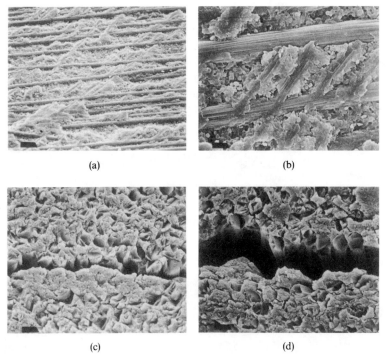

(a)                 (b)

(c)                 (d)

FIG. 9. Scanning electron micrographs of delamination damage. (a) View of delamination face. Unbroken fibres and the nature of the resin are clearly visible (original magnification × 500). (b) At higher resolution indentations left by fibres which have been pulled from the resin are clearly visible (original magnification × 2000). (c) Edge on view showing the resin-rich interfacial layer with the delamination along one side (original magnification × 1000). (d) Edge on view nearer centre of delamination (original magnification × 1000).

## 4. CONCLUSIONS

It is clear that the presence of damage within a composite material affects its engineering properties. From the point of view of the use of such properties to characterise failure it would appear that, based upon specimen observations, changes in damage area, stiffness, damping and natural frequency could all be used. The first two properties would give more dependable indications as they are easier to measure and changes are more significant. Fatigue performance could probably best be defined in terms of a percentage change in a given property over a fixed number of cycles at a given level of strain.

The observation that damage initiation occurs later as strain level is reduced indicates the possibility that a fatigue limit exists for the material. The lower the sinusoidal strain level, the later the onset of damage. After initiation, damage progresses at a uniform rate, regardless of strain level. However, this rate is different for autoclaved and platten-pressed specimens, it being greater for the former. This suggests that platten pressing is preferable if damage propagation is to be retarded.

## ACKNOWLEDGEMENTS

The authors would like to thank the RAE Farnborough for their sponsorship of this work via an unclassified research agreement. Considerable thanks are also due to Mr R. F. Mousley of the RAE for his guidance and encouragement throughout the programme of work.

## REFERENCE

1. PURSLOW, D., Matrix fractography of fibre-reinforced epoxy composites, *Composites*, **17** (1986), 289–303.

# 4

# The $T^*$-Integral and $S$ Criteria for the Analysis of Impact Damage at Fastener Holes in Graphite/Epoxy Laminates under Compression

T. E. TAY, J. F. WILLIAMS

*Department of Mechanical and Industrial Engineering,*
*University of Melbourne, Australia*

and

R. JONES

*Structures Division, ARL, Defence Science and Technology Organisation,*
*Box 4331, Melbourne, Victoria 3001, Australia*

## ABSTRACT

*This chapter illustrates the application of two fracture criteria, Atluri's $T^*$-integral and Sih's strain energy density factor S, for estimating the residual strength of an impact damaged fastener hole in a composite laminate. Finite element analyses are performed, and the magnitude and distribution of $T^*$ and S are determined around the delamination. It is found that the profiles of these distributions are extremely complex and sensitive to the modelling of the stress fields close to the delamination. The distribution of both $T^*$ and S is such that three local maxima occur and are situated at the same locations around the delamination. These locations approximately coincide with the points of maximum growth as revealed by ultrasonic C-scan of the damage growth of several specimens. The $T^*$-integral and S criteria may be successfully employed in the prediction of residual strength. However, S is better at predicting the direction of damage growth provided the stress fields near the delamination can be accurately modelled.*

## INTRODUCTION

Advanced laminated composite materials have been successfully and increasingly used in recent years for the construction of aircraft components. They offer many advantages over conventional materials in that they afford significant weight reductions, versatility, high stiffnesses and strengths. However, in order to improve and aid in the design and in-service maintenance of composite aircraft components, it is necessary that the damage tolerance and the failure mechanisms of these components be fully understood.[3-5] Low-energy impact by hard objects on the surface of these components can result in damage in the form of delaminations between the plies. Although the damage may not be detectable by the human eye (the only external indication is often a slight surface indentation), there can be a significant loss of compressive strength. This type of damage, called barely visible impact damage (BVID), has received considerable attention from several investigators.[3,6]

Fracture mechanics concepts which have been successfully used to predict failure for cracks in isotropic homogeneous metals must be used with caution when applied to composite structures. In this chapter, two fracture criteria, the strain energy density factor $S$ proposed by Sih[7] and the recent $T^*$-integral proposed by Atluri,[15] are applied to a specific problem of a composite laminated plate with an impact damaged fastener hole under compression.

## THE $T^*$ APPROACH

The path-independent $J$-integral of Rice[8,9] is denoted by

$$J = \int_{\Gamma} [Wn_1 - t_i u_{i,1}] \, ds \qquad (1)$$

where $\Gamma$ is any closed path surrounding the cracktip, $W$ is the strain energy defined as

$$W = \tfrac{1}{2}\sigma_{ij}\varepsilon_{ij} \qquad (2)$$

$n_i$ is the component of a unit normal to the path in the $x_i$ direction, $t_i$ is the traction vector defined as $t_i = n_j\sigma_{ij}$, and $u_i$ is the component of displacement. In this work we evaluate a modified form of the $J$-integral, which is defined as $T^*$ (Ref. 15) and where

$$T^* = \lim_{\varepsilon \to 0} \int_{\Gamma_\varepsilon} [Wn_1 - t_i u_{i,1}] \, ds \qquad (3)$$

Here the path $\Gamma_\varepsilon$ is a vanishingly small distance from the crackfront and in three dimensions must be normal to the crackfront. An important characteristic of $T^*$ is that it is evaluated near the cracktip while the traditional $J$ is really a far-field parameter. In low cycle fatigue analyses, this distinction becomes significant since reverse plastic deformation can occur upon unloading. Since $T^*$ is evaluated near the tip, it is claimed[15] that it is able to account for the near tip effects while $J$ cannot. In elastic analyses, $W$ is defined as in eqn. (2) for both $T^*$ and $J$. However, in the analyses of composite delaminations, it is important to be able to account for local fracture events along a three-dimensional crackfront. In elastic analyses, for a through-the-thickness crack propagating under mode $I$ fracture in a self-similar fashion, $J\,(=T^*)$ is equivalent to the classic energy release rate $G$, which is defined as

$$G = -\frac{\delta\pi}{\delta A} = -\frac{1}{B}\frac{\delta\pi}{\delta a} \tag{4}$$

where $\delta\pi$ is the change in potential energy per unit thickness of the system and $\delta a$ the increment of crack length. When measuring $G$ experimentally, eqn. (4) is often written as

$$G = \frac{P^2}{2B}\frac{dc}{da} \tag{5}$$

where $P$ is the applied load, $c$ the compliance as measured by movement of the loadpoints, and $B$ the specimen thickness. However, for three-dimensional delaminations, the growth is usually not self-similar and therefore eqn. (5) is invalid. Furthermore, eqns (4) and (5) are global quantities and do not provide information on near tip events such as local crack closure. For these cases, measurements of compliances do not yield true values of the cracktip energy release rates.

The assumption of self-similar crack growth is still inherent in the $T^*$-integral formulation. Although this assumption is incorrect for three-dimensional analyses of delamination growth, it can be said that for points of maximum crack growth along the crackfront, there is sometimes a 'local self-similarity', meaning that locally the crackfront remains parallel. Hence the $T^*$-integral may be useful in identifying points of maximum crack growth and in estimating the crack initiation load. At points other than the positions of local maximum crack growth, $T^*$ and other similar approaches cannot be said to give the local energy release rate.

## THE STRAIN ENERGY DENSITY METHOD

In the analysis of cracks in composite materials, the point stress failure criterion is often used.[16,17] In this approach, failure is said to occur when $\sigma$, the stress component perpendicular to the crack plane at a distance $r_0$ ahead of the cracktip, reaches the unnotched strength $\sigma_{oc}$ of the laminate. The distance $r_0$ is considered to be a material parameter and is determined experimentally. In this work, $r_0$ is taken to be 1 mm for the T300/5208 material system.[23] In delamination problems, however, it is necessary to consider the effects of interlaminar shear stress in addition to the peel stress. Since the matrix material is often said to only carry the shear and peel stresses, and since these stresses should be continuous across each ply interface,[10] we evaluate the quantity

$$\frac{\mathrm{d}\bar{W}}{\mathrm{d}V} = \frac{1}{2E_m}[\sigma_3^2 + 2(1 + v_m)(\tau_{13}^2 + \tau_{23}^2)] \tag{6}$$

which may be thought of as the strain energy density (i.e. $\mathrm{d}W/\mathrm{d}V$) of the matrix material. Here, $E_m$ and $v_m$ are the moduli and Poisson's ratio of the matrix material respectively. Equation (6) may be expressed in terms of the strain energy density factor, $S$, for the matrix material, where

$$S = r_0\left(\frac{\mathrm{d}\bar{W}}{\mathrm{d}V}\right) \tag{7}$$

This approximation will be discussed in more detail in a subsequent chapter. Using this interpretation for $S$, the strain energy density theory of failure as prepared by Sih[7] may then be used. The theory states that crack initiation occurs in a direction determined by the conditions

$$\frac{\delta S}{\delta \theta} = 0 \quad \text{and} \quad \frac{\delta^2 S}{\delta \theta^2} > 0 \quad \text{at } \theta = \theta_0 \tag{8}$$

i.e. in the direction of minimum strain energy density factor $S_{\min}$. Crack extension occurs when the minimum strain energy density factor reaches a critical value $S_c$, which is considered a material constant and a measure of its fracture toughness. Applied to delamination problems, the local minimum strain energy density factor $S_{\min}$ at $\theta = \theta_0$ must be determined for each position along the delamination front. The delamination is assumed to grow when the maximum of the minimum strain energy density factor, $S_{\min}^{\max}$, reaches the critical value $S_c$.

The purpose of this work is to apply both the $T^*$-integral and the strain energy density theory to a delamination problem in a composite laminate where the load to cause delamination growth is known, and to compare the predictions with available data.

## THE FINITE ELEMENT MODEL

Details of the finite element mesh used in this work are shown in Fig. 1. It is a model of an impact damaged laminate with a fastener hole under compression. The dimensions of the model are the same as those used in the experimental work of Lauraitis *et al.*[1] The specimen tested was a $[0/45/0_2/-45/0_2/45/0_2/-45/0]_s$ T300/5208 graphite–epoxy laminate and contained a centrally located hole at 9·5 mm diameter, surrounded by delamination damage due to impact and poor drilling. The elements used are mostly 20-node isoparametric elements with directionally reduced integration and $2 \times 2 \times 3$ Gaussian quadrature points, with the three points being taken through the ply thickness. Detailed description of the reduced integration scheme can be found in Ref. 16. The cracktip elements along the circular delamination are 15-node isoparametric wedge elements.

The initial damage around the fastener hole from Ref. 1 is modelled as a circular delamination of radius 13·75 mm between the second and third plies (i.e. between the 45° and 0° plies). This is an approximate simulation of the initial damage as shown in Fig. 2a. It can be seen from the ultrasonic C-scan that the initial delamination is nearly circular.

The two plies above and below the delamination are modelled separately with ordinary three-dimensional elements while the remaining 20 plies are modelled with super-elements with displacements varying quadratically in the local isoparametric coordinate system. The material properties used are those of Narmco-Thornel T300/5208.

It is important that in the FE model the faces of the delamination are prevented from overlapping. Otherwise non-physical solutions may be

TABLE 1
*Properties of matrix material*

| $E_m$(MPa) | $v_m$ |
|---|---|
| 3 500 | 0·3 |

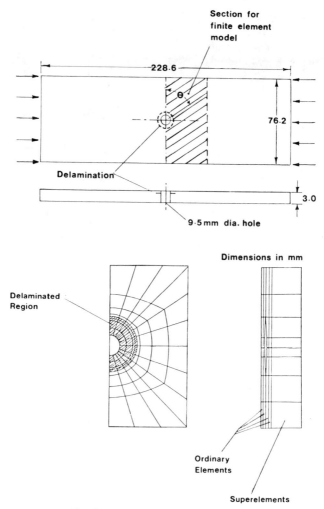

FIG. 1.    Details of finite element model.

obtained. By examining the solutions of the displacements, it is found that some parts of the delaminated faces have overlapped. Thus a series of constraint equations are applied to appropriate nodes to simulate local closure. A compressive strain of 0·006 is applied at the ends of the model. This value of strain represents the experimental measured, far-field failure strain for this specimen.[1]

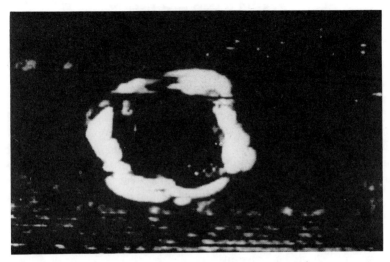

FIG. 2(a).   C-scan of initial damage from Ref. 1.

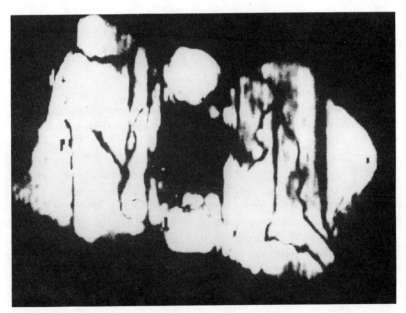

FIG. 2(b).   C-scan of final delamination growth after 20 000 cycles from Ref. 1.

## RESULTS AND DISCUSSION

An interesting characteristic of both the distribution of $S_{min}$ and $T^*$ in this particular problem is the existence of three peaks along the circular delamination front. The positions of these local peaks in the $T^*$ distribution coincide closely with corresponding peaks in the $S_{min}$ distribution (see Figs 3 and 4). The $T^*$ distribution has the largest peak located at $\theta = 60°$, where $\theta$ is defined in Fig. 1. The value of $T^*_{max}$ for the correct restraint condition is 70·8 J/m². This value compares favourably with the $G_{Ic}$ value of 76 J/m² for neat 5208 Narmco resin obtained by Bascom *et al.*[12] Double Cantilever Mode I tests provide values of 78–88 J/m² for the $G_{Ic}$ of the T300/5208 material system.[20,21]

The value of $S_{min}$ reaches its maximum peak at $\theta = 90°$. Here the value of $S^{max}_{min}$ is 11·64 J/m², as compared with a published value of critical strain energy density factor of 9·9 J/m² (Ref. 7). It should be noted that when failure is dominated by failure of the matrix material, the value of $S_c$ can be related to $G_{Ic}$ by the expression[19]

$$G_{Ic} = \frac{2(1 - v_m)}{(1 - 2v_v)} S_c$$

An $S_c$ of 11·64 J/m² substituted into the above expression (and $v_m$ taken as 0·3) would yield a value of $G_{Ic}$ of 128 J/m². Interestingly, this value is close to the value of $G_{Ic}$ of 137 J/m² for edge delamination tensile test specimens for the T300/5208 system.[11,22] A value of $G_{Ic}$ of 156 J/m² for a comparable matrix material was reported by Mohlin *et al.*[24] in an analysis of a delamination problem. It thus appears that the $S_c$ obtained seems to correlate well with fracture toughness values obtained from edge delamination tensile tests while the $T^*_{max}$ values correspond with those obtained through double cantilever tests.

Figure 2b shows the C-scan of the final size and shape of the damage after the specimen has undergone 20 000 fatigue cycles. It can be seen that the points of maximum delamination growth coincide approximately with the three positions of local maximum of both $T^*$ and $S_{min}$. It should be noted that only one delamination has been modelled in this finite element analysis while in reality impact produces multiple delaminations close to the impacted surface.[5] This will be the topic of another chapter. It has been stated that the value of $G_{23}$ can have a significant effect on the fracture calculations and should be more accurately determined. Despite these uncertainties and assumptions, the results seem to indicate that the $T^*$ and $S$ criteria may be useful parameters in predicting delamination behaviour.

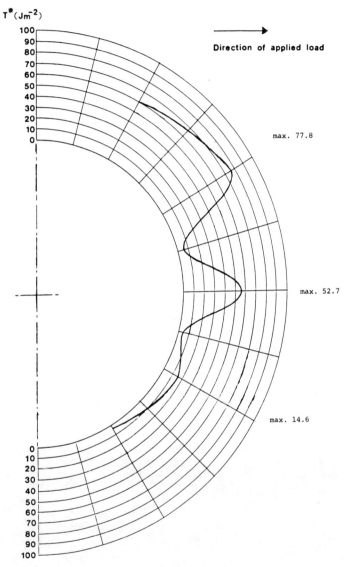

FIG. 3

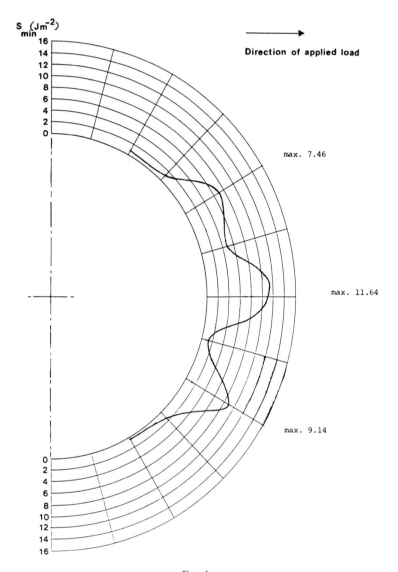

FIG. 4

## CONCLUSION

In the analysis of delaminations in impact damaged composite structures, it is important to consider the local fracture events along the delamination front since the global or far-field effects are inadequate for formulating a consistent failure criterion. The $T^*$-integral and $S$ theory are two fracture mechanics tools which may be used to characterize the neartip events. Their use, however, must be approached with caution since they rely on a reasonably accurate determination of stresses near the crackfront. The effects of simulating the stress singularity, the accurate determination of shear moduli and local crack closure are shown to be very important in the calculation $T^*$ and $S_{min}$.

The results of the analyses carried out in this work suggest that:

(a)  The $T^*$ and $S$ criteria may be useful fracture mechanics tools in the analysis of composite delaminations.

(b)  The effects of local crack closure (i.e. the closure of delamination faces) on the values of $T^*$ and $S_{min}$ are important; care must be taken in finite element analyses to avoid non-physical solutions through the use of appropriate restraints.

(c)  The peaks in the strain energy density approach correlate well with the directions of observed growth.

Further details concerning the effects of quarter/point shift, different interlaminar shear moduli and crack closure can be found in the companion paper.[25]

## ACKNOWLEDGEMENT

This work was undertaken as part of the Commonwealth Advisory Aeronautical Research Council Co-operative program on composite materials.

## REFERENCES

1.  LAURAITIS, K. N., RYDER, T. and PETTIT, D. E., Advanced residual strength degradation rate modelling for advanced composite structures, Vol. II, AFWAL-TR-79-3095, 1981.
2.  JANARDHANA, M. N., BROWN, K. C. and JONES, R., Designing for tolerance to impact damage at fastener holes in graphite/epoxy laminates under compression, *Theor. appl. Fract. Mech.*, **5** (1986), 51–55.

3. BAKER, A. A., JONES, R. and CALLINAN, R. J., Damage tolerance of graphite/ epoxy composites, *Composite Structures*, **4** (1985), 15–45.
4. JONES, R., PAUL, J. and BROUGHTON, W., On the effects of delamination damage in fibre composite laminates, ARL Structures Report 403, June, 1984.
5. KINLOCK, A. J., Workshop on damage tolerance of fibre-reinforced composites, Glasgow, Scotland, 12 September, 1985.
6. MOUSLEY, R. F., In-plane compression of damaged composite laminates, in: *Structural Impact and Crashworthiness 2* (J. Morton ed.), Amsterdam, Elsevier Applied Science, 1984.
7. SIH, C. C., *Mechanics of Fracture 6: Cracks in Composite Materials,* The Hague, Martinus Nijhoff, XVI-LX-XXI, 1981.
8. RICE, J. R., Mathematical analysis in the mechanics of fracture, in: *Fracture— An Advanced Treatise,* Vol. 2: *Mathematical Fundamentals* (Liebowitz ed.), New York, Academic Press, 1968.
9. RICE, J. R., A path independent integral and the approximate analysis of strain concentration by notches and cracks, *J. appl. Mech.,* June (1968), 379–387.
10. WANG, S. S., An analysis of delamination in angle-ply fibre-reinforced composites, *J. appl. Mech.,* **47**, March (1980), 64–70.
11. O'BRIEN, T. K., Interlaminar fracture of composites, *J. Aero. Soc. India,* **37**, No. 1 (1985), 61–69.
12. BASCOM, W. D., BITNER, J. L., MOULTON, R. J. and SIEBERT, A. R., The interlaminar fracture of organic-matrix, woven reinforcement composites, *Composites,* January (1980), 9–18.
13. WHITNEY, J. M. and BROWNING, C. E., Materials characterization for matrix-dominated failure modes, effects of defects in composite materials, ASTM STP 836, 1984, pp. 104–124.
14. RUSSELL, A. J. and STREET, K. M., The effect of matrix toughness on delamination: static and fatigue fracture under Mode II shear loading of graphite fibre composites. Presented at NASA/ASTM Symposium on Toughened Composites, Houston, 13–15 March, 1985.
15. BRUST, F. W., MCGOWAN, J. J. AND ATLURI, S. N., A combined numerical/ experimental study of ductile crack growth after a large unloading, using $T^*$, $J$ and CTOA criteria, *Engng Fract. Mech.,* **23**, No. 3 (1986), 537–550.
16. JONES, R., CALLINAN, R., TEH, K. K. and BROWN, K. C., Analysis of multilayer laminates using three-dimensional super elements, *Int. J. Numer. Methods Engng,* **18** (1984), 583–587.
17. NUISMER, R. J. and LABOR, J. D., Applications of the average stress failure criterion: Part I—Tension, *J. Comp. Mater.,* **12** (1978), 238–249.
18. NUISMER, R. J. and LABOR, J. D., Application of the average stress failure criterion: Part II—Compression, *J. Comp. Mater.,* **13** (1979), 49–60.
19. SIH, G. C., Analysis of defects and damage in composites, *Advances in Fracture Research, Vol. 1, Proc. 6th Int. Conf. on Fracture (ICF6),* New Delhi, India, Oxford, Pergamon Press, 1984, pp. 525–548.
20. WILKINS, D. J., EISENMANN, J. R., CAMIN, R. A., MARGOLIS, W. S. and BENSON, R. A., Characterizing delamination growth in graphite/epoxy, damage in composite materials, ASTM STP 775, 1982, pp. 168–183.
21. CHAI, H., The characterization of Mode I delamination failure in non-woven, multidirectional laminates, *Composites,* **15**, No. 4, October (1984), 277–290.

22. O'BRIEN, T. K., Analysis of local delaminations and their influence on composite laminate behaviour. *Delamination and debonding of materials*, ASTM STP 876, 1985, pp. 282–297.

23. DAVIS, M. J. and JONES, R., Damage tolerance of fibre composite laminates, *Proc. Int. Conf. on Fracture Mechanics Technology Applied to Material Evaluation and Structure Design*, Melbourne, Australia, 10–13 August, 1982, pp. 635–655.

24. MOHLIN, T., BLOM, A. F., CARLSSON, L. A. and GUSTAVSSON, A. I., Delamination growth in a notched graphite/epoxy laminate under compression fatigue loading. *Delamination and debonding of materials.*

25. TAY, T. E., WILLIAMS, J. F. and JONES, R., Application of the $T^*$ integral and $S$ criteria in the finite element analysis of impact damage at fastener holes in graphite/epoxy laminates under compression, *J. Comp. Struct.* (in press).

# 5

# Strengthening Mechanisms in Discontinuous SiC/Al Composites

R. J. ARSENAULT

*Engineering Materials Group, University of Maryland,
College Park, Maryland 20742-2105, USA*

## ABSTRACT

*The various potential factors which could affect the yield strength of discontinuous SiC/Al composites, such as load transfer mechanisms, residual elastic stresses, differences in texture, etc., were considered. It was found that the high dislocation density and small subgrain size generated as a result of the difference in thermal coefficient of expansion between the SiC and Al was the major contribution to the strengthening. The classical load transfer mechanism and the texture difference had no effect on the strength. The thermal elastic residual stress was on average in tension (for whisker composite) and it reduced the tensile yield stress of the whisker composite.*

## INTRODUCTION

The framework of a mechanism is slowly evolving to account for the strengthening due to the addition of SiC to an Al alloy matrix. However, it should be kept in mind that the number of detailed investigations of these composites is rather limited.

The basic strengthening mechanism is the high dislocation density, which is produced as a result of the differences in the thermal coefficients of expansion between SiC and Al, and the small subgrain size that results.

There could be other contributing causes to the strengthening, such as:

1. Residual elastic stresses.
2. Differences in texture.
3. Classical composite strengthening (load transfer).

2.70

The classical models based on continuum mechanics models will have to be considered in detail.

Two types of continuum models seem to have been used extensively, the shear lag type and the Eshelby type models. The former model, which was originally developed by Cox,[1] is simple and has been used for prediction of stiffness,[2] yield stress,[3] strength and creep strain rate.[4] In the case of continuous whisker composite, a shear lag type model was also applied to prediction of load concentration factor successfully.[5] However, it is known that the properties predicted by the shear lag type model will become a crude approximation when the aspect ratio of the short fiber ($l/d$) is small or the short fibers are misoriented. Nardone and Prewo[3] have recently proposed a variation of the shear lag type model to obtain a larger estimated tensile yield stress of a short whisker MMC with smaller values of $l/d$.

On the other hand, in the Eshelby type model, the short whisker is assumed to be a prolate ellipsoidal inhomogeneity. The analytical model to predict the thermal and mechanical properties of a composite was first developed by Eshelby, who considered a single ellipsoidal inclusion or inhomogeneity embedded in an infinite elastic body,[6] thus it is valid only for a small volume fraction of fiber $V_f$. Mori and Tanaka[7] modified the original Eshelby model for a finite volume fraction. The modified Eshelby type models are stiffness,[8] yield stress and work-hardening rate,[9-11] thermal expansion[12,13] and thermal conductivity.[14] The Eshelby type model has also been used to predict the thermal residual stress in a composite.[10,13,15] The detailed summary of the Eshelby type models is given in a book by Mura.[16]

The purpose of this investigation was to reconsider the load transfer mechanism of composite strengthening, and to consider elastic residual stress and texture differences which could exist on the strengthening of discontinuous composites.

## MODELS

### 1. Shear Lag Type Model

The original shear lag model, developed by Cox,[1] with its detailed derivation of the stiffness[2] and yield stress[3] has already been discussed elsewhere. Thus it is omitted here. The standard shear lag model has often been used for an aligned short fiber composite system where short fibers of the same size are assumed to be distributed in the matrix in a hexagonal array. The repeated cell (unit cell) is used for the detailed derivation.[17,18] The most important assumption in the shear lag type model is that load transfer occurs between a short whisker and matrix by means of shear stresses at the matrix–whisker interface. In the original shear lag model[1,2] the load transfer by the normal stress at the whisker ends and side surfaces was ignored. Nardone and Prewo[3] recently suggested that the load transfer at the fiber ends should be accounted for in predicting the yield stress, but still ignored the normal load at the side surface of fiber for the case of $l/d$ values. Below is the final formulae based on the shear lag type model to predict the stiffness[1,2] and the end result of a derivation of the tensile yield stress, including normal stresses on the end of the fiber of a short fiber composite. For the stiffness,

$$E_c/E_m = (1 - V_w) + (V_w E_w/E_m)[1 - (\tanh x)/x] \tag{1}$$
$$x = (l/d)\{(1 + v_m)(E_w/E_m)\ln[V_w]^{-1/2}\}^{-1/2}$$

For the yield stress,

$$\sigma_{yc}/\sigma_{ym} = 0.5V_w(2 + l/d) + (1 - V_w) \tag{2}$$

where $E_m$, $E_w$ and $E_c$ are Young's moduli of the matrix, whisker and composite, respectively; $V_w$ is the volume fraction of whiskers; $\sigma_{ym}$ and $\sigma_{yc}$ are the yield stresses of the matrix and composite, respectively; and $l/d$ is the whisker aspect ratio.

In the shear lag model, there exists an uncertainty regarding the relation between $l/d$ and $L/D$, which is usually found from observations of SEM photos. In deriving eqn. (1), we have used the same assumption as Kelly and Street,[4] i.e. $L = l$. This assumption would certainly induce errors in the analysis for the case of smaller whisker aspect ratios. This model is described in greater detail elsewhere.[17,18]

## 2. Eshelby Type Model

The original Eshelby model[6] is based on the assumption that an ellipsoidal inclusion with uniform nonelastic strain (eigenstrain) $e^*$ is embedded in an infinite elastic body. Eshelby[6] derived the formula to compute the stress field induced in and around an inclusion and also the associated strain energy of this system. Mori and Tanaka[7] modified the original Eshelby model to account for the interaction between inclusions, i.e. an aligned short fiber composite.

The Young's modulus along the fiber axis $E_c$ can be obtained as

$$E_c/E_m = 1/(1 + V_w(E_m/\sigma_0)e_{33}^*) \tag{3}$$

where $e_{33}^*$ is to be computed from formulation defined elsewhere,[17,18] and the results are expressed in terms of $c_0(\sigma_0/E_m)$ and $c_0$ is some numerical value.

The formula to predict the yield stress is finally reduced to[17,18]

$$\sigma_{yc} = c_1\sigma_{ym} + c_2e_p \tag{4}$$

where $c_1$ is a nondimensional parameter (the yield stress raiser), $c_2$ is the work-hardening rate and $e_p$ is the plastic strain in the matrix along the fiber axis.

## NUMERICAL RESULTS

In order to compare these models, the stiffness $(E_c)$ and the yield stress $(\sigma_{yc})$ of an aligned short whisker MMC are computed by using eqns (1) and (2) (shear lag type model) and eqns (3) and (4) (Eshelby type model). The target short whisker MMCs are spherical SiC $(SiC_s)$ and SiC whiskers $(SiC_w)/1100$ Al matrix composites.[18] The numerical results of $E_c/E_m$ based on the shear lag type and Eshelby type models are plotted as dashed and solid curves, respectively, as a function of whisker aspect ratio $l/d$ in Fig. 1. The experimental results[19] are also plotted as a circle in Fig. 1. The material constants used in this calculation are given in Ref. 18. It can be seen in Fig. 1

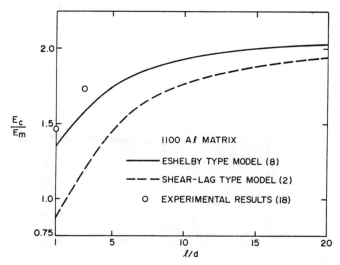

FIG. 1. The ratio of modulus of the composite to modulus of the matrix versus length to diameter ratio of fiber.

that the stiffnesses predicted by the shear lag type model are always less than those predicted by the Eshelby type model and that the shear lag model is a rather crude approximation for smaller $l/d$. A comparison with the experimental results reveals that the Eshelby type model gives a better prediction.

Next we have computed the values of $\sigma_{yc}/\sigma_{ym}$ based on eqns (2) and (4), the results of which are plotted as dashed and solid curves, respectively, in Fig. 2, where the closed circles denote the experimental results for a specific SiC$_w$/Al 6061-T6 composite[3] and the predicted values based on the modified Eshelby model, with residual stress accounted for, are plotted as a dash-dot curve. All predictions are plotted as a function of $l/d$. In the range of small $l/d$ the Eshelby type model gives an accurate prediction,

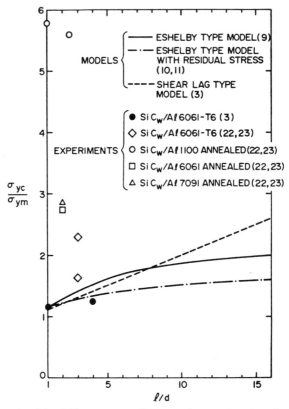

FIG. 2.    The ratio of the yield stress composite to matrix versus length to diameter ratio of fiber. For 20 V% SiC.

particularly at $l/d = 1$, while the shear lag model gives an underestimation. However, at larger $l/d$s both models predict basically the same order of composite tensile yield stress. In the same figure, our recent results for various $SiC_w/Al$ composite systems are also plotted as open symbols. Though the Eshelby type model gives reasonable predictions for the case of the T6-treated composite (●), the stress prediction of both models is low compared to the experimental results for most of the data (○, ◇, □ and △). The poorest prediction occurs for the case of an annealed 1100 aluminum composite (○).

The experimental $\sigma_{yc}$ and $\sigma_{ym}$ correspond to the stress at 0·2% offset, and $\sigma_{ym}$ is from a 0 V% material produced in the same manner as the composite. The reason that both of these models are not capable of predicting the observed strengthening is that these models assume that the matrix has the same strength as it has in the nonreinforced condition. In other words, these models assume that the addition of SiC does not change the strength of the matrix. It has been clearly shown that the dislocation density in the SiC/Al composites is much higher than in the nonreinforced Al.[20,21]

If the absolute magnitude of the increase in $\sigma_{yc}$ compared to $\sigma_{ym}$ is considered, then the apparent differences caused by the various matrix alloys is much less. The data[18] indicate that $\Delta\sigma$ ($\Delta\sigma = \sigma_{yc20V\%} - \sigma_{ym0V\%}$) is relatively independent of the composite matrix except for the T6 heat-treated case. The lack of a difference in $\Delta\sigma$ occurs because the thermal stress developed upon cooling is very large compared to the yield stress of the matrix in the annealed condition. The dislocations are generated in the initial cool down whether the sample is furnace cooled or quenched prior to aging. Therefore, the increased density of dislocations generated for all practical purposes is independent of the matrix.

Since the observed increase in the experimental $\sigma_{yc}$ is much greater than that predicted by the shear lag type model or the Eshelby type model, the increase in $\sigma_{yc}$ must be caused by the increased dislocation density in the composite matrix. This increase in dislocation density is the result of relaxation of a portion of the stresses developed upon cooling of the composite. The stresses arise from the localization of differences in coefficients of thermal expansion between the SiC and Al. The effect of this thermal expansion mismatch strain ($\Delta\alpha\Delta T$) has been considered in the Eshelby model, with residual stress accounted for (dot-dot curve in Fig. 2).[11,17] However, in the model $\Delta\alpha\Delta T$ is simulated by the equivalent surface dislocations[15] that are present at the matrix–whisker interfaces. In reality, the surface dislocations are more likely to relax by punching,[24] resulting in the localization of dislocations around a whisker.

If we now consider the proportional limit of the composite, an interesting correlation can be obtained. The proportional limits of the composites[22,25] are approximately equal to $\sigma_{ym}$. This correlation gives rise to two important points. First, the modulus $(E_c)$ that is predicted by the Eshelby type model, which is in agreement with the experimental data, is for the initial portion of the stress–strain curve, i.e. for stresses up to the proportional limit. Therefore, the Eshelby type model does operate up to the proportional limit, and the model predicts a very small increase in the proportional limit upon the addition of the reinforcement, again in agreement with the experimental data. Second, the increase in stress with strain in the stress region from the proportional limit to $\sigma_{yc}$ is caused by an exhaustion phenomena. In the stress–strain region, dislocation motion occurs in the lower dislocation density regions within the matrix. The increase in stress between the proportional limit and the $\sigma_{yc}$ is not caused by work-hardening, for there is no general increase in dislocation density.[26] There is an increase in dislocation immediately below the fracture surface, i.e. within 100 $\mu$m of the fracture surface. The dislocation density in the remainder of the sample is identical to that of the undeformed sample. However, only limited deformation can occur, because of the small volume of matrix. In order to have macro-deformation, i.e. to reach 0·2% offset strain, additional dislocation motion must occur in the higher stress regions of the matrix which are often quite localized in the specimen. Therefore, the macro-yielding of the composite is controlled by the inhomogeneous matrix which is a mixture of the high–low dislocation regions.

## RESIDUAL ELASTIC STRESS

An analytical model by Arsenault and Taya[11] based on an ellipsoidal-shaped SiC particle in Al matrix was developed, which predicts that a tensile thermal residual should exist in the matrix for a whisker of $l/d$ of 1·8 and the longitudinal residual stress should be higher than the transverse residual stress. The actual experimental data obtained from an X-ray analysis are shown in Table 1. The X-ray results do indicate that the matrix is in tension and that the longitudinal residual stress is higher than the transverse residual stress. However, the X-ray data indicate a higher value of residual stress than predicted.

The model of Arsenault and Taya[11] predicts that the yield stress in compression should equal the yield stress in tension if the SiC is in the form of spheres. The surprising result is that in the case of spherical SiC

TABLE 1
*Thermal residual stress (tensile) X-ray measurements*

| Material | Longitudinal (MPa) | Transverse (MPa) |
|---|---|---|
| 0 Vol. % whisker SiC 6061 matrix | 0·0 | 0·0 |
| 5 Vol. % whisker SiC 6061 matrix | 408 | 35 |
| 20 Vol. % whisker SiC 6061 matrix | 231 | 58 |
| Wrought 1100 Al | 0·0 | 0·0 |

composite, the tensile stress is higher than that of the compressive yield stress, whereas the difference predicted by the model is zero. However, the point to be made is that for the whisker case $\sigma_y^C > \sigma_y^T$. The model of Arsenault and Taya successfully predicts the differences in the tensile and compressive yield stress due to the thermal residual stresses. However, the model is completely incapable of predicting the absolute magnitude of the

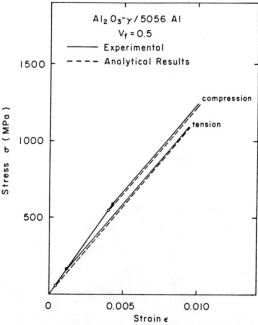

FIG. 3.   Tension and compression stress–strain curves of continuous $Al_2O_3$ fiber/5056 Al composite with $V_f = 0·5^y$. The experimental and theoretical results are denoted by solid and dashed curves, respectively. The solid and open circles denote the yield stress of the experimental and theoretical results, respectively.

increase in yield stress, for it does not have the capability of predicting the increase in the matrix strengthening.

In a case where the matrix strengthening is not such a predominant factor, i.e. continuous filament deposits, the model is excellent in predicting the differences between tension and compression, and the absolute values as shown in Fig. 3.

## TEXTURE

A texture investigation was undertaken and from a comparison of the data it is apparent that there is little difference in the texture of 99·99% Al, 0 V%, 6061 Al alloy and 20 V% 6061 Al alloy matrix composite, as shown in Fig. 4.

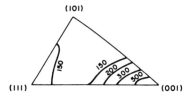

TEXTURE FOR EXTRUDED & ANNEALED
99.99% Al

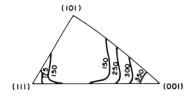

TEXTURE FOR EXTRUDED & ANNEALED
0V% SiC 1100 MATRIX

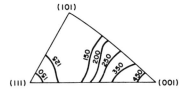

TEXTURE FOR EXTRUDED & ANNEALED
20V% SiC 1100 MATRIX

FIG. 4.    The texture of two control Al samples and a 20 V% SiC whisker composite in the annealed condition as determined by an X-ray technique.

## DISCUSSION

Some time ago, a listing of the components of the strengthening mechanisms was proposed:

$$\Delta\sigma_{yc} = \Delta\sigma_{disl} + \Delta\sigma_{sg} \pm \Delta\sigma_{res} + \Delta\sigma_{tex} + \Delta\sigma_{comp}$$

where $\Delta\sigma_{disl}$ is the increase in strengthening due to the increase in dislocation density resulting from the differences in thermal coefficients of expansion between the SiC and Al, $\Delta\sigma_{sg}$ is the increase in strengthening due to the reduced subgrain size, $\Delta\sigma_{res}$ is the difference in strengthening due to the thermal residual elastic stresses, $\Delta\sigma_{tex}$ is the strengthening due to differences in texture between the 0 volume % and the higher volume % composites, and $\Delta\sigma_{comp}$ is the strengthening due to classical composite strengthening, e.g. load transfer.

If the strengthening components are considered in detail, it is now necessary to consider the morphology of SiC. Table 2 is a listing of the

TABLE 2
*A listing of the strengthening components for the spherical SiC in 1100 Al matrix*

$$\Delta\sigma_{ys} = \Delta\sigma_{disl} + \Delta\sigma_{sg} \pm \Delta\sigma_{res} + \Delta\sigma_{tex} + \Delta\sigma_{comp}$$
$$\Delta\sigma_{disl} = 124\cdot2\,\text{MPa}$$
$$\Delta\sigma_{sg} = 55\cdot2\,\text{MPa}$$
$$\Delta\sigma_{res} = 0$$
$$\Delta\sigma_{tex} = 0$$
$$\Delta\sigma_{comp} = 0$$
$$\Delta\sigma_{ys} = 179\cdot4\,\text{MPa} \qquad \Delta\sigma_{ys\,exp} = 172\cdot5\,\text{MPa}$$

TABLE 3
*A listing of the strengthening components for the whisker SiC 1100 Al matrix*

$$\Delta\sigma_{yw} = \Delta\sigma_{disl} + \Delta\sigma_{sg} \pm \Delta\sigma_{res} + \Delta\sigma_{tex} + \Delta\sigma_{comp}$$
$$\Delta\sigma_{disl} = 124\cdot2\,\text{MPa}$$
$$\Delta\sigma_{sg} = 55\cdot2\,\text{MPa}$$
$$\Delta\sigma_{res} = -34\cdot5\,\text{MPa}$$
$$\Delta\sigma_{tex} = 0$$
$$\Delta\sigma_{comp} = 0$$
$$\Delta\sigma_{yw} = 144\cdot9\,\text{MPa} \qquad \Delta\sigma_{yw\,exp} = 144\cdot9\,\text{MPa}$$

strengthening components for the spherical SiC case. The strengthening due to $\Delta\sigma_{tex}$ and $\Delta\sigma_{comp}$ is equal to zero. The summation of the predicted strengthening and the observed strengthening are in very good agreement. Table 3 is a listing of the strengthening components for the whisker SiC case, The strengthening due to $\Delta\sigma_{disl}$ may be a little higher than the value given, however, again the agreement between the predicted and experimental results is very good.

## CONCLUSIONS

From the data presented in this paper, several conclusions can be drawn:

1. The data generated further support the concept that the strengthening mechanism is due to the difference in the thermal coefficient of expansion between SiC and Al, which results in a higher dislocation density and a small subgrain size.
2. The thermal residual stress, as measured by the X-ray technique, indicates that the matrix is in tension.
3. There is a difference between 6061 and 2124 in terms of the fracture of SiC whiskers and/or the bonding to SiC whiskers.
4. A model based on load transfer is completely incapable of explaining the increase in strength due to the addition of SiC to the Al alloy matrix.

## ACKNOWLEDGEMENT

This research was supported by the Office of Naval Research under grant No. N00014-85-K-007.

## REFERENCES

1. Cox, H. L. Jr, *J. appl. Phys.*, **3** (1952), 72.
2. Kelly, A., *Strong Solids*, Oxford, Clarendon Press, 1966.
3. Nardone, V. C. and Prewo, K. M., *Scripta Metall.*, **20** (1986), 43.
4. Kelly, A. and Street, K. M., *Proc. Roy. Soc. Lond.*, **A328** (1972), 283.
5. Hedgepeth, J. M. and Van Dyke, P., *J. Comp. Mater.*, **1** (1967), 294.
6. Eshelby, J. D., *Proc. Roy. Soc. Lond.*, **A241** (1957), 376.
7. Mori, T. and Tanaka, K., *Acta Metall.*, **21** (1973), 571.

8. TAYA, M. and MURA, T., *J. appl. Mech.*, **48** (1981), 361.
9. TANAKA, T., WAKASHIMA, K. and MORI, T., *J. Mech. Phys. Solids*, **21** (1973), 207.
10. WAKASHIMA, K., SUZUKI, S. and UMEKAWA, S., *J. Comp. Mater.*, **13** (1979), 391.
11. ARSENAULT, R. J. and TAYA, M., *Proc. ICCM-V* (Harrigan, W. C. Jr, Strife, J. and Dhingra, A. eds), TMS-AIME, Warrendale, PA, 1985, p. 21.
12. WAKASHIMA, K., OTSUKA, M. and UMEKAWA, S., *J. Comp. Mater.*, **8** (1974), 391.
13. TAKAO, Y. and TAYA, M., *J. appl. Mech.*, **107** (1985), 806.
14. HATTA, H. and TAYA, M., *J. appl. Phys.*, **58** (1985), 2478.
15. MURA, T. and TAYA, M., *Recent Advances in Composites in the United States and Japan*, ASTM STP 864 (Vinson, J. R. and Taya, M. eds), Philadelphia, PA, 1985, p. 209.
16. MURA, T., *Micromechanics of Defects of Solids*, The Hague, Martinus Nijhoff Publ., 1982.
17. ARSENAULT, R. J. and TAYA, M., to appear in *Acta Metall.*
18. TAYA, M. and ARSENAULT, R. J., to be published in *Scripta Metall.* (March 1987).
19. WOLFENDEN, A. and ARSENAULT, R. J., unpublished results.
20. VOGELSANG, M., ARSENAULT, R. J. and FISHER, R. M., *Met. Trans.*, **17A** (1986), 379.
21. ARSENAULT, R. J. and FISHER, R. M., *Scripta Metall.*, **17** (1983), 67.
22. ARSENAULT, R. J. and WU, B., submitted for publication.
23. ARSENAULT, R. J., *Proc. of Third Japan Conference, Composites '86. Recent Advances in Japan and United States* (Kauata *et al.* eds), in press.
24. ARSENAULT, R. J. and SHI, N., *Mater. Sci. Engng*, **81** (1986), 175.
25. ARSENAULT, R. J., *Mater. Sci. Engng*, **64** (1984), 171.
26. ARSENAULT, R. J. and FENG, C. R., to be submitted for publication.

# 6

# Considerations for Designing with Metal Matrix Composite Materials

JACQUES E. SCHOUTENS

MMCIAC, Kaman Tempo, 816 State Street,
Santa Barbara, California 93101, USA

## ABSTRACT

*This paper presents a methodology that is based upon the concepts of structural indices for the selection of metal matrix composite (MMC) materials. General design considerations of technical criteria and operational and cost criteria are discussed briefly. Relevant MMC material properties equations are given in a table. The structural indices that are used to calculate ratios between MMC and conventional material to compare weight to strength, impact resistance, efficiency of columns and plates, flexural rigidity, and structural efficiency indices for plates and shells are then discussed.*

## INTRODUCTION

Metal matrix composite (MMC) materials have distinct and, in some cases, overwhelming advantages in space and aircraft applications. This technology is still relatively immature compared to fiber reinforced resin composites. Hence, many problems of utilization with conventional materials remain to be solved.[1,2] As shown in Table 1, these advanced materials are expensive compared to unreinforced steel or aluminum (Al). The relative values shown in this table indicate that MMC materials do not appear to be competitive with unreinforced metals. However, a few words of caution are in order: cost comparison of raw materials does not really make sense. The cost of the function performed by a given component or subsystem must be considered. Such value analysis is quite complicated and

TABLE 1
*Approximate relative cost comparison between MMCs and conventional materials*

| Material | Relative cost |
|---|---|
| Steel plate (hot rolled) | 1 |
| Aluminum plates and castings | 1–4 |
| SiC/Al | 600 |
| Boron/Al | 1 800 |
| Graphite/Al, graphite/Mg | 4 800–20 000 |

must consider manhours required to produce the final product it replaces, the service life of the system, as well as many other considerations.[2]

In some applications, MMCs offer superior performance; for example, in high-temperature applications such as diesel engine pistons and connecting rods, and in turbomachinery. However, in some high-temperature applications, fiber reinforced ceramics appear to be a potentially strong competitor. MMC materials are finding numerous applications in designs where weight, decreased life cycle costs, wear resistance, and thermal stability are overwhelming considerations. Part of the resistance to broadening the applications of MMC materials is related to immaturity and insufficiency of reliable design data, particularly for safety critical applications. Consequently, the decision to utilize MMC materials must not be based on cost considerations alone, but on technical and operational criteria as well. Therefore, to help designers make such decisions and assist them in their choice of materials, a design methodology based on the concepts of structural indices is the main topic of this chapter.

## GENERAL CONSIDERATIONS

When considering the use of MMC materials in structural applications, it is not meaningful to consider each new structural concept for MMC applications through every stage of analytical details and reiteration of analysis. Such an approach is extremely tedious and may hide the implications of changes in strength, stiffness, density, and so on upon the performance of the structure. This would lead to the development of methodologies that become dependent upon the particular design under consideration. In this paper, a general methodology is discussed that is simple to use and adaptable to any structure. In considering structures in this manner, it is necessary to consider first the operational and cost criteria and the technical criteria that affect the design process. Examples of these criteria are shown in Table 2. These are not listed in order of their importance. There are interrelationships among operational and cost criteria and technical criteria. For example, the need for high strength may become a determining factor in a structural element or component that could override cost. Incidentally, the fracture toughness aspect of MMC material properties is not yet well developed or understood, which means that high risks in cost and safety would be involved in ignoring work of fracture where it is critical. Such trade-offs must be considered and evaluated during concept evaluation and design analysis. Table 2 shows a technical criterion known as stress density, defined here as follows. Quite often, structural designs require forces to be transmitted from one component to another through some confined space; for example, through a joint or attachment that must occupy a small volume arising from equipment or structural constraints. A high stress density is the applied force divided by the material cross-section of the force transmitter in a small volume. It may not be feasible to increase the material thickness of the force transmitter because of volume constraints. Consequently, to meet design requirements, high strength or high stiffness materials must be used to meet the required stress density. High stress density implies small spatial volume and, since it is a design variable, it cannot be defined as a material property.

TABLE 2
*Factors to be considered in assessing improvements in structures*

| Operational and cost criteria | Technical criteria |
|---|---|
| Weight | Strength |
| Indirect weight savings | Stiffness |
| Cost | Stress density |
| Maintainability | Critical crack length |
| Deployability | Work of fracture |
| Machinability | Corrosion resistance |
| Repair | Geometrical effects on structural stability |
| Performance | Weight |
| Reliability | Wear |
| Life cycle cost | Repairability |
| Simplicity | |
| Inspection | |
| Safety | |

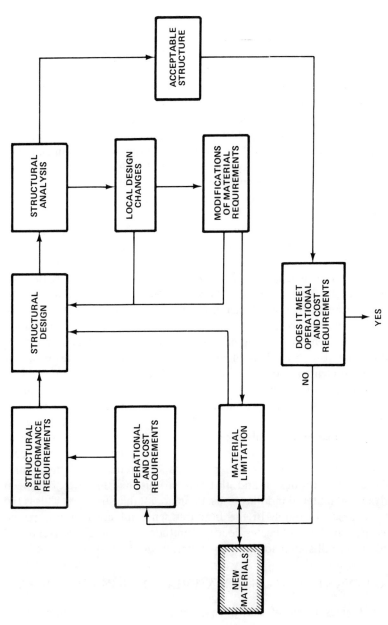

FIG. 1.    Schematic for material selection procedure.

The general design considerations consist in *qualitatively* analyzing elements and substructures or components of a design to uncover potential technical and operational weaknesses from the use of conventional materials, and then deciding whether or not these potential deficiencies can be overcome by the use of MMC materials.

A structure can be conceptually disassembled into its constituent elements where high stresses, wear, dynamic loads, and other critical factors may be identified, and each element can be represented with a free body diagram. These are standard steps in stress analysis, except that all information such as allowable maximum dimensions (constraints), limits on strain, and stress rates are specified, and time-dependent forces are indicated so that critical dynamical effects may be seen. Such diagrams then allow the designer to develop a complete physical and intuitive picture of the operation of the system.[1] Members loaded in torsion or compression need to be examined for possible loading limits set by buckling constraints of the material and design geometry. If cyclic loads exist, effects of long-term fatigue must be examined. MMC materials may be useful with better high-temperature mechanical properties than unreinforced metals. In some cases of cyclic loads, dynamic buckling analysis may be required. In addition, if such structural elements are impulse loaded, conventional static buckling analysis is inadequate since most materials exhibit some stress–strain behavior dependent upon strain rates.[3,4] Advanced composites may also be used to advantage with hybrid composites in which some components are cellular solids.[5–7] All loads on a structure can be reduced to tension, compression, flexure, torsion and shear, or combinations of these, with fracture, buckling, wear, corrosion, or plastic yielding failure states.

Figure 1 diagrammatically shows the analytical procedure discussed so far. The cycle begins with the definition of operational and cost requirements leading to structural performance requirements. This, in turn, leads to structural design and, through various iteration cycles between analysis and material modifications, to an acceptable structure. This is the usual course in design. In the broadest sense intended in this paper, structural analysis indicates load magnitudes for which conventional materials are ill-suited so that new materials are indicated.

## RELEVANT COMPOSITE MATERIAL DESIGN EQUATIONS

Table 3 gives a set of useful equations for preliminary design with composite materials.

TABLE 3
*Relevant composite design equations*[1]

Longitudinal elastic modulus (unidirectional fibers):

$$E_L = E_f V_f + (1 - V_f)E_m = \{(K_o - 1)V_f + 1\}E_m$$

Transverse elastic modulus (unidirectional fibers):

$$\frac{1}{E_T} = \frac{V_f}{E_f} + \frac{(1 - V_f)}{E_m} \quad \text{or} \quad E_T = \frac{E_m}{1 - \left(1 - \dfrac{1}{K_o}\right)V_f}$$

Strength along fiber direction:

$$\sigma_L = \sigma_{fu}\left\{\left(\frac{K_o - 1}{K_o} - \frac{S_c}{2S}\right)V_f + \frac{1}{K_o}\right\} \quad S \geq S_c$$

Critical $L/D$:

$$S_c = \frac{\sigma_{fu}}{2\tau_i} \qquad \tau_i \simeq \frac{\sigma_{my}}{2}$$

Strength for short fiber composites:

$$\sigma_L = V_f S \tau_i + (1 - V_f)\sigma_{my} \qquad S < S_c$$

Strength for 'infinitely' long fiber:

$$\sigma_L = \frac{\sigma_{fu}}{K_o}\{(K_o - 1)V_f + 1\} \qquad S \to \infty$$

Transverse strength (unidirectional fibers):

$$\sigma_T = \sigma_i V_f + (1 - V_f)\sigma_m \qquad \sigma_i \geq 2\tau_i \text{ or } \sigma_T \simeq \sigma_{mu}$$

Poisson ratio relations:

$$\nu_{LT} = -\frac{E_L}{E_T}\nu_{TL} = \{(K_o - 1)V_f + 1\}\left\{1 - \left(1 - \frac{1}{K_o}\right)V_f\right\}\nu_{TL}$$

or

$$\nu_{LT} \simeq \frac{1}{3}\left(1 - \frac{V_f}{4}\right) \quad \text{for most MMCs}$$

Thermal expansion—longitudinal:

$$\alpha_L = \frac{\alpha_f V_f K_o + \alpha_m(1 - V_f)}{(K_o - 1)V_f + 1}$$

Thermal expansion—transverse:

$$\alpha_T = \alpha_f V_f + (1 - V_f)\alpha_m + \frac{V_f 1 - V_f)(\nu_f - \nu_m K_o)}{(K_o - 1)V_f + 1}(\alpha_f - \alpha_m)$$

(*continued*)

TABLE 3—*contd*

Plate and shell buckling coefficients—MMC with isotropic reinforcement:

$$E_p = \frac{E}{1 - v^2} \quad \text{(plate)} \qquad E_s = \frac{E}{\sqrt{1 - v^2}} \quad \text{(shell)}$$

Plate and shell buckling coefficients—unidirectional fiber reinforced metals:

$$E_p = \frac{1}{2}\left\{ \frac{\sqrt{E_T E_L}}{1 - \sqrt{v_{TL} v_{LT}}} + 2G_{TL} \right\} \quad \text{(plate)}$$

$$E_{S1} = \left\{ \frac{2G_{TL}(E_T E_L)^{1/2}}{1 - \sqrt{v_{TL} v_{LT}}} \right\}^{1/2} \quad \text{(shell)}$$

or[a]

$$E_{S2} = \left\{ \frac{E_T E_L}{1 - v_{TL} v_{LT}} \right\}^{1/2} \quad \text{(shell)}$$

$$\frac{1}{G_{TL}} = \frac{V_f}{G_f} + \frac{(1 - V_f)}{G_m} \qquad G_{f,m} = \frac{E_{f,m}}{2(1 + v_{f,m})}$$

---

[a] Use $E_{S1}$ or $E_{S2}$, whichever has a smaller numerical value.
Definition of symbols: $E_f$, $E_m$, fiber and matrix modulus; $V_f$, fiber volume fraction; $K_o = E_f/E_m > 1$; $\sigma_{fu}, \sigma_m$, fiber ultimate tensile and matrix tensile strength; $S$, fiber length; $S_c$, critical fiber length; $\tau_i$, fiber/matrix interface shear strength; $\sigma_{my}$, matrix yield strength; $\sigma_{mu}$, matrix ultimate strength; $v_{LT}$, longitudinal transverse Poisson ratio; $E_L$, $E_T$, longitudinal and transverse elastic modulus of composite; $\alpha_L, \alpha_T$, longitudinal and transverse coefficient of thermal expansion; $\alpha_f, \alpha_m$, coefficient of thermal expansion for fiber and matrix; $v_f, v_m$, fiber and matrix Poisson ratio; $G_{TL}$, composite shear modulus; $G_f, G_m$, fiber and matrix shear modulus.

## STRUCTURAL INDICES

One of the most useful tools in the study of optimum structural design is the application of the principles of dimensional similarity; in structural engineering, this is accomplished by using structural indices. A structural index can be considered as a measure of the loading intensity, i.e. a comparison between the magnitude of the carried load and the distance over which this load is transmitted. The structural index is to structural design what coefficients are in aero- and hydrodynamics, except that it has dimensional form.[8] To render structural indices to dimensionless form, it is only necessary to divide them by a stress. One of the great advantages of this approach is that design proportions that are optimum (minimum weight) for a particular structure are also optimum for structures of any size, provided they all have the same structural index.

In determining optimum design on a weight/strength basis, the variables are the *material* and the *configuration* of the structure. For simple tension, optimum configuration is a straight line, and the proportions of cross-sections have no direct effect on the weight, Consequently, the weight/ strength factor is determined entirely by material properties. For compression, the optimum configuration is also a straight line, but the size and shape of the cross-section, as well as the constraints (fixed ends, lateral support, etc.), play an important part in the buckling strength of the member. In this case, the material properties and configuration are interrelated. Therefore, for any structural index, it is possible to find a combination of material and configuration that will result in the lightest structure. Use of structural indices illustrates the consequence of choosing some parameters in favor of others and the potential effects of these choices upon structural design. It is recognized that weight reduction, improved fracture toughness, better area weight-to-stiffness or strength ratios, and improved buckling properties can be understood for a broad class of designs. Consequently, various structural indices can be used to test for the effects of changing geometry or material properties to improve the design by lowering critical stresses.

### Weight-strength

If weight is an important consideration, then two important numerical factors in choosing between materials are the specific strength, $\sigma/\rho$, and the specific stiffness, $E/\rho$, where $\sigma$ is the strength, $E$ is the elastic modulus, and $\rho$ is the material density. These factors can be related to a characteristic material thickness, $t$, to obtain $\rho t/\sigma$ and $\rho t/E$, where these factors have the dimensions of length. The characteristic thickness can be shell or plate thickness of any other critical dimension that characterizes supporting a load. A material with the highest strength or stiffness efficiency would have the lowest factors.

### Impact Resistance

Another parameter relevant to structural design is the ratio of impact resistance to the dynamic load, particularly where dynamic loads are concerned. This is a fairly complex phenomena.[1] For simple calculations, one can consider the ratio of the impact resistance to the dynamic energy delivered, both terms being generally difficult to determine. The impact resistance is a material property that is not well understood for MMC at the present time. The delivered energy is the time integral of the force, which is not always easily determined for a real system. The force may be periodic and, mathematically speaking, well behaved, in which case it may be represented by a sine or cosine wave. If the force is impulsive, it can be approximated by a narrow square pulse or a step function, depending on the nature of the problem. In the case of shock loading, Fourier representation of actual force versus time function may be used. If an airblast from an explosion produces the impulse loading, then the force function may be represented by an exponential pulse.[3,9] For example, a force function due to an airblast can be represented by

$$F(t) = AP_0 t \exp(-t/t_m)$$

where $A$ is the exposed area, $t$ is time, $t_m$ is the time to maximum pressure, and $P_0 = 0.368 P_{max}/t_m$, where $P_{max}$ is the peak pressure at $t_m$.[3] Sometimes, for analytical convenience, a triangular pulse can be used to approximate the actual pulse. In that case, the energy delivered is $E \simeq AP_{max}\tau/2$, where $\tau$ is the maximum duration of the triangular pulse.

## Efficiency of Columns and Plates

Gordon[10] and Shanley[8] have shown that the efficiency of plates and columns to support compressive loads is given by

$$e_c = C_0 \left( \frac{\sqrt{E}}{\rho} \right) \left( \frac{\sqrt{F}}{L^2} \right) \quad \text{(column)} \tag{1}$$

$$e_p = C_1 \left( \frac{\sqrt[3]{E}}{\rho} \right) \left( \frac{F^{2/3}}{L^{5/3}} \right) \quad \text{(plate)} \tag{2}$$

where $C_0$ and $C_1$ are constants; $E$ and $\rho$ are the elastic modulus and material density, respectively; $L$ is the length of the plate or column along the loading direction; and $F$ is the load. When comparing the efficiency of two plates or columns made from different materials, the constants in eqns (1) and (2) cancel. The elastic modulus applying to conventional metals or composites (Table 3) must be used. The material density is that for conventional materials or the rule of mixture (see Table 3) for composites.

The terms $\sqrt{E}/\rho$ and $\sqrt[3]{E}/\rho$ in eqns (1) and (2) are called material parameters or material efficiency criteria.[10] The terms $\sqrt{F}/L^2$ and $F^{2/3}/L^{5/3}$ are called the structure loading coefficients.[10] These parameters account for the weakening in compression loaded members due to buckling (Euler effect). The critical stress for buckling is always less than the material compressive yield strength because buckling is a geometric effect. These results are useful indicator numbers because they point to the direction in which improvements are most easily made: either geometric or material.

The structure loading coefficients indicate that weight savings are achieved with the highest value of these coefficients, that is, by using the most compact and highly loaded devices, cases where MMCs have distinct advantages over conventional materials. However, not much can be done to raise the structure loading coefficients themselves because of fundamental limitations. The effects of low structure loading coefficients can be offset to some extent by the material efficiency criteria ($\sqrt{E}/\rho$ for a column or $\sqrt[3]{E}/\rho$ for a plate) by using material with high specific stiffness values such as composites.

A simple example can be used to compare, say, steel with graphite/aluminum (Gr/Al). For Gr/Al, $E_c = 390\,\text{GPa}$, with $V_f = 0.5$, and $E = 212\,\text{GPa}$ for steel. The densities for Al and steel are 2.7 and 7.8 g/cm³, respectively. Then using eqns (1) and (2) to compare Gr/Al and steel for a column and a plate gives

$$\frac{(\sqrt{E}/\rho)_{\text{Gr/Al}}}{(\sqrt{E}/\rho)_{\text{steel}}} = \frac{\sqrt{390}/2.7}{\sqrt{212}/7.8} = 1.36(2.89) = 3.92$$

for a column and, for a plate,

$$\frac{(\sqrt[3]{E}/\rho)_{\text{Gr/Al}}}{(\sqrt[3]{E}/\rho)_{\text{steel}}} = \frac{\sqrt[3]{390}/2.7}{\sqrt[3]{212}/7.8} = 1.23(2.89) = 3.54$$

It can be seen from these results that an increase in stiffness by a factor of 1.84 (composite stiffness over unreinforced material stiffness) results in an improved stiffness by a factor of 1.36 for a column and 1.23 for a plate. A decrease in the density alone nearly triples (2.89) the plate and column efficiency. Thus, net improvements in efficiency of 3.5 for a plate and nearly 4 for a column are possible by simply changing materials to improve the design. A similar calculation comparing unreinforced Al to steel results in a net improvement of only 1.7 for a plate and 2 for a column, the dominant contribution being from density reduction alone, which somewhat offsets the negative effects of the lower stiffness of Al. In the above example, the geometrical properties or design ($F$ or $L$) have not been changed.

**Flexural Rigidity**

In the simple bending of beams, the elastic deformation results in a curvature of radius $R$. If the bending forces result in a moment $M$, then

$MR = EI$, where $EI$ is the flexural rigidity with $I$ the moment of inertia about the neutral axis. A beam with great stiffness has a high value of $EI$ and, since for a beam of height $h$ and unit width, $I = h^3/12$, so $EI = Eh^3/12$. The weight of the beam per unit length and width is $\rho h$. Then comparing two materials, a composite with a conventional material (designated by the subscripts c and o) for the same flexural rigidity, they will have their thickness related by $E_o h_o^3 = E_c h_c^3$, and the ratio of their weight is $W_o / W_c = \rho_o h_o / \rho_c h_c$, so that $W_o / W_c = (\rho_o / E_o^{1/3}) / (\rho_c / E_c^{1/3})$. The most efficient material in terms of stiffness is one having the highest value of $E^{1/3}/\rho$.

### Efficiency Structural Indices for Plates and Shells

We now discuss plate and shell efficiency structural indices when these structural elements are loaded in axial compression. A simple analysis shows that the weight index $W/b$ or $W/R$ versus the load index $N_x/b$ or $N_x/R$ for plates and shells, respectively, exhibits a knee as shown in Figs 2 and 3.[11] $N_x$ is the axial compressive load, $b$ is the plate width, $R$ is the shell radius and $W$ is the weight. In Figs 2 and 3, the right-hand sides of these curves have slopes of 45° representing the region in which the structure is limited by material strength. The left-hand sides of these curves correspond to the elastic stability limited region, that is, the region limited by plate or shell buckling as is shown by the sketches in these figures. The knee is a discontinuity corresponding to the point at which failure is, hypothetically, simultaneously a material and a stability failure.[11] For shells under lateral pressure and under combined load, the corresponding slopes of the elastic stability region of the curve vary between those for shells and plates under axial load decreasing from $E_s^{1/2}$ for shells under axial load to $E_s^{1/2.75}$ for shells under axial compression and lateral pressure as shown in Fig. 4.[11]. This result is approximate because the length of the shell must also be considered. It can be shown[1] that below the knee

$$\frac{W}{b} = \rho\left(\frac{t}{b}\right) = \rho\sqrt{\frac{N_x/b}{\pi^2 E_p/3}} \quad \text{(plate)} \tag{3}$$

and

$$\frac{W}{R} = \rho\left(\frac{t}{R}\right) = \rho\sqrt[3]{\frac{N_x/R}{E_s/2\sqrt{3}}} \quad \text{(shell)} \tag{4}$$

In eqns (3) and (4), $E_p$ is the plate buckling modulus and $E_s$ is the shell buckling modulus (Table 3) for unreinforced or composite materials.

In general, the designer has no control over the values of $N_x/b$ or $N_x/R$ to be satisfied. If the design value of the load index is below the knee, simply

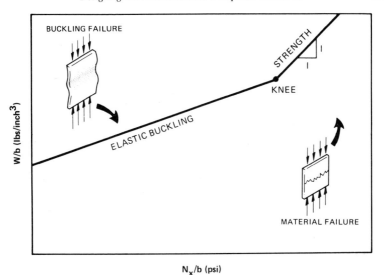

FIG. 2.  Plot for evaluating the efficiency of flat plates in axial compression.

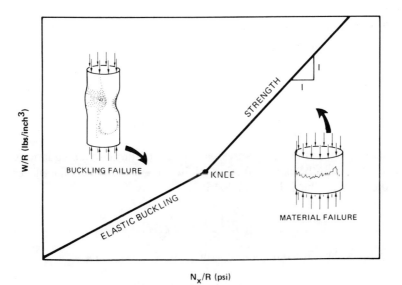

FIG. 3.  Plot for evaluating the efficiency of circular shells in axial compression.

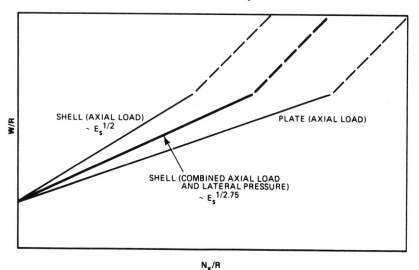

FIG. 4.  Diagram showing the slope of shells loaded in axial compression and lateral pressure relative to axial compression alone for shells and plates.

increasing the strength of the material will not reduce the weight. If the design value is above the knee, simply increasing the buckling efficiency is equally useless. Consequently, the knee itself is an important index in efficient design.[11] The efficiency index for a plate in axial compression loading is[1]

$$e_p^* = \left\{ \left( \frac{\sqrt[3]{E_p}}{\rho} \right) \left( \frac{\sigma_{cu}}{\rho} \right) \right\}^{1/2} \quad \text{(plate)} \tag{5}$$

and

$$e_s^* = \left\{ \frac{E_s \sigma_{cu}}{\rho^3} \right\}^{1/3} \quad \text{(shell)} \tag{6}$$

where $\sigma_{cu}$ is the ultimate compressive strength of the material. Equations (5) and (6) can now be used to obtain a loading intensity at which the efficiency indices are attained. For a plate,[1] it is

$$\left( \frac{N_x}{b} \right)_{e_p^*} = \frac{\sigma_{cu}^{3/2}}{\sqrt{\pi^2 E_p/3}} \tag{7}$$

and for a shell in axial compression, it is

$$\left( \frac{N_x}{R} \right)_{e_s^*} = 2\sqrt{3} \left( \frac{\sigma_{cu}^2}{E_s} \right) \tag{8}$$

In the above analysis, the composite density can be obtained from the rule of mixture, that is, $\rho_c = \rho_f V_f + \rho_m(1 - V_f)$, where $\rho_c$, $\rho_f$ and $\rho_m$ are the composite, the fiber and the matrix densities, respectively.

## Design Data

A variety of design data are available in the literature (see, for example, Ref. 1). These data are useful for conceptual design and preliminary analyses. For final design, actual test data must be used, and scale model tests should be performed for design verification, particularly in cases where public safety is concerned. At the present time, there are no design data available for MMC from which commercial structures may be safely built directly without resorting to material or demonstration model testing. Consequently, until such time as design handbooks are published which contain MMC data that can be applied with confidence, designers and engineers will have to develop their own database in particular design instances.

## DESIGN PRACTICE

Good structural design is a compromise among design requirements and constraints. Criteria established by design and analysis determine the degree of emphasis to be placed upon each factor considered. As discussed in this chapter, the designer now has at hand an additional class of materials

TABLE 4
*Recommendations for good composite design practice (after Ref. 14)*

1. Fibers should be in the direction of the principal stresses
2. Load must be transferred to the composite through shear
3. Holes should not be cut in highly loaded regions of fiber reinforced composites
4. Isotropic unreinforced or reinforced metals[a] should be used in complex stress areas
5. The highest payoffs are in areas where the load path is unidirectional and well defined
6. Shear effects must be considered in compression stability analysis
7. The matrix material, fiber and fiber volume fraction should be chosen to optimize desired properties
8. Allowance should be made for reasonable analysis capability

[a]An example of an isotropic reinforced metal is silicon carbide particulate or whisker reinforced aluminum.

that are very stiff and strong in a single direction. Moreover, these materials can be prepared so that different fiber directions can be exploited, thus producing tailorable component properties. This requires that the load paths be well defined. Also, reliability is an important consideration, and the structure utilizing new materials must have a reliability at least as great as the structure it replaces. Moreover, considerations of the design factors discussed in this paper, combined with experience and common sense, will result in designs that demonstrate the potential of fiber reinforced composite materials. However, it will also become necessary for designers and engineers to learn to 'think composites' in design in order to take full advantage of these new materials.[12-14]

To assist the designer to focus on the principles of good design with composites and to promote design concepts based on the philosophy of 'thinking composites', we conclude by presenting a set of design guidelines, adapted from June and Lager,[14] in Table 4.

## REFERENCES

1. SCHOUTENS, J. E., Structural indices in design optimization with metal matrix composite materials, *Composites*, **17** (1986), 188–204.
2. SCHOUTENS, J. E., Metal matrix composite materials today—invited commentary, *J. Metals*, June 1985, 43.
3. HUDSON, C. C. and SCHOUTENS, J. E., Analysis of dynamic stresses and motion of pressure gages in alluvium, *ISA Trans.*, **20** (1982), 13.
4. LINDBERG, H. E. and FLORENCE, A. L., *Dynamic Pulse Buckling—Theory and Experiment*, Handbook No. DNA 6503H, SRI International, Menlo Park, CA, February 1983.
5. SCHOUTENS, J. E., The elastic and compressive properties of hybrid metal matrix composites/hexagonal prismatic cellular solids, *Composite Structures*, **4** (1985), 267–291.
6. SCHOUTENS, J. E., Direct measurements of non-linear stress–strain curves and elastic properties of metal matrix composite sandwich beams with any core material, *J. Mater. Sci.*, **20** (1985), 4421–4430.
7. SCHOUTENS, J. E. and HIGA, D. I., Linear stress relations for a metal matrix composite sandwich beam with any core material, *J. Mater. Sci.*, **21** (1986), 1943–1946.
8. SHANLEY, F. R., *Weight-strength Analysis of Aircraft Structures*, New York, McGraw-Hill, 1952.
9. SCHOUTENS, J. E., unpublished data, 1980.
10. GORDON, J. E., *Structures—Or Why Things Don't Fall Down*, London, A Decapo Paperback, 1982.
11. DOW, N. F. and DERBY, E., Personal communication, 1984.

12. LOCKETT, F. J., Development of design analysis methods for new materials, in: *Modern Vehicle Design Analysis*, London, International Association for Vehicle Design, June 1983.

13. OBRAZTSOV, I. F., Some problems of composite mechanics, in: *Mechanics of Composites* (Obraztsov, I. F. and Vasil'ev, V. V., eds), Moscow, MIR Publishers, 1982.

14. JUNE, R. R. and LAGER, J. R., Commercial aircraft, in: *Applications of Composite Materials*, ASTM STP 524, American Society for Testing and Materials, 1973, pp. 1–42.

# 7

# The Mechanical Properties of Carbon Fibre Reinforced Zinc Base Alloy Composite

Wang Chengfu, Ying Meifang and Tao Jie

*Department of Metallic Material Engineering, Hefei Polytechnical University, Anhui, People's Republic of China*

## ABSTRACT

*The composite has been prepared by the infiltration of carbon fibres with melting zinc and the zinc alloy. The properties of longitudinal tensile, impact and lubricated wear of the composite were investigated respectively. It was shown that the tensile strength and wear resistance of the zinc alloy were improved considerably after the incorporation of carbon fibres, and these properties were influenced by the volume percentage, distribution and orientations of carbon fibres.*

## INTRODUCTION

Casting zinc alloys have found wide application in latter years since they have a series of advantages. They have become the effective competitor of cast irons, copper alloys and aluminium alloys in some areas.[1] This Zn–Al system alloy has a higher tensile strength and a good casting property. Therefore, zinc alloys have widespread use in the production of low melting point dies and bearings. Zinc alloy dies, compared with steel dies, give a better performance without deformation after heat treatment. In addition, they can be remelted and used again. However, the zinc alloy chosen has some defects such as large thermal coefficients of expansion and inferior wear resistance. Attempts were made to use high strength carbon fibres to reinforce zinc alloys in order to improve their characteristics.

## COMPOSITE FABRICATION

The carbon fibre used in this work was about 7 $\mu$m diameter and its typical ultimate tensile strength was 2940 Pa. In order to improve the compatibility of carbon fibre with zinc or its alloys at elevated temperature, the coating technique of electroless deposition was adopted, which was capable of coating a $3 \times 10^3$ filament tow and provided each of the fibres within the tow with a uniform covering of nickel. The coated fibres were washed and vacuum dried at 110°C before the fabrication of the composite.

The matrixes chosen were industrial pure zinc and zinc alloy (4% Al, 1% Cu, 0·05% Mg). 120 mm lengths of the coated fibre were laid longitudinally in a steel mould covered with a layer of graphite and the assembly heated to a temperature of about 500°C in vacuum. The composites were prepared by infiltration of carbon fibres with melting zinc and its alloy.

## MECHANICAL TESTING

### 1. Tensile Strength

Tensile specimens with a cross-section of $2 \times 20\,mm^2$ were stressed to failure in an TJL-10 testing machine at a crosshead speed of 2 mm/min.

Longitudinal tensile strength data for the carbon fibre–zinc base alloy composite are shown in Table 1.

The strength of a tow of carbon fibres is approximately 2940 MPa, which decreased to 2548–2646 MPa after the fibres were coated with nickel and the composite was fabricated. Using the value of 2646 MPa as the strength

TABLE 1
*Tensile strength of the composite*

| Material | Fibre volume (%) | Average tensile strength $c$ (MPa) | Rule-of-mixtures value $\bar{c}$ (MPa) | $c/\bar{c}$ (%) |
|---|---|---|---|---|
| Zinc | 0 | 35·87 | | |
| Zinc | 6·0 | 165·33 | 192·47 | 85·9 |
| Zinc | 10·8 | 187·87 | 317·72 | 59·1 |
| Zinc | 12·7 | 236·77 | 367·40 | 64·5 |
| Zinc alloy | 0 | 90·45 | | |
| Zinc alloy | 6·0 | 179·24 | 243·82 | 73·5 |
| Zinc alloy | 10·0 | 198·84 | 346·04 | 57·5 |
| Zinc alloy | 14·6 | 262·05 | 463·54 | 56·5 |

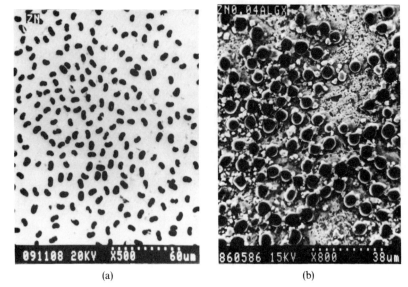

(a)                                    (b)

FIG. 1.    (a) Distribution of carbon fibres coated nickel in pure zinc. (b) Cross-section of the
zinc alloy composite.

of the carbon fibre, the rule-of-mixtures values for the longitudinal tensile
strength of the composite are calculated respectively according to different
volumes of carbon fibre. It can be seen in Table 1 that the tensile strength of
zinc or zinc alloy was appreciably increased after the incorporation of
carbon fibres.

Microstructure examination of the composite showed a quite even
distribution of fibres (Fig. 1a, b).

## 2. Impact Resistance

An impact bending test of the composite was performed. The impact
specimens were $10 \times 10 \times 55$ mm without notch. The volume percentage of
carbon fibres was approximately 20%.

Table 2 shows the results of zinc and zinc alloy with and without carbon
fibres. The impact resistance of the zinc composite was higher than that of
the zinc alone, but the zinc alloy composite showed quite the opposite.

## 3. Lubricated Wear

The material used was the zinc alloy composite with 17% volume of fibre.
The cross-sectional area of the specimens was $12\cdot35 \times 12\cdot35$ mm$^2$ and the

TABLE 2
*Longitudinal impact resistance*

| Material | Impact resistance (*J*) | Number of specimens |
|---|---|---|
| Pure zinc | 5·596 | 5 |
| Zinc composite | 7·350 | 5 |
| Zinc alloy | 22·393 | 5 |
| Zinc alloy composite | 7·473 | 5 |

normal force employed was 78·5 N. The wear specimens were pressed against a GCr15 steel ring about 36 mm in diameter. Oil and water were used as lubricants respectively.

The specimens of the composite were tested in three different directions. (a) Orientation I: carbon fibres perpendicular to the counterface. (b) Orientation II: carbon fibres parallel to the counterface and perpendicular to sliding direction. (c) Orientation III: carbon fibres parallel to both the counterface and sliding direction.

The results are summarized in Tables 3 and 4. It was found that the wear rate of the composite was only about 1/20 that of the zinc alloy, and that the wear rate and friction coefficient were lower when the fibres were perpendicular rather than parallel to the counterface.

TABLE 3
*Effect of sliding distance on wearing volume (mm³)*

| Material | revs 2 | revs 29 | revs 121 | revs 400 | revs 601 |
|---|---|---|---|---|---|
| Zinc alloy | 0·051 4 | 0·239 4 | 1·149 7 | 3·791 3 | 5·152 2 |
| Orientation I | $1·763 \times 10^{-3}$ | 0·053 9 | 0·057 9 | 0·150 8 | |
| Orientation II | $2·099 \times 10^{-3}$ | 0·138 9 | 0·153 5 | 0·158 4 | |

TABLE 4
*Friction coefficients of the composite*

| Type | 300 rpm | 400 rpm | 500 rpm | 600 rpm | 700 rpm | 800 rpm |
|---|---|---|---|---|---|---|
| Orientation I | 0·056 74 | 0·071 85 | 0·072 00 | 0·069 04 | 0·058 15 | 0·063 91 |
| Orientation II | 0·070 81 | 0·071 62 | 0·080 27 | 0·071 42 | 0·068 39 | 0·038 62 |
| Orientation III | | 0·067 74 | 0·079 46 | 0·079 27 | 0·079 27 | 0·086 99 |

## DISCUSSION

There are several elements which influence the strength of metal matrix composites, such as the bond force between fibre and matrix, the volume percentage of fibre and the mechanical properties of the matrix itself. In addition, the distribution of fibres in the matrix and the impairing of fibres during composite fabrication also affect the strength of the composites. The flowing property of matrices and the wettability of matrices by fibres, as is very essential when composites are fabricated by liquid infiltration techniques, govern both the interfacial bond force and the distribution of fibres.

It is easy for zinc or zinc alloy to be oxidized at high temperature. During the fabrication of the composite, an oxide of zinc or zinc alloy will form on the surface of the melting zinc or zinc alloy. This layer of oxide prevents the liquid matrix materials from immersing the fibres, which impairs the flowing property of the liquid matrix and the wettability of the matrixes by carbon fibres, so that the interfacial bond between the matrixes and fibres is poor. Meanwhile, because carbon fibres in the composite are not uniform in distribution, it will cause stress concentrations so that the composite fails at a lower stress. Therefore, a proper pressure must be exerted in order to eliminate the deteriorative effect of the oxide.

(a)           (b)

FIG. 2. Friction surface of the zinc alloy composite.

Investigation of the microstructure of pure zinc showed the presence of a very large grain size. The incorporation of the carbon fibres greatly refined the coarse grain structure, which may be the main cause of improvement of impact resistance of pure zinc composite. In the case of zinc alloy composite, however, the fibre/matrix interfacial bonding may influence its impact value to a great extent; further work is in progress on this.

Carbon fibre is not only potentially capable of reducing friction and wear by acting as a solid lubricant, but can also increase strength and thermal conductivity, and reduce the coefficient of thermal expansion. Because of the effect of the solid film and the chopped fibres, the composite has a good wear resistance (Fig. 2a, b).

The matrix material had low plasticity. Cracks formed easily at and below the friction surface owing to the fact that both tangential and normal loads were applied. This could facilitate debonding at the matrix/carbon fibre interface.[2] Subsequently, debris began to come off, which were smaller than those of the matrix material alone. So far as the friction coefficients of three orientations of carbon fibre are concerned, it is suggested[3] that extra work is necessary to bend and break the fibres when the fibres are oriented perpendicular to the counterface as compared to when the fibres are parallel to the sliding surface.

## CONCLUSIONS

(a)    The carbon fibre coated nickel has a uniform distribution in the zinc or the zinc alloy.

(b)    The zinc or the zinc alloy is reinforced after incorporation of carbon fibres. The strength of the composite is controlled by the properties of interfacial bonding, volume percentage and distribution of fibres in the composite.

(c)    The zinc alloy composite has a lower wear rate than the matrix alloy. The effect is different with different fibre orientations.

## ACKNOWLEDGEMENTS

The authors thank Mr Xu Renbiao, Mr Lu Xinghua and Mr Wang Yingliang for their assistance with the majority of this work.

# REFERENCES

1. KANICKI, D., Zinc in the foundry industry, where does it fit, *Modern Casting*, **60**, No. 3 (1979), 49–56.
2. ELIWZER, Z., KHANNA, V. D. and AMATEAU, M. F., Wear mechanism in composites: a qualitative model, *Wear*, **51** (1978), 169–179.
3. ELIWZER, Z., KHANNA, V. D. and AMATEAU, M. F., On the effect of fibre orientation on the wear of composite materials, *Wear*, **53** (1979), 387–389.

# 8

# The Boundary Between the Concepts of Material and Structure in the Plate and Shell Theories for Laminated Media

F. M. Brito

*Laboratório Nacional de Engenharia Civil,
Av. do Brasil 101, 1799 Lisboa, Portugal*

## ABSTRACT

*From the viewpoint of the mechanics of heterogeneous media each ply in laminates shall be considered to be a different material.*

*As a result, the analysis of the laminate structures should be made following three-dimensional models of heterogeneous bodies, the internal boundary conditions being incorporated in the interfaces.*

*On account of the high complexity involved in such a concept, ordinarily the analysis of laminated plates and shells follows two-dimensional models. The hypotheses of the theories behind such models in general allow us to establish constitutive equations that correlate generalized stresses and strains. Hence the laminate is implicitly dealt with as a hypothetical two-dimensional material.*

*In the present research work the general consistence of this type of theories is discussed, particular emphasis being laid on their mutual relationships and on the universality of the constants that they foresee.*

## NOTATION AND CONVENTIONS

| | |
|---|---|
| $x_1, x_2, x_3$ | System of coordinates with $x_3$ always perpendicular to the laminate surface |
| $W_3, W_2$ | Three-dimensional and two-dimensional strain energies |
| $u_i$ | Components of the field of displacements |
| $\varepsilon_{ij}$ | Components of the strain tensor |

| | |
|---|---|
| $\sigma_{ij}$ | Components of the stress tensor |
| $C_{ijkl}$ | Components of the tensor of the elastic constants in each ply |
| $L_{ijkl}^{(m,n)}$ | Components of matrix of order $(m,n)$ of the global constants of the laminate |
| $N_{ij}^{(n)}$ | Generalized stresses of order $n$ |
| $\varepsilon_{ij}^{(n)}$ | Generalized strains of order $n$ |
| $K_\alpha$ | Principal curvatures of the middle surface of the laminate |
| $A_\alpha$ | Square root of the diagonal elements of the metric tensor of the middle surface of the laminate |
| $H_i$ | Square root of the diagonal elements of metric tensor of the three-dimensional system associated with laminated shells |
| $\lambda_\alpha = H_\alpha/A_\alpha$ $= 1 + K_\alpha x_3$ | Auxiliary curvature parameters. |

**Superscripts**

Superscripts between parentheses are applied to the generalized variables and identify the coefficient of the term of the same order in the polynomial expansion in $x_3$ of the corresponding three-dimensional variables; in the double index that appears in the matrices of the laminate constants, the two values respectively refer to the order of generalized stresses and generalized strains correlated by each matrix

Superscript 0 outside parentheses identifies residual fields.

**Subscripts**

Latin indices can take three distinct values $(i, j, k, l = 1, 2, 3)$

Greek indices can take two distinct values $(\alpha, \beta = 1, 2)$

The repetition of an index in the same term equals a summation extended to all values of this index, unless otherwise indicated.

## 1.   INTRODUCTION

In the mechanics of heterogeneous media the concept of material is associated with the possibility of its being described as macroscopically homogeneous, and thus with the existence of a representative volume element (RVE), very small as compared to the body dimensions *in all directions*, that keeps the properties of any larger-size region of the media.

Of the two levels of heterogeneity observed in ordinary fibre-reinforced composite structures (matrix/fibre inside each ply and between plies forming the laminate), only the first may be given a macroscopically homogeneous description, since the composition of the second cannot be represented but for the complete thickness of laminate. Therefore laminates are heterogeneous media formed of superposed plies of different materials.

With the increasing availability of powerful computers the structural analysis of homogeneous media is more and more based on numerical solutions of three-dimensional models. There is not such a tendency in the case of laminated media since the high-stress gradients throughout the thickness that result from difference of properties in the plies would require an enormous discretization of the structures. This, together with the fact that most practical applications concern small-thickness laminates, prompt the structural analysis to be made by using two-dimensional models of plates and shells.

The classical laminated plate theory (CLPT) is only the simplest approximation (first order together with the additional assumption of the Kirchhoff hypothesis) in the infinity of two-dimensional laminated plate theories. A similar diversity can be found for shells, although this case is much more complex as its mathematical formulation must use reference bases that consider the intrinsic geometry of its middle surface.

Actually even the methods of two-dimensional theory generation are numerous; they can be grouped in two fundamental families:

(a)  autonomous generation through consideration of the form of the two-dimensional strain energy;
(b)  generation by introducing simplifications in the three-dimensional model which is considered an exact pattern in this context.

The former family comprises the theories generally designated by 'multipole' or 'microelastic', the simplest of which starts from the identification of the medium as a Cosserat continuum. Their complexity would rather place them in the field of theoretical mechanics, though some essential conclusions must be retained, namely the *autonomous consistency of two-dimensional models*. Kunin[1] even holds that the type of simplification required for concentrating on a surface the properties of a three-dimensional medium is of the same order of magnitude as that for creating macroscopically homogeneous three-dimensional models on heterogeneous solids where a RVE can be defined. From the theoretical point of view it makes sense to describe a laminate as an ideal two-dimensional material, the properties of whose surface element are representative of the corresponding prismatic volume element with the total laminate thickness.

## 2. FORMULATION OF THE PROBLEM

The present work is only concerned with the practical aspects of the issue, and so will consider only the current use theories, which all pertain to the latter family (b). In this context, the autonomous consistency of the two-dimensional model depends on the possibility of defining *constitutive equations* in which the *relationship* between *generalized stresses* (stress resultants, couple resultants and high-order moments) and *generalized strains* (middle surface strains, curvatures and higher order terms) can be established through *universal constants*. Such universality should comprise the two following conditions:

(1)   constants should only be a function of the properties of laminate plies and geometry, and specifically they should not depend on the surface shape;

(2)   the higher order theories should contain the constants of the lower order ones, new constants only being added.

The passage of the three-dimensional model to an approximate two-dimensional model will implicitly or explicitly involve an assumption as to the evolution of the stress fields or displacement fields throughout the thickness. In laminated media the continuity of one of such field functions necessarily implies the discontinuity of the other. As it does not make sense to build an elastic model with a discontinuous field of displacements, theories are always based on an approximation to this field, a polynomial function being admitted as a rule:

$$u_i = u_i^{(0)} + u_i^{(1)}x_3 + u_i^{(2)}x_3 + \cdots + u_i^{(N)}x_3^N \qquad (1)$$

with $u_i^{(n)} = u_i^{(n)}(x_1, x_2)$.

The different theories only diverge as regards the surface shape, the degree of the polynomial and possible further simplifications. One may obtain the possible forms of constitutive equations by calculating the two-dimensional strain energy through the integration of the three-dimensional strain energy, as exemplified for the case of Cartesian metrics:

$$W_2(\varepsilon_{ij}^{(n)}) = \int_{-h/2}^{h/2} W_3[\varepsilon_{ij}(\varepsilon_{ij}^{(n)})]\,\mathrm{d}x_3 \qquad (2)$$

and determining their derivatives with reference to generalized strains $\varepsilon_{ij}^{(n)}$:

$$N_{ij}^{(n)} = \frac{\partial W_2}{\partial \varepsilon_{ij}^{(n)}} \qquad (3)$$

Usually the strain energy is considered to vanish when all stresses are null. On the other hand, from the point of view of structural analysis it is convenient to consider that strains vanish simultaneously to displacements in the natural state of the structure (lack of external forces). In these conditions and assuming the physical and geometrical linearity of the materials forming all plies, the three-dimensional strain energy is given by

$$W_3 = \tfrac{1}{2}C_{ijkl}\varepsilon_{ij}\varepsilon_{kl} + \sigma_{ij}^0\varepsilon_{ij} + \tfrac{1}{2}\sigma_{ij}^0\varepsilon_{ij}^0 \tag{4}$$

where $\sigma_{ij}^0$ and $\varepsilon_{ij}$ are residual stresses and strains presented by the structure in the natural state, calculated against an initial state in which they are supposed to vanish simultaneously.

The above equations define a general methodology for building up plate and shell theories based on a polynomial approximation to the field of displacements. Nevertheless, its development will differ from one case to another, which will be analysed separately.

## 3. ANALYSIS OF CONSTITUTIVE EQUATIONS IN LAMINATED PLATE THEORIES

Let us consider a laminated plate with constant thickness, in which a Cartesian orthogonal reference base is supposedly set out as Fig. 1 shows.

As the analysis is made against the natural state of the structure and assuming infinitesimal strains,

$$\varepsilon_{ij} = \frac{1}{2}\left(\frac{\partial u_i}{\partial x_j} + \frac{\partial u_j}{\partial x_i}\right) \tag{5}$$

On account of the form of eqn. (4), development of eqns (1) and (3) will only lead to constitutive equations such as desired if a polynomial expansion of strains similar to that of displacements is possible, i.e. of type

$$\varepsilon_{ij} = \varepsilon_{ij}^{(0)} + \varepsilon_{ij}^{(1)}x_3 + \varepsilon_{ij}^{(2)}x_3^2 + \cdots + \varepsilon_{ij}^{(N')}x_3^{N'} \tag{6}$$

with $\varepsilon_{ij}^{(n)} = \varepsilon_{ij}^{(n)}(x_1, x_2)$.

Substituting eqn. (1) into eqn. (5) and collecting the resulting terms of the same degree, we can obtain an expression of type (6) for all components of strain except for transverse shear strain. For the latter this will not be possible unless the order of the polynomial of approximation to $u_3$ is one unit lower than that of the other two components of displacement. This condition only generalizes to theories of any order the method followed by Lo et al.[2] in the development of a third-order theory.

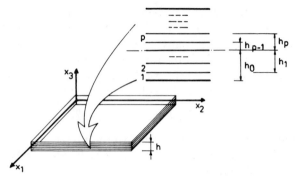

Fig. 1.    Orthogonal Cartesian coordinate system used in plate analysis.

In these conditions the substitution of (1) into (5) leads to

$$\varepsilon_{\alpha\alpha} = \sum_{n=0}^{N} \left[ \frac{\partial u_\alpha^{(n)}}{\partial x_\alpha} \right] x_3^n \quad \text{(no summation refers to } \alpha\text{)}$$

$$\varepsilon_{\alpha\beta} = \frac{1}{2} \sum_{n=0}^{N} \left[ \frac{\partial u_\alpha^{(n)}}{\partial x_\beta} + \frac{\partial u_\beta^{(n)}}{\partial x_\alpha} \right] x_3^n$$

$$\varepsilon_{\alpha3} = \frac{1}{2} \sum_{n=0}^{N-1} \left[ (n+1)u_\alpha^{(n+1)} + \frac{\partial u_3^{(n)}}{\partial x_\alpha} \right] x_3^n \tag{7}$$

$$\varepsilon_{33} = \sum_{n=0}^{N-2} \left[ (n+1)u_3^{(n+1)} \right] x_3^n$$

Hence the following relations between generalized strains and displacements:

$$\varepsilon_{\alpha\alpha}^{(n)} = \frac{\partial u_\alpha^{(n)}}{\partial x_\alpha} \quad \text{(no summation)} \qquad (n = 0, 1, 2, \ldots, N)$$

$$\varepsilon_{\alpha\beta}^{(n)} = \frac{1}{2} \left( \frac{\partial u_\alpha^{(n)}}{\partial x_\beta} + \frac{\partial u_\beta^{(n)}}{\partial x_\alpha} \right) \quad (\beta \neq \alpha) \quad (n = 0, 1, 2, \ldots, N) \tag{8}$$

$$\varepsilon_{\alpha3}^{(n)} = \frac{1}{2} \frac{\partial u_3^{(n)}}{\partial x_\alpha} + (n+1)u_\alpha^{(n+1)} \qquad (n = 0, 1, 2, \ldots, N-1)$$

$$\varepsilon_{33}^{(n)} = (n+1)u_3^{(n+1)} \qquad (n = 0, 1, 2, \ldots, N-2)$$

System (8) comprises $6N + 2$ independent equations that relate $6N + 2$ generalized strains to $3N + 2$ generalized displacements. The number of generalized stresses will also be $6N + 2$.

By developing eqn. (4) as a function of generalized strains, the following expression is obtained in which terms derived from those of eqn. (4) are put into brackets for clarity:

$$W_3 = \{[\tfrac{1}{2}C_{ijkl}(\varepsilon_{ij}^{(0)}x_3 + \varepsilon_{ij}^{(2)}x_3^2 + \cdots)(\varepsilon_{kl}^{(0)} + \varepsilon_{kl}^{(1)}x_3 + \varepsilon_{kl}^{(2)}x_3^2 + \cdots)]$$
$$+ [\sigma_{ij}^0(\varepsilon_{ij}^{(0)}x_3 + \varepsilon_{ij}^{(2)}x_3^2 + \cdots)] + [\tfrac{1}{2}\sigma_{ij}^0\varepsilon_{ij}^0]\} \tag{9}$$

The summations indicated will lead to terms of different order, depending on the strain component concerned. Nevertheless, it is easier to assume that all polynomials are of degree $N$, and cancel out terms in excess at the end:

$$W_3 = \{[\tfrac{1}{2}C_{ijkl}(\varepsilon_{ij}^{(0)}\varepsilon_{kl}^{(0)} + \cdots + \varepsilon_{ij}^{(0)}\varepsilon_{kl}^{(N)}x_3^N + \varepsilon_{ij}^{(1)}\varepsilon_{kl}^{(0)}x_3 + \cdots$$
$$+ \varepsilon_{ij}^{(1)}\varepsilon_{kl}^{(N)}x_3^{N+1} + \cdots + \varepsilon_{ij}^{(N)}\varepsilon_{kl}^{(0)}x_3^N + \cdots + \varepsilon_{ij}^{(N)}\varepsilon_{kl}^{(N)}x_3^{2N})]$$
$$+ [\sigma_{ij}^0\varepsilon_{ij}^{(0)} + \sigma_{ij}^0\varepsilon_{ij}^{(1)}x_3 + \cdots + \sigma_{ij}^0\varepsilon_{ij}^{(N)}x_3^N] + [\tfrac{1}{2}\sigma_{ij}^0\varepsilon_{ij}^0]\} \tag{10}$$

For the integration of (10) intended to obtain $W_2$, each term should be analysed separately.

As generalized strains do not depend on $x_3$, quantities within the first parentheses can be represented by

$$\tfrac{1}{2}\varepsilon_{ij}^{(n)}\varepsilon_{kl}^{(m)} \int_{-h/2}^{h/2} C_{ijkl}x_3^{n+m}\,dx_3 \quad \text{(no summation)} \tag{11}$$

or in a more condensed way

$$\tfrac{1}{2}L_{ijkl}^{(n,m)}\varepsilon_{ij}^{(n)}\varepsilon_{kl}^{(m)} \quad \text{(no summation)} \tag{12}$$

with

$$L_{ijkl}^{(n,m)} = \int_{-h/2}^{h/2} C_{ijkl}x_3^{n+m} \tag{13}$$

$C_{ijkl}$ being constant within each ply, the integral can be replaced by a summation referred to the total number of plies:

$$L_{ijkl}^{(n,m)} = \sum_p C_{ijkl}^p \left[\frac{1}{n+m+1}(h_p^{n+m+1} - h_{p-1}^{n+m+1})\right] \tag{14}$$

The operators $L_{ijkl}^{(n,m)}$ only depend on the elastic properties of each ply and of the sequence of lamination. The total number of distinct operators is

$[(N+1)(N+2)]/2$ due to symmetry as regards superscripts between parentheses. Each operator has just the same symmetries as the tensor of elastic constants:

$$L_{ijkl}^{(n,m)} = L_{klij}^{(n,m)} = L_{ijlk}^{(n,m)} = L_{jikl}^{(n,m)} \tag{15}$$

The quantities within the second parentheses of $W_2$ will have the following general form:

$$\varepsilon_{ij}^{(n)} \int_{-h/2}^{h/2} \sigma_{ij}^0 x_3^n \, dx_3$$

and the term within the third parentheses does not depend on generalized strains, and disappears in subsequent derivations:

$$W_2 = \left\{ \left[ \tfrac{1}{2}(L_{ijkl}^{(0,0)}\varepsilon_{ij}^{(0)}\varepsilon_{kl}^{(0)} + \cdots + L_{ijkl}^{(0,N)}\varepsilon_{ij}^{(0)}\varepsilon_{kl}^{(N)} + L_{ijkl}^{(1,0)}\varepsilon_{ij}^{(1)}\varepsilon_{kl}^{(0)} + \cdots \right. \right.$$
$$\left. + L_{ijkl}^{(1,N)}\varepsilon_{ij}^{(1)}\varepsilon_{kl}^{(N)} + \cdots + L_{ijkl}^{(N,0)}\varepsilon_{ij}^{(N)}\varepsilon_{kl}^{(0)} + \cdots + L_{ijkl}^{(N,N)}\varepsilon_{ij}^{(N)}\varepsilon_{kl}^{(N)} \right]$$
$$\left. + \left[ \varepsilon_{ij}^{(0)} \int_{-h/2}^{h/2} \sigma_{ij}^0 \, dx_3 + \cdots + \varepsilon_{ij}^{(N)} \int_{-h/2}^{h/2} \sigma_{ij}^0 x_3^N \, dx_3 \right] + [C^{te}] \right\} \tag{16}$$

Substituting the value of $W_2$ in eqn. (3),

$$N_{ij}^{(n)} = [L_{ijkl}^{(0,n)}\varepsilon_{kl}^{(0)} + L_{ijkl}^{(1,n)}\varepsilon_{kl}^{(1)} + \cdots + L_{ijkl}^{(N,m)}\varepsilon_{kl}^{(N)}] + \left[ \int_{-h/2}^{h/2} \sigma_{ij}^0 x_3^n \, dx_3 \right] \tag{17}$$

Obviously the second term is the moment of order $n$ in the distribution of residual effective stresses. It can be easily shown that the overall expression represents the moment of order $n$ of the distribution of total effective stresses. The constitutive equations in condensed notation can be written under the following form:

$$N_{ij}^{(n)} = \sum_{m=0}^{N} L_{ijkl}^{(m,n)}\varepsilon_{kl}^{(m)} + N_{ij}^{0(n)} \tag{18}$$

The operators $L_{ijkl}^{(m,n)}$ only depend on the properties of the plies and composition of the laminate, thus playing the role of true elastic constants of the two-dimensional models. Besides, for increasing degrees of the polynomial $N$ of approximation to displacements, the lower order theory constants still remain, and new constants are added.

TABLE 1

| Operator | | Components which will cancel | |
|---|---|---|---|
| $L^{(n,N-1)}$ | $(n=0,1,2,\ldots,N)$ | $L_{ij33}$ | $(i,j=1,2,3)$ |
| $L^{(n,N)}$ | $(n=0,1,2,\ldots,N)$ | $L_{ij33}, L_{ij\alpha3}$ | $(i,j=1,2,3)$ $(\alpha=1,2)$ |

In the most general case, each operator should have 21 independent components, since its symmetry is such as that of tensor $C_{ijkl}$. Nevertheless, as said in the note before eqn. (10), some operators will contain components which will cancel as a result of the polynomials referred to, $\varepsilon_{33}$ and $\varepsilon_{\alpha3}$ being respectively of degrees $N-2$ and $N-1$ (Table 1).

## 4. ANALYSIS OF CONSTITUTIVE EQUATIONS IN LAMINATED SHELL THEORIES

Let us consider a shell of whatever shape, limited by two continuous and parallel surfaces, in which a curvilinear reference base is assumed to be set, with the following characteristics (see Fig. 2):

— the origin lies in the middle surface, parallel and at the same distance to the faces;
— two of the axes, $x_1$ and $x_2$, are tangent to the principal lines of curvature of the middle surface, i.e. lines whose normal curvature radii take extreme values at all points;
— the third axis, $x_3$, is perpendicular to the middle surface.

Axes $x_1$ and $x_2$ form a two-dimensional orthogonal reference system associated with the intrinsic geometry of the middle surface, the first quadratic form being given by

$$ds^2 = A_1^2(dx_1)^2 + A_2^2(dx_2)^2 \qquad \text{com } A_\alpha = A_\alpha(x_1,x_2) \qquad (19)$$

As $x_3$ is perpendicular to the middle surface, the complete reference system is still orthogonal, and its metrics are defined by the quadratic form

$$ds^2 = H_1^2(dx_1)^2 + H_2^2(dx_2)^2 + H_3^2(dx_3)^2 \qquad (20)$$

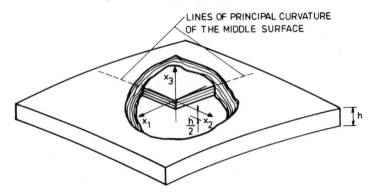

FIG. 2.   Orthogonal curvilinear coordinate system used in shell analysis.

Owing to the special form in which the system is defined, the following relations hold:

$$H_1 = A_1(1 + k_1 x_3) \qquad H_2 = A_2(1 + k_2 x_3) \qquad H_3 = 1 \qquad (20')$$

where $k_1, k_2$ are the principal curvatures of the middle surface and depend on $x_1$ and $x_2$ only. For facility, auxiliary parameters are introduced:

$$\lambda_\alpha = \frac{H_\alpha}{A_\alpha} = 1 + K_\alpha x_3 \qquad (21)$$

In the reference system so defined, the strain–displacement relations are given by

$$\varepsilon_{\alpha\alpha} = \frac{1}{\lambda_\alpha}\left(\frac{1}{A_\alpha}\frac{\partial u_\alpha}{\partial x_\alpha} + \frac{1}{A_\alpha A_\beta}\frac{\partial A_\beta}{\partial x_\alpha}u_\beta + K_\alpha u_3\right)$$

$$\varepsilon_{\alpha\beta} = \frac{1}{2}\left[\frac{1}{\lambda_\alpha}\left(\frac{1}{A_\alpha}\frac{\partial u_\beta}{\partial x_\alpha} - \frac{1}{A_\alpha A_\beta}\frac{\partial A_\alpha}{\partial x_\beta}\right) + \frac{1}{\lambda_\beta}\left(\frac{1}{A_\beta}\frac{\partial u_\alpha}{\partial x_\beta} - \frac{1}{A_\alpha A_\beta}\frac{\partial A_\beta}{\partial x_\alpha}\right)\right]$$

$$\text{(no summation; } \beta \neq \alpha) \quad (22)$$

$$\varepsilon_{\alpha 3} = \frac{1}{2}\left[\frac{\partial u_\alpha}{\partial x_3} + \frac{1}{\lambda_\alpha}\left(\frac{1}{A_\alpha}\frac{\partial u_3}{\partial x_\alpha} - K_\alpha u_\alpha\right)\right]$$

$$\varepsilon_{33} = \frac{\partial u_3}{\partial x_3}$$

By introducing expression (1) into (22) and following a process of reorganization of terms similar to those used for plates, we obtain

$$\varepsilon_{\alpha\alpha} = \frac{1}{\lambda_\alpha} \left\{ \left[ \left( \frac{1}{A_\alpha} \frac{\partial u_\alpha^{(0)}}{\partial x_\alpha} + \frac{1}{A_\alpha A_\beta} \frac{\partial A_\beta}{\partial x_\alpha} + K_\alpha u_3^{(0)} \right) \right] \right.$$
$$\left. + \sum_{n=1}^{N} \left[ \left( \frac{1}{A_\alpha} \frac{\partial u_\alpha^{(n)}}{\partial x_\alpha} + K_\alpha u_3^{(n)} \right) \right] x_3^n \right\}$$

$$\varepsilon_{\alpha\beta} = \frac{1}{2} \left[ \frac{1}{\lambda_\alpha} \left( \frac{1}{A_\alpha} \frac{\partial u_\beta^{(0)}}{\partial x_\alpha} - \frac{1}{A_\alpha A_\beta} \frac{\partial A_\alpha}{\partial x_\beta} \right) + \frac{1}{\lambda_\beta} \left( \frac{1}{A_\beta} \frac{\partial u_\alpha^{(0)}}{\partial x_\beta} - \frac{1}{A_\alpha A_\beta} \frac{\partial A_\beta}{\partial x_\alpha} \right) \right]$$
$$+ \frac{1}{2} \sum_{n=1}^{N} \left[ \frac{1}{\lambda_\alpha} \left( \frac{1}{A_\alpha} \frac{\partial u_\beta^{(n)}}{\partial x_\alpha} \right) + \frac{1}{\lambda_\beta} \left( \frac{1}{A_\beta} \frac{\partial u_\alpha^{(n)}}{\partial x_\beta} \right) \right] x_3^n \quad (23)$$

(no summation; $\beta \neq \alpha$)

$$\varepsilon_{\alpha 3} = \frac{1}{2} \frac{1}{\lambda_\alpha} \left[ \left( \frac{1}{A_\alpha} \frac{\partial u_3^{(0)}}{\partial x_\alpha} - K_\alpha u_\alpha^{(0)} \right) + (u_\alpha^{(1)}) \right]$$
$$+ \frac{1}{2} \sum_{n=1}^{N-1} \left[ \frac{1}{\lambda_\alpha} \left( \frac{1}{A_\alpha} \frac{\partial u_3^{(n)}}{\partial x_\alpha} - K_\alpha u_\alpha^{(n)} \right) + (n+1)(u_\alpha^{(n+1)}) \right] x_3^n$$

$$\varepsilon_{33} = \sum_{n=0}^{N-2} [(n+1)(u_3^{(n+1)})] x_3^n$$

Equations (23) show that it is not possible to obtain coefficients depending on $x_1$ and $x_2$ only, once parameters $\lambda_\alpha$ are a function of $x_3$. They measure the variation of the metric parameters in successive surfaces coordinated to $x_3 = C^{te}$ as a result of curvature.

This, however, is not the sole obstacle to the establishment of general constitutive equations for shells. In fact, both the definition of the two-dimensional strain energy density and that of moments of order $n$ in the distribution of effective stresses must be modified on account of the non-Cartesian metrics associated with the system of coordinates. Expression (2) should be replaced by

$$W_2(\varepsilon_{ij}^{(n)}) = \int_{-h/2}^{h/2} W_3[\varepsilon_{ij}(\varepsilon_{ij}^{(n)})] \lambda_1 \lambda_2 \, dx_3 \quad (24)$$

and those moments have the following definition:

$$N_{\alpha\alpha}^{(n)} = \int_{-h/2}^{h/2} \sigma_{\alpha\alpha}\lambda_\beta x_3^n \, dx_3$$

$$N_{\alpha\beta}^{(n)} = \int_{-h/2}^{h/2} \sigma_{\alpha\beta}\lambda_\beta x_3^n \, dx_3 \quad \text{(no summation; } \beta \neq \alpha) \tag{25}$$

$$N_{\alpha 3}^{(n)} = \int_{-h/2}^{h/2} \sigma_{\alpha 3}\lambda_\beta x_3^n \, dx_3$$

$$N_{33}^{(n)} = \int_{-h/2}^{h/2} \sigma_{33}\lambda_\alpha\lambda_\beta x^n \, dx_3 \tag{26}$$

Nevertheless, they are not completely independent due to the symmetry of the stress tensor, and the following relations can be established:

$$N_{\alpha\beta}^{(n-1)} = N_{\beta\alpha}^{(n-1)} - K_\alpha N_{\alpha\beta}^{(n)} + K_\beta N_{\beta\alpha}^{(n)}$$
$$N_{\alpha 3}^{(n-1)} = N_{3\alpha}^{(n-1)} - K_\alpha N_{\alpha 3}^{(n)} \qquad (\beta \neq \alpha; \, n > 0) \tag{27}$$

The number of generalized stresses and strains and so the number of constitutive equations is higher than that established for plates. Equations (23)–(26) show that for shells with whatever thickness one cannot obtain constitutive equations based on operators that only depend on properties of the plies and lamination sequence, unless for an eventual and very special case in which a set of geometric singularities may remove the obstacles referred to. It does not seem of any interest to investigate whether this most improbable special case might exist for two reasons:

— the origin of the obstacles lies in the variation of the metric characteristics of space with coordinate $x_3$ owing to the curvature, which remains in successive particularizations leading to simpler and more regular geometries—revolution shells, cylindrical shells, circular cylindrical shells—notice that in the last two cases the Gaussian curvature is null, the surfaces being isometric to the Cartesian plane ($A_\alpha = \text{constant}$);
— even in this improbable case, the operators $L_{ijkl}^{(m,n)}$ would never coincide with those determined for plates, i.e. they could not represent the same symmetries.

The only circumstance in which constitutive equations of the type desired could be defined relies on the validity for the *thin shell* assumption in such as way as

$$\lambda_\alpha \simeq 1 \Leftrightarrow K_\alpha |x_3| = \frac{|x_3|}{R_\alpha} \leq \frac{h}{2R_\alpha} \ll 1 \tag{28}$$

If such a hypothesis is assumed, the coefficients of polynomial expansion in $x_3$ of the strains (23) are only a function of $x_1$ and $x_2$, and so the redefinition of concepts through eqns (24) and (25) is no longer meaningful. In these conditions constitutive equations identical to eqns (18) can be deduced, taking the above coefficients as generalized strains (these expressions obviously differ from those in plates (7), into which they degenerate with $\lambda_\alpha = 1$ and $K_\alpha = 0$).

Nevertheless, so that operators $L_{ijkl}^{(n,m)}$ may be constant along the surface and coincide with those obtained for plates, it is also necessary that the symmetry characteristic directions of the laminate along its surface do have constant orientation with reference to the lines of principal curvatures of that surface. Otherwise, although the mathematical formalism remains valid, the values of the operator components depend on the variable angle between those directions. This limitation results from the specificity of the system of axes used and from the definition of operators (13).

## 5. CONCLUSIONS

The main conclusions drawn from this research work can be summed up as follows.

Any theory based on a polynomial approximation of any order to the field of displacements that may be applied to plates or shells whose lines of principal curvatures are at constant orientation, with reference to the directions of symmetry of the laminate, makes it possible to define identical constitutive equations (18) by relating generalized stresses and strains through a set of linear operators depending only on the composition of the laminate.

As a corollary, the laminates in these conditions can be dealt with as hypothetical two-dimensional materials.

For thick shells these conclusions no longer hold. At most, in some cases it would be possible to do a formal separation of the influence of the laminate composition from the influence of curvature (as, for instance, Bert[3] does on the basis of the Kirchhoff–Love hypothesis); however, the impossibility of obtaining constitutive equations of the type required still remains.

## REFERENCES

1. KUNIN, I. A., On foundations of the theory of elastic media with microstructure, *Int. J. Engng Sci.*, **22** (1984), 969.
2. LO, N. H., CHRISTENSEN, R. M. and WU, E. M., A high-order theory of plate deformation, Parts 1 and 2, *J. appl. Mech.*, **44** (1977), 663–676.
3. BERT, C. W., Structural theory for laminated anisotropic elastic shells, *J. Comp. Mater.*, **1** (1967), 414–423.

# 9

# A Causal Approach to Effective Dynamic Moduli of Random Composites

ABRAHAM I. BELTZER and NEIMA BRAUNER

*Holon Institute for Technological Education, Affiliated with Tel-Aviv University, P.O.B. 305, Holon 58102, Israel*

## ABSTRACT

*Random elastic composites behave in a 'viscoelastic' way when subject to dynamic loading. They exhibit a spatial decay of running waves and, consequently, large-scale structures made of these materials possess the absorption capability because of energy transfer to incoherent motions. This chapter presents some of the recently established theories of these phenomena and their implications to predictions of effective dynamic moduli. It is shown that a homogeneous viscoelastic medium may serve as an effective model for random elastic mixtures. The concept of shock modulus is discussed also.*

## 1. INTRODUCTION

The effective stiffness of random elastic composites may show strong frequency dependence. Furthermore, under dynamic loading these elastic mixtures surprisingly behave in a 'viscoelastic' way, exhibiting a spatial decay of progressive disturbances or temporary decay of free vibrations. Scattering by randomly dispersed inclusions induces these effects, which render the static theories of the effective moduli inapplicable to a dynamic case.

The well-known theory of the effective dynamic response is that of configurational averaging, which was pioneered by Foldy and further extended by Lax, Waterman and Twersky (see the survey by Willis[1]). This theory provides a reliable prediction for the static and low-frequency behavior. However, it has been noted by Beltzer and Brauner[2-4] and recently confirmed by Weaver[5] that some of the widely-spread versions of this theory may provide a causal prediction for the wave speed. As far as one is concerned with low-frequency behavior, the above drawback might be regarded as minor, but it may become fatal in the case of moderate and high-frequency waves and, consequently, in that of transient disturbances.

In fact, any causal linear passive system must obey the so-called Kramers–Kronig (K–K) relations, which indicate that the wave speed, $c(\omega)$, and attenuation, $\alpha(\omega)$, are Hilbert transforms of each other. In particular,

$$\frac{c_0}{c(\omega)} = 1 + \frac{2\omega^2 c_0}{\pi} P \int_0^\infty \frac{\alpha(\omega')\,d\omega'}{\omega'^2(\omega'^2 - \omega^2)} \tag{1}$$

where $P$ denotes the Cauchy principal value and

$$c_0 = \lim_{\omega \to 0} c(\omega) \tag{2}$$

It has been shown[2-5] that some of the versions of the configurational averaging technique explicitly violate the K–K relations, which may result in substantial errors.

More complicated versions of the ensemble averaging technique make use of the quasicrystalline approximation (QCA) to break the hierarchy of an infinite chain of equations. While the QCA may be consistent with causality, as shown by Weaver,[5] it is yet unclear whether this approximation complies with the K–K relations. Also, with increasing frequency, the final determinantal equation yields multiple roots for the coherent wave number, which makes the analysis difficult.

The correct analysis of the effective high-frequency response of random composites is not only of academic interest. The destructive influence of irregularities on wave propagation is similar to that of viscoelastic damping (see, for example, Hodges[6]). In the case of vibrations of large-scale structures made of these materials, the absorption capability is described by the logarithmic decrement, $\delta$, which is given by

$$\delta = 2\pi\alpha(\omega)c(\omega)/\omega \tag{3}$$

The above expression holds for viscoelastic waves, too. Moreover, the knowledge of $c(\omega)$ and $\alpha(\omega)$ for the entire frequency interval, $0 \le \omega < \infty$, appears necessary for analysis of a transient response.

Finally, the high-frequency limit

$$c_\infty = \lim_{\omega \to \infty} c(\omega) \tag{4}$$

can be shown[7] to represent the velocity of a small amplitude shock and be remarkably related to the static limit, $c_0$, by

$$\frac{\pi(1 - c_0/c_\infty)}{2c_0} = \int_0^\infty \alpha(\omega)\omega^{-2}\,d\omega \tag{5}$$

which has interesting theoretical and applied implications. For example, since it is difficult to accurately measure attenuation for a wide frequency interval, the left-hand side may yield a useful estimate, which can be used in non-destructive evaluations of materials.

In what follows, we present an approach which yields the dispersion curves, $c(\omega)$ and $\alpha(\omega)$, for the entire frequency interval and enforces the K–K relations. The approach is solely based on simple energy considerations and an appeal to such physically meaningful properties of the effective medium as its causality, linearity and passivity.

## 2. CAUSAL THEORIES OF EFFECTIVE RESPONSE

It should be pointed out that because of the limited nature of the information available on the composite microstructure any attempt at the exact solution seems unreasonable. In fact, the known theories are essentially approximate. This is why the theory should incorporate as many of the *a priori* given features as possible.

To this end, consider an elastic homogeneous isotropic material containing randomly dispersed inclusions. An incident plane wave first gives rise to primary scattered waves, which then cause secondary scattered waves, and so forth.

A natural way to start the analysis of this tangled situation is to neglect rescattering. This is a reasonable approximation if the mismatch of elastic moduli and that of the densities are slight and the volume fraction, $\phi$, is small. Then we get for attenuation the simple expression

$$\alpha(\omega) = N\gamma/2 \tag{6}$$

where $N$ is the number of inclusions per unit volume and $\gamma$ the total scattering cross-section of a single inclusion.

Equation (6) imposes no frequency limitation. Therefore, upon substituting eqn. (6) into eqn. (1) we immediately get the wave velocity, $c(\omega)$. For evaluation of $c_0$ appearing in eqn. (1) we can use any suitable result from the static analysis, since

$$c_0^2 = E/\rho \tag{7}$$

where $E$ is the relevant static modulus and $\rho$ the effective density:

$$\rho = (1 - \phi)\rho_1 + \phi\rho_2 \qquad (8)$$

Here, 1 refers to matrix material and 2 to inclusion.

This simple approach of independent scatterers invokes the obvious physical properties of the effective medium, namely, its causality, passivity and linearity, which enables one to resort to the K–K relations. However, it neglects the phenomena of multiple scattering, which may substantially modify the effective response.

To this end, we apply the concept of differential effective media, which is well known in the statics of composites (see Norris[8] for a survey). We extend it to the dynamic case, making use of the theory of independent scatterers as a building block.

The concept of differential effective media is based on an observation that a particular way of producing a material is not reflected in the governing equations, which, under the usual statement, account for the final phase geometry only. The solution for 'dense' systems may therefore be approximated by superimposing that for a 'tenuous' medium.

Consider the example of a two-phase mixture with a lossless homogeneous background medium and randomly dispersed identical lossless inclusions at the volume concentration $\phi$. We begin the construction process with a matrix of phase, say, 1 and imbed inclusions of phase 2 in dilute concentration, which is chosen so as to insure slight rescattering. In the dynamic case, care also must be taken to avoid overlapping and a consequent occurrence of inclusion-size distribution. Next, a homogenization is carried out by employing a theory of the effective dynamic response, and then new inclusions are imbedded into the 'homogeneous' matrix. By repeating this build-up procedure one arrives at the volume concentration prescribed.

Let $\{I\}$ be a set of parameters which identify relevant physical properties of inclusions and $\{M\}$ that of a background medium. Clearly, in the above construction process $\{I\}$ remains constant, while $\{M\}$ takes current values defined by homogenization. This differs from the static differential scheme, for which $\{M\}$ consists of the static moduli solely. In the sequel subscript 1 refers to matrix, 2 to inclusion and $\sim$ denotes the current value.

The increase in the total absolute concentration, $d\phi$, and the current increase, $d\tilde{\phi}$, are related by

$$d\phi = (1 - \phi)\,d\tilde{\phi} \qquad (9)$$

with initial condition $\phi = 0$ at $\tilde{\phi} = 0$, while the evolution of the static effective moduli, $\mathbf{L}$, is given by

$$\frac{d\tilde{\mathbf{L}}_1}{d\phi} = \frac{(L_2 - \tilde{\mathbf{L}}_1)T}{(1 - \phi)} \tag{10}$$

with initial condition $\tilde{\mathbf{L}} = L_1$ at $\phi = 0$. Here $T$ is a tensor describing the matrix–inclusion interaction, the explicit form of which is known for several particular cases (Norris[8]).

In the dynamic case, the set $\{M\}$ contains also the mass density, $\tilde{\rho}_1$, attenuation, $\tilde{\alpha}_1(\omega)$, and velocity, $\tilde{c}_1(\omega)$. For the case of solid composite media, the effective density, $\rho$, is linear in $\phi$ and is given by eqn. (8), which provides

$$\tilde{\rho}_1 = (1 - \phi)\rho_1 + \phi\rho_2 \tag{11}$$

It is to be noted that this expression should be viewed as a reasonable approximation, in particular, when no statistical information of higher order, such as a correlation function, is available. As other effective entities, it is associated with the representative volume element and may not be applicable to smaller volumes. Recently, Willis[9] treated the effective density as a non-local operator and derived the associated bounds.

Now we turn to the attenuation, $\alpha(\omega)$, which in the absence of rescattering is given by eqn. (6). This approximate additive rule may be readily adapted in the framework of the differential scheme. We apply the well-known incremental removal-replacement process leading to a realizable effective medium. Namely, at each infinitesimal step the volume of 'homogeneous' material is removed by $\Delta\tilde{\phi}$ and replaced with the same volume of inclusion material. Thus

$$\tilde{\alpha}(\phi + \Delta\phi) = \tilde{\alpha}(\phi)(1 - \Delta\tilde{\phi}) + \Delta\tilde{\phi}\tilde{\gamma}/(2V) \tag{12}$$

where $\tilde{\gamma}$ is the current scattering cross-section of a single inclusion in homogenized matrix and $\Delta\tilde{\phi}$ the increase in current concentration. Since the homogenized matrix is a dispersive and attenuative medium, the concept of scattering cross-section, which is usually applied to elastic waves, should be properly reformulated. However, if the attenuation is not too large, a satisfactory estimate may be obtained from the elastic scattering cross-section, $\gamma$, by replacing the matrix parameters with the effective current parameters. In this case, the effective wave number, $\tilde{q}_1$, is frequency-dependent.

The first term in the right-hand side of eqn. (12) describes the 'accumulated' attenuation of the matrix, while the second describes the scattering losses due to the increase, $\Delta\tilde{\phi}$. Equations (12) and (9) provide

$$\frac{d\tilde{\alpha}}{d\phi} \approx \frac{\tilde{\gamma}/(2V) - \tilde{\alpha}}{1 - \phi} \tag{13}$$

with initial condition $\tilde{\alpha} = 0$ at $\phi = 0$.

It remains to find the phase velocity, $\tilde{c}(\omega)$. This is given by

$$\tilde{c}(\omega) = \tilde{c}_0 \left[ 1 + \frac{2\omega^2}{\pi} \tilde{c}_0 P \int_0^\infty \frac{\tilde{\alpha}(\omega')\,d\omega'}{\omega'^2(\omega'^2 - \omega^2)} \right]^{-1} \tag{14}$$

Having determined $c(\omega)$ and $\alpha(\omega)$ by the above procedure one may define the dynamic modulus, $E(i\omega)$, for the full frequency interval, $0 \le \omega < \infty$, via the usual relations of viscoelasticity:

$$\text{Re } E(i\omega) = (C_r^2 - C_i^2)\rho \tag{15}$$

$$\text{Im } E(i\omega) = 2C_r C_i \rho \tag{16}$$

where

$$C_r = \text{Re } C(i\omega) = c\omega^2/(\omega^2 + \alpha^2 c^2) \tag{17}$$

$$C_i = \text{Im } C(i\omega) = -\alpha\omega c^2/(\omega^2 + \alpha^2 c^2) \tag{18}$$

Here and in what follows we drop the ~ to denote the values associated with the final geometry of the composite.

Taking into account that $\alpha(\omega) \sim \text{const}$ as $\omega \to \infty$ and $\alpha(\omega) \sim 0(\omega^4)$ as $\omega \to 0$ ($\alpha(\omega) \sim 0(\omega^3)$, as $\omega \to 0$ for two-dimensional problems) we get for the geometric modulus

$$E_\infty = \lim_{\omega \to \infty} \text{Re } E(i\omega) = c_\infty^2 \rho \tag{19}$$

and for the static modulus

$$E_0 = \lim_{\omega \to 0} \text{Re } E(i\omega) = c_0^2 \rho \tag{20}$$

Since[2]

$$c_\infty > c_0 \tag{21}$$

this implies

$$E_\infty > E_0 \tag{22}$$

This is similar to the case of homogeneous viscoelastic media, which serve, in fact, as an effective model for random elastic mixtures.

The above limiting values of $E(i\omega)$ for $\omega \to 0$ and for $\omega \to \infty$ satisfy the conditions imposed on the complex moduli in the theory of viscoelasticity (Christensen[10]). This modulus may thus be ascribed to an imaginary homogeneous medium, which implies that $c_\infty$ is the speed of propagation of small-amplitude shocks.[7] Unlike the viscoelasticity, which interprets the basic variables as local values, the modulus given by eqns (15) and (16) is an overall effective entity. It may be inapplicable to volumes smaller than the effective representative volume, similarly to the effective density, $\rho$.

## 3. NUMERICAL RESULTS AND CONCLUSIONS

Multiple interactions among inclusions consist of mutual influence of their acoustic fields and of purely geometric correlations of their positions. The latter effect manifests for appreciable volume concentrations only, while the relevance of acoustic interactions is governed by the intensity of scattered fields and may be significant even for dilute mixtures.

Like its static precursor, the model of causal differential effective media (CDEM) disregards the geometrical correlations and involves therefore no higher-order information concerning the phase geometry. The above build-up procedure can be fulfilled in the two ways which lead either to 'overlapping' or 'stirred together' inclusions. In the latter version care is taken to place new inclusions so as to avoid mutual 'penetration' of scatterers. Equations describing the static behavior hold in either case, for they incorporate no information concerning the inclusion size or a correlation function. However, in the dynamic case the response does show sensitivity to the size of inclusions. Thus, to arrive at the composite under consideration, which contains identical inclusions, care must be taken to place new scatterers so as to avoid overlapping and strong rescattering. This may put a limitation on the admissible concentration, $\phi$, a restriction mentioned previously. Hence, this method is particularly suitable for modeling dilute random composites with strong acoustic interactions, which may occur in materials with a considerable mismatch of elastic properties and densities.

Computations have been made for the case of acoustic media containing randomly dispersed spherical inclusions[11] and for the case of axial shear waves in fiber-reinforced composites.[12] It was found that multiple acoustic interactions taken into account by the CDEM model may substantially decrease the resonance overshoot, which is available in the case of intensive scattering, and shift it to a higher frequency. However, the dispersion curves, $c(\omega)$ and $\alpha(\omega)$, preserve, in general, their typical pattern. Namely, when frequency increases from zero, $c(\omega)$ decreases at first, then rises to a larger value and then approaches $c_\infty$, which is greater than $c_0$. The attenuation, $\alpha(\omega)$, increases at first and may then show a resonance overshoot, but decreases thereafter and approaches its geometric limit as $\omega \to \infty$.

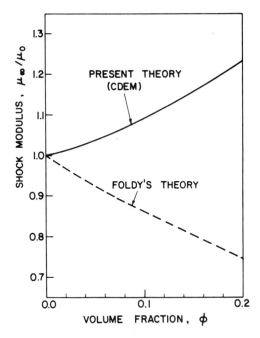

FIG. 1.    Ratio of the effective shock and static moduli, $\mu_\infty/\mu_0$, versus volume fractions, $\phi$.

The capability of predicting the response for the entire frequency interval can be used for evaluating the shock modulus, associated with propagation of a wide-band pulse. Figure 1 shows the ratio of the shock modulus, $\mu_\infty$, to the static modulus, $\mu_0$, as a function of the volume fraction, $\phi$, for boron–aluminum fiber composite subjected to axial shear waves. The ratio of densities is 1·075 and that of moduli is 0·155. Thus, the present theory predicts that the composite is much stiffer in the case of transient (shock) loading than in the static case, unlike Foldy's theory.[4]

# REFERENCES

1. WILLIS, J. R., Variational and related methods for the overall properties of composites, in: *Advances in Applied Dynamics* (C. S. Yih, ed.), Vol. 21, New York, Academic Press, 1981, pp. 1–78.

2. BELTZER, A. I. and BRAUNER, N., SH-waves of an arbitrary frequency in random fibrous composites via the Kramers–Kronig relations, *J. Mech. Phys. Solids*, 33 (1985), 471–487.

3. BRAUNER, N. and BELTZER, A. I., The Kramers–Kronig relations method and wave propagation in porous elastic media, *Int. J. Engng Sci.*, 23 (1985), 1151–1162.

4. BELTZER, A. I. and BRAUNER, N., The causal effective field approximation—application to elastic waves in fibrous composites, *Mech. Mater.*, 5 (1986), 161–170.

5. WEAVER, R. L., Causality and theories of multiple scattering in random media, *Wave Motion*, 8 (1986), 473–483.

6. HODGES, C. H., Confinement of vibration by structural irregularity, *J. Sound Vib.*, 82 (1982), 411–424.

7. BELTZER, A. I. and BRAUNER, N., Shear waves in polycrystalline media and modifications of the Keller approximation, *Int. J. Solids Struct.* (in press).

8. NORRIS, A. N., A differential scheme for the effective moduli of composites, *Mech. Mater.*, 4 (1985), 1–16.

9. WILLIS, J. R., The nonlocal influence of density variations in a composite, *Int. J. Solids Struct.*, 21 (1985), 805–817.

10. CHRISTENSEN, R. M., *Theory of Viscoelasticity*, New York, Academic Press, 1971.

11. BELTZER, A. I. and BRAUNER, N., Acoustic waves in random discrete media via a differential scheme, *J. appl. Phys.*, 60 (1987), 538–540.

12. BELTZER, A. I. and BRAUNER, N., Elastic waves in random composites via a differential scheme, *Mech. Res. Communs* (in press).

# 10

# The Effects of Combined Shear and Direct Loads on the Damping Behaviour of Unidirectional FRPs*

T. A. WILLWAY

*Department of Civil Engineering, Brighton Polytechnic, Moulsecoomb, Brighton BN2 4GJ, UK*

## ABSTRACT

*This chapter describes a theoretical and experimental study of the effects of both applied and induced combined shear and direct loads on the damping of unidirectionally reinforced FRPs.*

*A theoretical model based on the energy being dissipated in the resin only is derived using equivalent strain analyses previously developed for isotropic materials. The model was applied to two different combined loading situations and predictions compared with experimental results.*

## INTRODUCTION

Previous investigations of the dynamic properties of FRPs have been limited to purely direct or shear loading situations. Whilst such data are an essential first phase of the assessment of material properties, practical loading situations inevitably involve combined shear and direct forces. This chapter describes an experimental and theoretical investigation of the effect of both applied and induced combined dynamic shear and direct loads on the loss factors of unidirectional short and continuous CRFPs via

---

* Work carried out by author at the Institute of Sound and Vibration Research, University of Southampton, UK.

consideration of the energy dissipated and stored in the separate phases of the composite.

## THEORY

### General Theory of Damping under Combined Stresses (for Isotropic Materials)

The fundamental assumption of the equivalent stress analysis of damping under combined direct and shear stresses is that the energy dissipated per unit volume per cycle in a material is the same for shear stress ($\tau$) as for direct stress ($\sigma$), provided the shear stress amplitude is $K$ times the direct stress amplitude, $K$ may be evaluated from the loss factors and moduli of the material in shear and flexure as follows:

$$K = \left[ \frac{G}{E} \frac{\eta}{\beta} \right]^{1/2} \tag{1}$$

An analogous equivalent strain approach may also be adopted when $K^1$ (equivalent strain constant) may be found from

$$K^1 = \left[ \frac{E}{G} \frac{\eta}{\beta} \right]^{1/2} \tag{2}$$

where $G$ is shear modulus, $E$ is Young's modulus, $\eta$ is axial or flexural loss factor, $\beta$ is shear loss factor.

If the equation $\tau = K\sigma$ is used to indicate that for pure shear stress $\tau$ numerically equal to $K\sigma$, some chosen physical quantity has the same value as in direct stress $\sigma$, Table 1 can be prepared. If $K$ is found from (1) to coincide with $K$ for one of the physical properties of Table 1, it is possible that the damping of the material may be governed by that parameter.

TABLE 1
*K values in $\tau = K\sigma'$*

| Physical quantity | K | Comment |
|---|---|---|
| Damping capacity | 0·48–0·60 | Experimental result |
| Maximum shear stress (or strain) | 0·55 | |
| Distortion strain energy | 0·577 | |
| Total strain energy | 0·632 | Evaluated for $v = 0·25$ |
| Principal strain | 0·80 | Evaluated for $v = 0·25$ |
| Principal stress | 1·0 | |

Hooker[1] derived an equivalent strain expression, assuming the energy dissipated due to distortion and dilation to be independent:

$$\varepsilon_e = \left[ \varepsilon_x^2 + \frac{dG}{E} \gamma xy^2 \right]^{1/2} \tag{3}$$

where $\varepsilon_x$ is direct strain, $\varepsilon_e$ is equivalent direct strain, $\gamma_{xy}$ is shear strain, $d = $ a factor accounting for the governing energy dissipation parameter, determined from $K$ or $K^1$ values.

The value of $d = 1$ corresponds to dissipation dependent on total strain energy.

## The Relationship between the Energy Dissipated in Shear and Flexure for FRPs

Values of $K$ determined from both previous experimental and theoretical measurements of the shear and flexural dynamic properties of FRPs were so dissimilar to those in Table 1, as to suggest that the energy dissipated in a unidirectional FRP is not related to any of those parameters in Table 1.

For isotropic materials it would then be reasonable to conclude that the energies dissipated in shear and flexure are independent. However, as most energy dissipated in the FRP is dissipated in the resin, whilst most of the energy is stored in the fibres, perhaps a more accurate representation of the damping under combined stresses would result from consideration of the energy dissipated and stored in the separate phases of the composite.

### Energy Dissipated in a Viscoelastic Resin

For a viscoelastic resin, the shear loss factor $\beta$ and flexural loss factor $\eta$ are very similar in magnitude. Therefore

$$K_r = \sqrt{\frac{G_r}{E_r}}$$

where subscript r denotes resin for $v = 0.25$, $K_r = 0.632$, i.e. the energy dissipated in a resin is dependent on the total strain energy in the resin (Table 1) and from (3)

$$\varepsilon_{er} = \left[ \varepsilon_{xr}^2 + \frac{G_r}{E_r} \gamma_{xy} r^2 \right]^{1/2} \tag{4}$$

Substitution of this expression into the fundamental equations for material loss factor reveals the resin loss factor to be independent of mode.

**Energy Dissipated in Unidirectional FRPs under Combined Loading**

If energy dissipation in a FRP (perfectly bonded) occurs only in the resin, then the energy dissipated in the FRP under combined stresses must be governed by the mechanism of energy dissipation in the resin, i.e. the total strain energy in the resin matrix.

Therefore

$$\Delta U_{cc} = J_r E_r^2 \int_{v_r} \varepsilon_{er}^2 \, dv \qquad (5)$$

where $\Delta U_{cc}$ = energy dissipated in the FRP under combined loading, $J_r$ = resin damping constant,[3] $v_r$ = volume of matrix, $\varepsilon_{er}$ = equivalent strain distribution in the resin.

**Energy Stored in Unidirectional FRPs under Combined Loading**

Energy is stored in both the fibres and matrix of FRPs. Under combined shear and flexural stresses, the total energy stored is the sum of the flexural and shear strain energies.

In flexure, the strain in the fibres $\varepsilon_F$ equals the strain in the matrix $\varepsilon_r$ equals the strain in the FRP, $\varepsilon$. The flexural strain energy $U_f$ may then be evaluated from

$$U_f = \int_v \frac{1}{2} E_c \varepsilon^2 \, dv \qquad (6)$$

However, in shear, the stress in the fibres $(\tau_f)$ and matrix $(\tau_m)$ are equal $(\tau)$. Then the strain energy as a result of torsion of the rod $U_s$

$$U_s = \int_v \frac{\tau^2}{2G_c} \, dv \qquad (7)$$

The total energy stored in the unidirectional FRP rod subjected to combined shear and flexure is then equal to $U_T$:

$$U_T = U_s + U_F \qquad (8)$$

**Loss Factor of Unidirectional FRPs under Combined Loading**

The loss factor of the unidirectional FRPs under combined loading may then be found by substituting eqns. (5) and (8) into the fundamental equation for loss factor,[2-4] as follows:

$$\eta = \frac{\Delta U_{cc}}{2U_T} \qquad (9)$$

(a) For a rod with combined applied torsion and a linear bending moment, equation (9) may be shown to yield[2]

$$\eta_{cc} = (1 - v_f)\eta_r \frac{\left[ E_r\left(\dfrac{\varepsilon_{max}}{\gamma_{maxr}}\right)^2 + 6G_r \right]}{\left[ E_c\left(\dfrac{\varepsilon_{max}}{\gamma_{maxr}}\right)^2 + \dfrac{6G_r^2}{G_c} \right]} \tag{10}$$

where $V_f$ = volume fraction of fibres, $\varepsilon_{max}$ = maximum flexural strain in resin/composite, $\gamma_{maxr}$ = maximum shear strain in resin on rod perimeter.

Although no exact theoretical analysis of the loss factor of short FRPs under combined stresses was made, (10) may be assumed as an approximation.[2]

(b) For a rod under combined applied torsion and induced axial strains. When a rod having longitudinally restrained ends is subjected in torsion to a shear strain $\gamma$ an axial strain $\varepsilon_a$ is induced. The total strain (TS) of a combined nature can be shown to be:[2]

$$TS = \varepsilon_a + \gamma \tag{11}$$

where

$$\gamma_{r,c} = \frac{r}{R_0}\gamma_{maxr,c} \tag{12}$$

$$\varepsilon_{ar,c} = \frac{1}{2}\left(\frac{r}{R_0}\gamma_{maxr,c}\right)^2 \tag{13}$$

subscripts r, c refer to resin and composite respectively.

Substituting eqns (12) and (13) into eqns (4)–(9) yields the following equation for the loss factor of a unidirectional FRP in torsion including the effects of induced axial strains[2] ($\eta_{hc}$).

$$\eta_{hc} = (1 - V_f)\eta_r \frac{(E_r\gamma_{maxr}^2 + 6G_r)}{\left( E_c\gamma_{max}^2 + 6\dfrac{G_r^2}{G_c} \right)} \tag{14}$$

Note: if $V_f = 0$ (a resin rod), $\eta_{hc} = \eta_r$ independent of strain amplitude as expected.

Using eqns (10) and (14) the loss factors of the CFRP specimens described in Table 2 under combined shear and direct dynamic loads were predicted from the properties of their resin, fibres and composites and compared with experimental results.

TABLE 2
*Description of specimens*

| | Description | Dimensions (mm) | | $E$ (GN/m$^2$) | $G$ (GN/m$^2$) |
|---|---|---|---|---|---|
| | | clamped-clamped | clamped-free | | |
| CRN1 | RB6 + 60% vf aligned CCF | | $l = 171$ $\varphi = 7$ | 207 | 0·95 |
| CRN4 | RB8 + 60% vf aligned CCF | | $l = 171$ $\varphi = 7$ | 207 | 1·54 |
| CRN10 CRN11 | RB13 + 50% vf aligned CCF | $l = 342$ $\varphi = 7$ | $l = 171$ $\varphi = 7$ | 174 | 2·08 |
| SCRN1 SCRN2 | RB13 + 60% vf aligned short CF (length = 2·25 mm) | $l = 342$ $\varphi = 7$ | $l = 171$ $\varphi = 7$ | 204 | 2·57 |
| RR1 | Pure resin-rod | $l = 342$ $\varphi = 7$ | $l = 171$ $\varphi = 7$ | 2·2 | 0·80 |

CCF = continuous EHm-s carbon fibre.
Short CF = short EHm-s carbon fibre.

## Experiments

The effects of combined dynamic stresses on the damping of the rods described in Table 2 were studied using an especially developed apparatus shown in Fig. 1 and fully described in Ref. 2.

The loss factors of both clamped-clamped and cantilevered solid rods undergoing steady state torsional vibration were calculated from measurements of energy supplied and strain energy stored in the specimens at resonance.[1,2] This technique makes no assumptions as to the linearity of specimen damping. Loss factors were also estimated using the linear free decay method[2,3] to indicate the sensitivity of that model to deviations from simple strain distributions.

Comparisons of experimental and theoretical results for each specimen for fully clamped and cantilevered ends enabled conclusions to be drawn concerning the effects of high amplitude torsional vibrations and induced axial strains on the shear damping behaviour of rods and the accuracy of developed theories.

Alternatively, the coupled bending–torsion vibrations of the cantilevered specimens could be obtained by adding weights to the inertia bar. Harmonic excitation by the coil and magnet units then produced coupled rotational and translational motion of the inertia bar and consequently

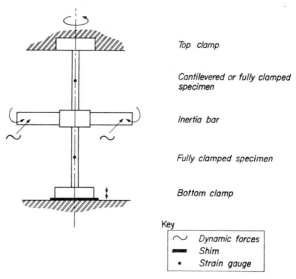

Top clamp

Cantilevered or fully clamped specimen

Inertia bar

Fully clamped specimen

Bottom clamp

Key

| | |
|---|---|
| $\sim$ | Dynamic forces |
| ▬ | Shim |
| • | Strain gauge |

FIG. 1.    Combined loading apparatus.

combined torsion and bending in the specimen. Using the inertia bar and two additional masses at one attachment point, six different natural frequencies and mode shapes were found.

Extraneous losses in the apparatus were measured via tests on duralumin specimens. No significant variation in extraneous losses were found with end constraint. Although the losses were low, even at high strain amplitudes, they were corrected for as detailed in Refs 1 and 2.

## DISCUSSION OF RESULTS

### (a) Combined Torsion and Linear Bending Moment

As predicted theoretically, the loss factor of the visco-elastic resin rod RR1 was found to remain constant for all modes (Table 3) confirming the energy dissipated in the visco-elastic resin to be dependent on the total strain energy in the resin as assumed in the theory.

From the theoretical curves (10) of loss factor versus $\varepsilon_{max}/\gamma_{maxr}$ for CRN1, CRN4, CRN10 and SCRN1 (see Fig. 2) it can be concluded that the loss factors of the unidirectional CFRPs decreased monotonically as the ratio of flexure to shear in the mode increased, from a maximum in pure flexure.

TABLE 3
*Combined mode loss factors found using logarithmic decrement method*

| Mode | CRN1 | CRN4 | CRN10 | SCRN1 | RR1 |
|------|------|------|-------|-------|-----|
| 1 | 0·466 7 | 0·088 7 | 0·017 0 | $7·5 \times 10^{-3}$ | 0·029 0 |
| 2 | 0·035 7 | 0·010 5 | 0·002 7 | $1·7 \times 10^{-3}$ | 0·026 8 |
| 3 | — | 0·086 9 | 0·016 6 | $7·4 \times 10^{-3}$ | 0·026 6 |
| 4 | 0·037 9 | 0·010 2 | 0·002 6 | $1·6 \times 10^{-3}$ | 0·026 5 |
| 5 | — | 0·084 1 | 0·016 5 | $6·9 \times 10^{-3}$ | 0·026 4 |
| 6 | 0·038 5 | 0·011 8 | 0·002 9 | $2·1 \times 10^{-3}$ | 0·028 8 |

Experimental results for the unidirectional CFRPs showed similar behaviour to that predicted theoretically, particularly those determined using the logarithmic decrement method (see Fig. 2 and Table 3). However, agreement was qualitative only, actual values being similar only in the near torsional modes of CRN4 and CRN10. The apparent underestimate of near flexural mode loss factors may be attributed to transverse shear deformation which is significant in members such as FRPs having a high ratio of flexural to shear moduli.[2,5]

If the fundamental assumption regarding energy dissipation in FRPs is correct and the assumed expressions for the strain distribution in the

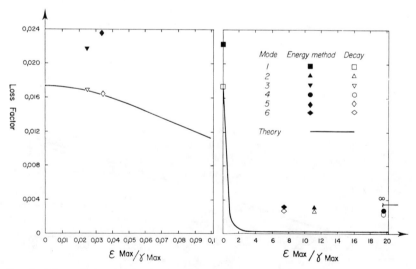

FIG. 2. Theoretical and experimental loss factor versus $\varepsilon_{max}/\gamma_{max}$ CRN10 (at principal strain = $100 \, \mu$s).

TABLE 4
*Resin and fibre properties*

|  | E (GN/m²) | η | G (GN/m²) | ν |
|---|---|---|---|---|
| RB6 | 1·07 | 0·3500 | 0·39 | 0·385 |
| RB8 | 1·75 | 0·0774 | 0·63 | — |
| RB13 | 2·98 | 0·0180 | 1·08 | — |
| RR1 | 2·20 | 0·028 | 0·8 | — |
| EHM-s fibres | 345 (longitudinal) 33·5 (transverse) | — | 33·5 | 0·24 |

composite and resin accurate, graphs of energy stored versus energy dissipated in the resins of the FRPs should reveal all modal loss factors occurring on a single straight line, the gradient of which should be $2\pi \times$ the component resin loss factor given in Table 4. However, such graphs plotted from values interpolated from experimental results as described in Ref. 2 comprise separate strain line plots for each mode (e.g. see Fig. 3). The gradient of the lines (apparent resin modulus) increases with the ratio of the

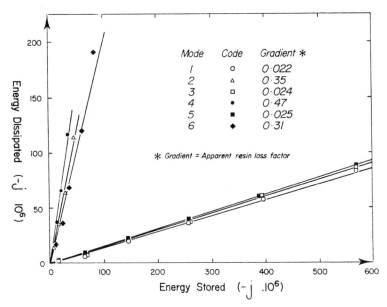

FIG. 3.   Energy dissipated and stored in resin of CRN10.

flexural to shear strain in the mode for each specimen. Furthermore, the increase in apparent loss factor appears to be proportional to the ratio of the flexural to shear moduli of the specimens (or inversely to the resin moduli). This behaviour, apparently contradicting the theory, may also be explained with reference to shear deformation.[2,5]

If the energy stored in the resin and CFRP as a result of transverse shear deformation had been included in the model, the theoretical and experimental results would have shown better agreement for modes including flexure.

Agreement between experimental values of loss factor found using the energy and free decay methods was poor, the latter being more consistent and generally smaller. It was concluded that inconsistencies in rod manufacture may lead to strain gauge readings being quantitatively misrepresentative and consequently errors in the calculation of loss factor using the energy method. Measurements found using the free decay technique are however, representative of the rod as a whole although no indications of variation through the volume are provided.

## (b) Combined Torsion and Induced Axial Strains

As the energy dissipation in both resins and CFRPs is dependent on the dissipation in the resin for which the damping exponent $n = 2$, the loss factor of cantilevered resin and CFRP rods in torsion (with no induced axial strains), theoretically should remain constant with varying shear strain amplitude.

Furthermore, according to (14) the loss factor of a resin rod is independent of strain distribution and consequently should also remain constant for all strain amplitudes for clamped-clamped boundary conditions when axial strains are induced.

However, for clamped-clamped unidirectional CFRP rods, (eqn. 14) predicts a very slight decrease in loss factor as the shear strain amplitude increases (e.g. see Fig. 4). Although the magnitude of the decrease in loss factor is effectively negligible, more significant is the fact that the loss factor of a clamped-clamped rod in torsion should not, theoretically, increase at high shear strain amplitudes as a result of the induced axial strain alone. Therefore, theoretically high amplitude shear vibrations and induced axial strains do not effect the loss factor of resin or unidirectional CFRP rods in torsion.

The experimentally measured loss factors of resin rod RR1 were found to be essentially independent of shear strain amplitude measurement technique and boundary conditions as predicted theoretically.

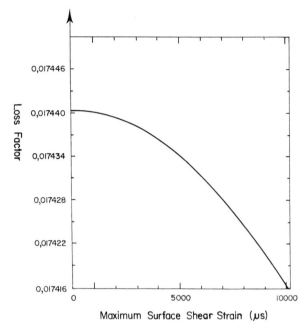

FIG. 4.  Theoretical variation of loss factor with shear strain amplitude, clamped-clamped CRN11 (including the effects of induced axial strains).

Therefore, the assumption that the energy dissipated in a resin is dependent on the total strain in the resin, inherent in the theory, has been verified. Furthermore, it appears that for a resin, the free decay technique is adequate for measuring the loss factor of rods having both applied shear and induced axial strains.

The loss factor of CRN11, an aligned continuous CFRP rod (see Fig. 5) remained constant up to a surface shear strain amplitude of approximately 900 $\mu$s, thereafter increasing slightly with increasing strain thus contradicting theoretical predictions. This behaviour was exhibited by the cantilevered and clamped-clamped rods using both the energy input at resonance and the free decay techniques of loss factor measurement. The increasing loss factors could not therefore be a result of reduced axial strains.

Considering the gradient (Fig. 5) it can be seen that for the CRN11 the increase in loss factor with increasing strain above 900 $\mu$s is very slightly greater for clamped-clamped than clamped-free end constraints. This effect, observed using both the energy input at resonance and the logarithmic

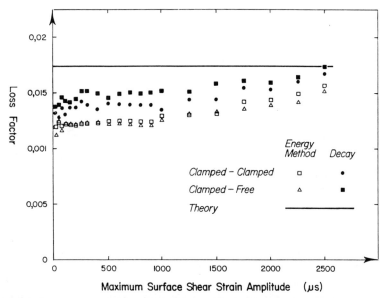

Fɪɢ. 5.  Experimental variation of loss factor with shear strain amplitude CRN11.

decrement methods, suggests that induced axial strains resulted in a small increase in the loss factor of the CFRP rod, contrary to theoretical predictions.

The loss factor of SCRN2 (an aligned short fibre CFRP rod) increased steadily from a constant value at low strain amplitudes, with further increasing shear strain amplitudes, once more contradicting theoretical predictions.

However, the loss factor of the cantilevered rod was found to be greater than that of the clamped-clamped rod using both experimental methods. The increases in loss factor with increasing shear strain measured using the free decay method were similar for both clamped-clamped and clamped-free end constraints. Thus, as predicted, theoretically, the induced axial strains had a negligible effect on the CFRP rod loss factor. However, experimental results according to the energy method suggested that the induced axial strain significantly decreased the loss factor of the rod.

Differences in results taken using the logarithmic decrement and energy input at resonance techniques may be explained once again by the non-uniformity of the rod materials through their volumes, which would particularly affect the latter results. Such material irregularities would also cause the variations in loss factors observed between the fully clamped and cantilevered specimens at low strain amplitudes when induced axial strains are negligible.

## CONCLUSIONS

For a rod subjected to torsion with a linear bending moment, the loss factor was found, experimentally and theoretically, to increase from a minimum in pure flexure to a maximum pure shear with the increasing ratio of shear to flexural strain in the mode.

Comparisons between theoretical and experimental results revealed shear deformation to be the most significant source of energy dissipation in CFRPs having a high ratio of flexural to shear moduli, for all modes having a flexural component. It was concluded that although there was some evidence of other mechanisms of energy dissipation, a model based upon the energy dissipated in the resin as (10), but including the effects of shear deformation would reasonably represent the damping of unidirectional CFRPs under combined dynamic shear and flexure.

When the model was applied to resin and unidirectional CFRPs undergoing high amplitude torsional vibrations, with and without induced axial strains, the structural loss factor of the resin rod was predicted to be independent of the shear strain amplitude and induced axial strains, whilst that of the CFRP was predicted to decrease by a negligible amount with increasing strain amplitudes.

However, results of experiments on resin and CFRP rods with clamped-clamped and clamped-free ends suggested other mechanisms not included in the model to cause the loss factors of both rods to increase with increasing shear strain amplitude. However, as predicted theoretically, the effects of induced axial strains on the shear loss factors were negligible.

## REFERENCES

1. HOOKER, R. J., High damping metals, Ph.D. thesis, University of Southampton, 1975.
2. WILLWAY, T. A., Stiff light, highly damped CFRPs, and the effect of complex loads on damping, Ph.D. thesis, University of Southampton, 1986.
3. LAZAN, B. J., *Damping of Materials and Members in Structural Mechanics*, Oxford, Pergamon Press, 1968.
4. NASHIF, A. D., JONES, D. J. G. and HENDERSON, J. P., *Vibration Damping*, New York, Interscience, 1985.
5. HEYBEY, O. and KARASZ, F. E., Experimental study of flexural vibrations of thick beams, *Journal of Applied Physics*, **47**, No. 7 (1976), 3232.

# 11

# Effects of Shear Deformation on the Onset of Flexural Failure in Symmetric Cross-ply Laminated Rectangular Plates

G. J. TURVEY

*Department of Engineering, University of Lancaster,
Bailrigg, Lancaster LA1 4YR, UK*

## ABSTRACT

*An initial flexural failure analysis for symmetric cross-ply laminated simply supported rectangular plates is described. The analysis is based on a simple higher-order shear deformation plate theory (HSDT) and simple lamina and interlaminar shear failure criteria. Initial failure pressures and associated plate centre deflections are computed for CFRP plates. The computed values are compared with corresponding values derived from classical laminated plate theory (CPT). The comparison demonstrates that the neglection of the inherent shear flexibility of a laminated plate leads to an unconservative estimation of the failure pressure and a conservative estimation of the associated centre deflection. Even for moderately slender plates these under/over-estimations may be significant, e.g. for a three-layer square plate with a width–thickness ratio of 10 the difference amounts to about 10%.*

## NOTATION

| | |
|---|---|
| $a, b$ | Plate width and length |
| $A_{ij}, B_{ij}, D_{ij}, E_{ij}, F_{ij}, H_{ij}\,(i,j = 1, 2, 6)$ | Extensional–flexural stiffnesses |
| $A_{ij}, D_{ij}, F_{ij}\,(i,j = 4, 5)$ | Transverse shear stiffnesses |
| $D_{\mathrm{T}}\,(= E_{\mathrm{T}} h_{\mathrm{o}}^3 (1 - v_{\mathrm{LT}} v_{\mathrm{TL}})^{-1}/12)$ | Reference plate stiffness |
| $e_1^{\mathrm{o}}, e_2^{\mathrm{o}}, e_6^{\mathrm{o}}$ | Mid-plane strain components |
| $e_4^{\mathrm{o}}, e_5^{\mathrm{o}}$ | Transverse shear strain components |

2.141

| | |
|---|---|
| $e_L, e_T, e_{LT}$ | In-plane strain components with respect to lamina principal material axes |
| $e_{LZ}, e_{TZ}$ | Transverse shear strain components with respect to lamina principal material axes |
| $E_L, E_T$ | Lamina moduli parallel and transverse to the fibre direction |
| $G_{LT}, G_{LZ}, G_{TZ}$ | Lamina in-plane and transverse shear moduli |
| $h_o$ | Plate thickness |
| $k_1^o, k_2^o, k_6^o$ | Mid-plane curvature components |
| $k_1^2, k_2^2, k_6^2$ | Second-order curvature components |
| $k_4^2, k_5^2$ | Second-order transverse shear strain components |
| $M_1, M_2, M_6$ | Stress couples |
| $N_1, N_2, N_6$ | Stress resultants |
| $N_L$ | Number of laminae |
| $P_1, P_2, P_6$ | Second-order stress resultants |
| $q, q_o$ | Normal pressure |
| $\bar{q}_{T-H} (= q_o a^2 s_T^{t-1} h_o^{-2}), \bar{q}_{ILS}^L, \bar{q}_{ILS}^T$ | Dimensionless initial failure pressures for in-plane and interlaminar shear failure modes |
| $Q_1, Q_2$ | Transverse shear stress resultants |
| $R(= ba^{-1})$ | Plate aspect ratio |
| $R_1, R_2$ | Second-order transverse shear stress resultants |
| $s_L^t, s_L^c$ | Lamina fibre direction tensile and compressive strengths |
| $s_T^t, s_T^c$ | Lamina tensile and compressive strengths transverse to the fibre direction |
| $s_{LT}$ | Lamina in-plane shear strength |
| $s_{LZ}, s_{TZ}$ | Lamina transverse shear strengths parallel and transverse to the fibre direction |
| $u, v$ | In-plane displacement components |
| $w$ | Plate deflection |
| $\bar{w}_c^{T-H} (= w_c D_T a^{-2} s_T^{t-1} h_o^{-2})$ | Dimensionless associated plate centre deflection |

| | |
|---|---|
| $x, y, z$ | Cartesian coordinates |
| $\bar{x}(=xa^{-1}), \bar{y}(=yb^{-1})$ | Dimensionless coordinates |
| $\lambda(=ah_o^{-1})$ | Plate width–thickness ratio |
| $v_{LT}, v_{TL}, v_{LZ}$ | Lamina Poisson ratios |
| $\sigma_L, \sigma_T$ | Lamina direct stresses parallel and transverse to the fibre direction |
| $\sigma_{LT}$ | Lamina in-plane shear stress |
| $\sigma_{LZ}, \sigma_{TZ}$ | Lamina transverse shear stress components parallel and normal to the fibre direction |
| $\phi_1, \phi_2, \phi_3$ | Initial failure functions for in-plane and interlaminar shear failure modes |
| $\phi_x, \phi_y$ | Normal rotation components at the plate mid-plane |

**Mnemonics**

CPT    Classical laminated plate theory or value of variable derived from classical laminated plate theory

HSDT    Higher-order shear deformation laminated plate theory or value of variable derived from higher-order laminated plate theory

ILS    Value of variable derived from the interlaminar shear failure criterion

T–H    Value of variable derived from the Tsai–Hill failure criterion

## 1.  INTRODUCTION

The transverse shear stiffnesses of fibre-reinforced composite plastics materials such as carbon fibre reinforced plastics (CFRP) are typically of the order of 4% of the tensile stiffness in the fibre direction. For isotropic materials, such as mild steel, the corresponding figure is 38%. It is, therefore, not surprising that shear deformation effects are more significant in structures made of the former material. This 'fact of life' is becoming more widely appreciated as evidenced by the increasing use of Reissner–Mindlin-type first-order shear deformation theory for the analysis/design of laminated fibre-reinforced plate structures (see, for example, Refs 1–5). These first-order theories account for shear deformation in an average sense and, therefore, require the specification of so-called 'shear correction factors'. For isotropic plates a single factor suffices,[6,7] though there has been a long-running debate as to what the precise value of this factor should be (see Refs 8 and 9). For laminated plates the situation is, as expected, more complex. In general, three shear correction factors have to be specified. Moreover, it is far from clear that the suggested methods[10,11] for their computation are both valid and accurate for arbitrary lay-ups. Thus, the need to compute several shear correction factors combined with the uncertainties associated with their computation must be viewed as a major weakness of the Reissner–Mindlin-type first-order shear deformation laminated plate theories.

It has been shown[12] that first-order shear deformation plate theory is superior to classical laminated plate theory, particularly in respect of deflection predictions. Nevertheless, neither theory provides accurate stress distribution predictions when compared with the distributions obtained from a full three-dimensional analysis. This latter characteristic may be regarded as the *other* major shortcoming of first-order theory, especially as there are many practical situations in which strength rather than stiffness design criteria predominate.

The two major weaknesses of first-order shear deformation laminated plate theory outlined above may be overcome by recourse to the use of higher-order shear deformation plate theory. In recent years, several such theories have been advanced. Perhaps the first, and probably the most complex, is that due to Lo *et al.*[13] Other theories have since been developed by Murthy,[14] Reddy[15,16] and Bhimaraddi and Stephens.[17] The linear higher-order theories due to Reddy[15] and Bhimaraddi and Stephens[17] are, in fact, identical. Several of these higher-order theories are neither simple nor consistent. Here, we have made use of Reddy's theory,[15] because of its relative simplicity and also because it satisfies the requirement that the shear stresses vanish at the plate surfaces. Reddy's theory does not require shear correction factors and has been shown to provide accurate stress distribution predictions.[15] Moreover, exact solutions may be derived from this theory for specific types of lay-up, plate geometry and edge conditions. Thus, these solutions provide a means of accurately quantifying the effect of shear deformation on a variety of response characteristics.

As far as the present paper is concerned, attention is focused on assessing the effect of shear deformation on the initial failure pressure of uniformly loaded rectangular plates. Previous studies to determine the initial failure pressure of laminated strips and plates have been reported by the author[18-24] and others,[25] but all have been based on classical laminated plate theory. The author has also conducted an analogous study[26] on *isotropic* circular plates to assess the effect of shear deformation on the onset of flexural first yield. However, the latter study utilised Mindlin first-order shear deformation plate theory.

Exact higher-order theory solutions[15] for uniformly loaded symmetric cross-ply laminated simply supported plates are used to compute the stress field within each lamina. These stresses are then scaled to satisfy appropriate failure criteria at one or more points within the plate and in so doing the magnitude of the transverse pressure initiating failure is established. This procedure is used to study the effects of thickness–width ratio, aspect ratio and number of laminae on the initial failure pressure and associated centre deflections of rectangular CFRP plates. It is shown that, except for unrealistically thick plates, initial failure is flexure rather than interlaminar shear dominated. Moreover, it is also shown that, compared with initial failure pressures predicted by classical laminated plate theory, shear deformation reduces the initial failure pressure and the percentage reduction increases as the width–thickness ratio and number of laminae decrease.

## 2. INITIAL FLEXURAL FAILURE ANALYSIS

The essential components of an initial failure analysis are two-fold, viz. the solution of the governing higher-order plate equations and an appropriate failure criterion (or criteria). It is convenient to introduce each component separately below before describing how they are combined to determine the initial failure load.

### 2.1. Higher-order Flat Rectangular Plate Theory

The derivation of the governing plate equations has been fully described in Ref. 15. Here, we shall merely restate them, but in a non-dimensional and, therefore, marginally simpler format.

#### 2.1.1. Equilibrium equations

The five equations of static equilibrium (three translational and two rotational) are as follows:

$$N_1^{\cdot} + N_6' = 0$$
$$N_6^{\cdot} + N_2' = 0$$
$$\lambda^2(Q_1^{\cdot} + Q_2') + q + 4\lambda^2(R_1^{\cdot} + R_2') + (4/3)(P_1^{\cdot\cdot} + 2P_6^{\cdot\prime} + P_2'') = 0 \qquad (1)$$
$$M_1^{\cdot} + M_6' - \lambda^2 Q_1 + 4\lambda^2 R_1 - (4/3)(P_1^{\cdot} + P_6') = 0$$
$$M_6^{\cdot} + M_2' - \lambda^2 Q_2 + 4\lambda^2 R_2 - (4/3)(P_6^{\cdot} + P_2') = 0$$

### 2.1.2. Compatibility equations

The relationships between the strains, curvatures and higher-order curvatures are

$$e_1^o = u^{\cdot} \qquad e_2^o = v' \qquad e_6^o = u' + v^{\cdot} \qquad e_4^o = \phi_y + w' \qquad e_5^o = \phi_x + w^{\cdot} \quad (2)$$

and

$$k_1^o = \phi_x^{\cdot} \qquad k_2^o = \phi_y' \qquad k_6^o = \phi_x' + \phi_y^{\cdot} \quad (3)$$

and

$$\begin{aligned}
k_1^2 &= -(4/3)(\phi_x^{\cdot} + w^{\cdot\cdot}) \\
k_2^2 &= -(4/3)(\phi_y' + w'') \\
k_6^2 &= -(4/3)(\phi_x' + \phi_y^{\cdot} + 2w^{\cdot'}) \\
k_4^2 &= -4(\phi_y + w') \\
k_5^2 &= -4(\phi_x + w^{\cdot})
\end{aligned} \quad (4)$$

### 2.1.3. Constitutive equations

The stress resultants, etc., of eqns (1) and the strains, etc., of eqns (2)–(4) are related via the constitutive equations, which are conveniently expressed in matrix form as

$$
\begin{bmatrix} N_1 \\ N_2 \\ N_6 \end{bmatrix} =
\begin{bmatrix} A_{11} & A_{12} & A_{16} \\ & A_{22} & A_{26} \\ \text{symm.} & & A_{66} \end{bmatrix}
\begin{bmatrix} e_1^o \\ e_2^o \\ e_6^o \end{bmatrix} +
\begin{bmatrix} B_{11} & B_{12} & B_{16} \\ & B_{22} & B_{26} \\ \text{symm.} & & B_{66} \end{bmatrix}
\begin{bmatrix} k_1^o \\ k_2^o \\ k_6^o \end{bmatrix}
$$

$$
+ \begin{bmatrix} E_{11} & E_{12} & E_{16} \\ & E_{22} & E_{26} \\ \text{symm.} & & E_{66} \end{bmatrix}
\begin{bmatrix} k_1^2 \\ k_2^2 \\ k_6^2 \end{bmatrix}
$$

$$
\begin{bmatrix} M_1 \\ M_2 \\ M_6 \end{bmatrix} =
\begin{bmatrix} B_{11} & B_{12} & B_{16} \\ & B_{22} & B_{26} \\ \text{symm.} & & B_{66} \end{bmatrix}
\begin{bmatrix} e_1^o \\ e_2^o \\ e_6^o \end{bmatrix} +
\begin{bmatrix} D_{11} & D_{12} & D_{16} \\ & D_{22} & D_{26} \\ \text{symm.} & & D_{66} \end{bmatrix}
\begin{bmatrix} k_1^o \\ k_2^o \\ k_6^o \end{bmatrix}
$$

$$
+ \begin{bmatrix} F_{11} & F_{12} & F_{16} \\ & F_{22} & F_{26} \\ \text{symm.} & & F_{66} \end{bmatrix}
\begin{bmatrix} k_1^2 \\ k_2^2 \\ k_6^2 \end{bmatrix} \quad (5)
$$

$$
\begin{bmatrix} P_1 \\ P_2 \\ P_6 \end{bmatrix} =
\begin{bmatrix} E_{11} & E_{12} & E_{16} \\ & E_{22} & E_{26} \\ \text{symm.} & & E_{66} \end{bmatrix}
\begin{bmatrix} e_1^o \\ e_2^o \\ e_6^o \end{bmatrix} +
\begin{bmatrix} F_{11} & F_{12} & F_{16} \\ & F_{22} & F_{26} \\ \text{symm.} & & F_{66} \end{bmatrix}
\begin{bmatrix} k_1^o \\ k_2^o \\ k_6^o \end{bmatrix}
$$

$$
+ \begin{bmatrix} H_{11} & H_{12} & H_{16} \\ & H_{22} & H_{26} \\ \text{symm.} & & H_{66} \end{bmatrix}
\begin{bmatrix} k_1^2 \\ k_2^2 \\ k_6^2 \end{bmatrix}
$$

$$\begin{bmatrix} Q_2 \\ Q_1 \end{bmatrix} = \begin{bmatrix} A_{44} & A_{45} \\ A_{45} & A_{55} \end{bmatrix} \begin{bmatrix} e_4^o \\ e_5^o \end{bmatrix} + \begin{bmatrix} D_{44} & D_{45} \\ D_{45} & D_{55} \end{bmatrix} \begin{bmatrix} k_4^2 \\ k_5^2 \end{bmatrix}$$

$$\begin{bmatrix} R_2 \\ R_1 \end{bmatrix} = \begin{bmatrix} D_{44} & D_{45} \\ D_{45} & D_{55} \end{bmatrix} \begin{bmatrix} e_4^o \\ e_5^o \end{bmatrix} + \begin{bmatrix} F_{44} & F_{45} \\ F_{45} & F_{55} \end{bmatrix} \begin{bmatrix} k_4^2 \\ k_5^2 \end{bmatrix}$$

$$(6)$$

The stiffnesses in eqns (5)–(6) are computed by summing the contribution from each lamina through the plate thickness and the stiffness expressions are

$$[A_{ij}, B_{ij}, D_{ij}, E_{ij}, F_{ij}, H_{ij}] = \sum_{m=1}^{N} \int_{z_m}^{z_{m+1}} \bar{Q}_{ij}^m [1, z, z^2, z^3, z^4, z^6] \, dz \qquad (7)$$

and

$$[A_{ij}, D_{ij}, F_{ij}] = \sum_{m=1}^{N} \int_{z_m}^{z_{m+1}} \bar{Q}_{ij}^m [1, z^2, z^4] \, dz \qquad (8)$$

in which $i, j = 1, 2, 6$ and $i, j = 4, 5$ for eqns (7) and (8), respectively. For symmetric cross-ply lay-ups, which are the only types considered here, many of the stiffnesses vanish, i.e.

$$A_{16} = A_{26} = A_{45} = B_{16} = B_{26} = D_{16} = D_{26} = D_{45} = E_{16} = E_{26}$$
$$= F_{16} = F_{26} = F_{45} = H_{16} = H_{26} = 0$$

### 2.1.4. Boundary conditions

It appears that exact solutions for the foregoing complex system of plate equations may be obtained for the following special type of simply supported boundary conditions:

$$w = \phi_y = M_1 = P_1 = 0 \quad [\text{along } x = 0, a] \qquad (9a)$$

$$w = \phi_x = M_2 = P_2 = 0 \quad [\text{along } y = 0, b] \qquad (9b)$$

## 2.2. Solution of Governing Equations

A Navier-type solution of the governing equations is possible for the simply supported boundary conditions defined by eqns (9). Thus, the displacement components are defined in double series form as

$$u = \sum_{m=1}^{\infty} \sum_{n=1}^{\infty} u_{mn} \cos \alpha x \sin \beta y$$

$$v = \sum_{m=1}^{\infty} \sum_{n=1}^{\infty} v_{mn} \sin \alpha x \cos \beta y$$

$$w = \sum_{m=1}^{\infty} \sum_{n=1}^{\infty} w_{mn} \sin \alpha x \sin \beta y \tag{10}$$

$$\phi_x = \sum_{m=1}^{\infty} \sum_{n=1}^{\infty} \phi_{x_{mn}} \cos \alpha x \sin \beta y$$

$$\phi_y = \sum_{m=1}^{\infty} \sum_{n=1}^{\infty} \phi_{y_{mn}} \sin \alpha x \cos \beta y$$

in which $\alpha = m\pi a^{-1}$ and $\beta = n\pi b^{-1}$. In a similar manner, the loading on the plate is defined as

$$q = \sum_{m=1}^{\infty} \sum_{n=1}^{\infty} q_{mn} \sin \alpha x \sin \beta y \tag{11}$$

and for a uniformly distributed load—the only case considered here—the pressure coefficients are given by

$$q_{mn} = 16\pi^{-2} m^{-1} n^{-1} q_{\text{o}} \qquad (m, n = 1, 3, \ldots) \tag{12}$$

The unknown displacement amplitudes of eqns (10) are determined by first forming the appropriate displacement derivatives for substitution in eqns (2)–(4), which in turn are substituted into eqns (5) and (6) to yield stress resultants, etc., in terms of displacement amplitudes. The appropriate stress resultant, etc., derivatives are then formed and together with the displacement derivatives are substituted into eqns (1) which may be expressed as five simultaneous equations in terms of the unknown displacement amplitudes as follows:

$$\begin{bmatrix} a_{11} & a_{12} & a_{13} & a_{14} & a_{15} \\ & a_{22} & a_{23} & a_{24} & a_{25} \\ & & a_{33} & a_{34} & a_{35} \\ & \text{symm.} & & a_{44} & a_{45} \\ & & & & a_{55} \end{bmatrix} \begin{bmatrix} u_{mn} \\ v_{mn} \\ w_{mn} \\ \phi_{xmn} \\ \phi_{ymn} \end{bmatrix} = \begin{bmatrix} 0 \\ 0 \\ -q_{mn} \\ 0 \\ 0 \end{bmatrix} \tag{13}$$

Details of the coefficients, $a_{ij}$, in eqns (13) are given in the Appendix. Inversion of eqns (13) leads to the definition of the displacement amplitudes in terms of the known load coefficients and, hence, the complete solution of the higher-order plate equations.

## 2.3. Failure Criteria

As mentioned at the beginning of this section, the failure criterion represents the second principal component of the initial failure analysis. The choice of criterion is clearly very important. A variety of criteria are available to choose from and they range from the very simple to the rather complex. None, as yet, has been found to be entirely satisfactory. Moreover, for our purposes, we shall tacitly assume that lamina failure in flexure (matrix cracking, fibre fracture, etc.) and interlaminar shear failure are independent modes. It is further assumed that the former failure mode is governed by the Tsai–Hill criterion, which may be expressed as

$$\delta_{t,c}^2 \sigma_L (\sigma_L - \sigma_T) + \eta_{t,c}^2 \sigma_T^2 + \zeta^2 \sigma_{LT}^2 = 1 \tag{14}$$

in which $\left. \begin{array}{l} \delta_t = s_L^{t-1} \\ \eta_t = 1 \end{array} \right\}$ when the lamina is in tension

and $\left. \begin{array}{l} \delta_c = s_L^{c-1} \\ \eta_c = s_T^{c-1} \end{array} \right\}$ when the lamina is in compression

and $\zeta = s_{LT}^{-1}$

and that interlaminar shear failure arises whenever either of the following two conditions is satisfied:

$$\sigma_{LZ} s_{LZ}^{-1} = 1 \tag{15a}$$

$$\sigma_{TZ} s_{TZ}^{-1} = 1 \tag{15b}$$

The lamina failure strengths, $s_L^t$, etc., which are used in the later CFRP plate initial failure studies are listed in Table 1, together with the elastic constants.

TABLE 1
*(a) Elastic modular ratios of uni-directional CFRP*

| $E_L/E_T$ | $G_{LT}/E_T$ | $G_{LZ}/E_T$ | $G_{TZ}/E_T$ | $\nu_{LT}$ | $\nu_{LZ}$ |
|---|---|---|---|---|---|
| 14·444 | 0·533 33 | 0·533 33 | 0·533 33 | 0·28 | 0·28[a] |

*(b) Strength ratios of uni-directional CFRP*

| $s_L^t/s_T^t$ | $s_L^c s_T^t$ | $s_T^c/s_T^t$ | $s_{LT}/s_T^t$ | $s_{LZ}/s_T^t$ | $s_{TZ}/s_T^t$ |
|---|---|---|---|---|---|
| 31·25 | 24·25 | 5·25 | 1·875 | 1·875 | 1·875[a] |

[a] Assumed ratios.

## 2.4. Evaluation of the Initial Failure Pressure

Once the elastic solution to the higher-order shear deformation plate bending problem has been obtained, i.e. the coefficients, $u_{mn}$, etc., have been determined, it may be used to establish the strain field throughout the plate. First, the strain components in any lamina are determined from

$$
\begin{aligned}
e_1 &= \dot{u} + z\{\dot{\phi}_x - (4/3)z^2(\dot{\phi}_x + \ddot{w})\} \\
e_2 &= v' + z\{\phi'_y - (4/3)z^2(\phi'_y + w'')\} \\
e_6 &= u' + \dot{v} + z\{(\phi'_x + \dot{\phi}_y) - (4/3)z^2(\phi'_x + \dot{\phi}_y + 2\dot{w}')\} \\
e_4 &= \phi_y + w' - 4z^2(\phi_y + w') \\
e_5 &= \phi_x + \dot{w} - 4z^2(\phi_x + \dot{w})
\end{aligned}
\tag{16}
$$

in which $z_{k-1} \le z \le z_k$. These strain components are then transformed to material principal axes as follows:

$$
\begin{bmatrix} e_L \\ e_T \\ e_{LT} \end{bmatrix} =
\begin{bmatrix} c^2 & s^2 & sc \\ s^2 & c^2 & -sc \\ -2sc & 2sc & (c^2 - s^2) \end{bmatrix}
\begin{bmatrix} e_1 \\ e_2 \\ e_6 \end{bmatrix}
\tag{17a}
$$

and

$$
\begin{bmatrix} e_{LZ} \\ e_{TZ} \end{bmatrix} =
\begin{bmatrix} c^2 & s^2 \\ s^2 & c^2 \end{bmatrix}
\begin{bmatrix} e_4 \\ e_5 \end{bmatrix}
\tag{17b}
$$

in which $c = \cos\theta$ and $s = \sin\theta$.

The lamina stresses, which are required for checking in the failure criteria, are determined from the lamina constitutive equations as follows:

$$\begin{bmatrix} \sigma_L \\ \sigma_T \\ \sigma_{LT} \end{bmatrix} = \begin{bmatrix} Q_{11} & Q_{12} & 0 \\ & Q_{22} & 0 \\ \text{symm.} & & Q_{66} \end{bmatrix} \begin{bmatrix} e_L \\ e_T \\ e_{LT} \end{bmatrix} \qquad (18a)$$

and

$$\begin{bmatrix} \sigma_{LZ} \\ \sigma_{TZ} \end{bmatrix} = \begin{bmatrix} Q_{44} & 0 \\ 0 & Q_{55} \end{bmatrix} \begin{bmatrix} e_{LZ} \\ e_{TZ} \end{bmatrix} \qquad (18b)$$

Now substituting the stresses from eqns (18) into eqns (14) and (15) three failure pressures may be defined. They are

$$\bar{q}_{\text{T-H}} = \phi_1^{-1/2} \qquad \bar{q}_{\text{ILS}}^L = \lambda^{-1}\phi_2^{-1} \qquad \bar{q}_{\text{ILS}}^T = \lambda^{-1}\phi_3^{-1} \qquad (19)$$

Clearly, the initial failure functions, $\phi_1$, $\phi_2$ and $\phi_3$ (which are complex functions of the lamina stresses defined by eqns (14) and (15)), will assume their maximum values at one or more points within the plate. These maximum values, as is clear from eqns (19), correspond to minimum pressure values. In general, the pressure minima are not equal in magnitude and the smallest value is taken as the initial failure pressure, i.e.

$$\min\left(\bar{q}_{\text{T-H}}, \bar{q}_{\text{ILS}}^L, \bar{q}_{\text{ILS}}^T\right)$$

## 3. COMPUTATION OF INITIAL FAILURE FUNCTION MAXIMA

A simple, but effective, procedure was used to compute the maximum values of the initial failure functions, $\phi_1$, etc. It consists of evaluating each function at the nodes of a rectangular grid stretching over one-quarter of the plate (symmetry is exploited). The individual $\phi$-values are scanned to isolate zones in which the maxima are likely to be found. A finer grid is then defined over each local zone and the local nodal values of the functions are re-computed. These values are then re-scanned to yield more accurate $\phi$-maxima. In practice, it was often unnecessary to proceed to the local zone re-computation stage, especially as the first stage coarse grid often included the node(s) at which the initial failure functions assumed their maximum values. Where this was not the case, only small changes (less than 1%) in these maxima were usually determined on proceeding to the local zone re-computation stage, because the initial failure functions were slowly varying and did not exhibit local peaks.

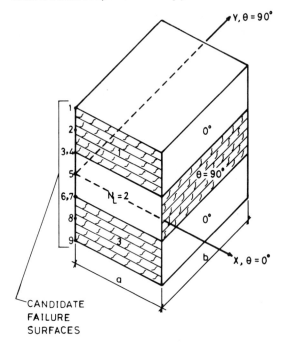

FIG. 1.    Three-layer cross-ply CFRP plate (vertical scale exaggerated).

Because it was not known *a priori* in which lamina failure would initiate, it was necessary to carry out the foregoing initial failure function computations at intervals throughout the plate thickness. These corresponded to the upper and lower faces and the mid-plane of each lamina. Thus, for example, in a three-layer plate the initial failure functions were evaluated at nine levels through the plate thickness (see Fig. 1).

## 4.    DISCUSSION OF COMPUTED FAILURE PRESSURES FOR CFRP PLATES

Two numerical studies for simply supported, symmetric cross-ply laminated, CFRP plates subjected to uniform pressure loading were conducted using the shear deformable plate theory outlined under Section 2. The details of each study are presented separately below.

In the first numerical study the plate aspect ratio was purposely fixed at unity. By so doing the effects of the stacking sequence as well as aspect ratio could be eliminated and the influence of both the number of laminae, $N_L$, and the width–thickness ratio, $\lambda$, on the computed initial failure pressures could be more thoroughly examined. Failure pressures and associated plate centre deflections were computed for three-, five-, seven- and nine-layer plates with $\lambda$-values varying from 1 (thick end of the range) to 100 (thin end of the range). The results for $\lambda = 1$ are, of course, not really meaningful (they correspond to a cube rather than a plate), but have been included for the sake of completeness. The three computed failure pressures (one for lamina failure and two for interlaminar shear failure) for each $\lambda$-value are listed for each plate in Tables 2a–2d. There are several common features in these tables. First, the lowest failure pressure, i.e. the initial flexural failure pressure, appears to be associated with lamina failure (cracking, fibre fracture, etc.) rather than interlaminar shear failure. Only in unrealistically thick plates does it appear possible for interlaminar shear failure to occur first. For this reason only the associated plate deflections for lamina failure are listed in Tables 2a–2d. A second common feature is that flexural failure always initiates in the highest numbered lamina at or near its centre. It is also evident that the *non-dimensional* initial failure pressure reduces as the width–thickness ratio, $\lambda$, decreases. This feature is more pronounced in plates with few laminae (cf. corresponding results in Tables 2a and 2d). The implication of these latter two comments is *not* that the absolute magnitude of the failure pressure decreases, but rather that for thicker plates the use of classical laminated plate theory (CPT) instead of higher-order shear deformation theory (HSDT) leads to unconservative estimates of the failure pressure. To illustrate the point more vividly the initial failure pressures, $\bar{q}_{T-H}$, in Tables 2a–2d have been expressed as percentages of the corresponding pressure for $\lambda = 100$ (equivalent to the pressure that would be computed using CPT) and plotted against $\lambda$ in Fig. 2a. This figure clearly shows that even when the plate is only moderately thick, $\lambda = 10$, and comprises only a few laminae, $N_L = 3$, CPT leads to an over-estimation of the failure pressure of about 10% compared with the HSDT predicted pressure.

The associated plate centre deflections at failure for these plates are plotted against the width–thickness ratio in Fig. 2b. It is evident that as the value of $\lambda$ decreases, i.e. the plates become thicker, the deflection at the plate centre is progressively under-estimated by CPT. For the case $N_L = 3$ and $\lambda = 10$, the under-estimation is around 9%.

The results of the first study showed that failure pressure predictions made with HSDT and CPT differed significantly even when the plate was only moderately thick ($\lambda = 10$) and the number of laminae was small

TABLE 2

*Dimensionless initial failure pressures and associated centre deflections for uniformly loaded, square, simply supported, symmetric cross-ply laminated CFRP plates*

(a) $N_L = 3$

| $\lambda$ | $\bar{q}_{T-H}$ | $\bar{q}_{ILS}^L$ | $\bar{q}_{ILS}^T$ | $\bar{w}_c^{T-H}$ ($\times 10^{-1}$) |
|---|---|---|---|---|
| 1 | 3·323 1 | 5·037 2 | 5·405 8 | 0·468 08 |
| | $(0·5, 0·3)^a$ | $(0·0, 0·5)$ | $(0·5, 0·0)$ | |
| | $9^b$ | 5 | 5 | |
| 2 | 5·901 9 | 7·383 4 | 8·306 4 | 0·292 64 |
| | $(0·5, 0·5)$ | $(0·0, 0·5)$ | $(0·0, 0·5)$ | |
| | 9 | 5 | 3 | |
| 5 | 11·260 | 14·133 | 15·900 | 0·186 70 |
| | $(0·5, 0·5)$ | $(0·0, 0·5)$ | $(0·0, 0·5)$ | |
| | 9 | 5 | 3 | |
| 7·5 | 13·556 | 20·085 | 22·596 | 0·168 06 |
| | $(0·5, 0·3)$ | $(0·0, 0·5)$ | $(0·0, 0·5)$ | |
| | 9 | 5 | 3 | |
| 10 | 14·634 | 26·193 | 29·467 | 0·158 98 |
| | $(0·5, 0·3)$ | $(0·0, 0·5)$ | $(0·0, 0·5)$ | |
| | 9 | 5 | 3 | |
| 15 | 15·566 | 38·593 | 43·417 | 0·151 63 |
| | $(0·5, 0·3)$ | $(0·0, 0·5)$ | $(0·0, 0·5)$ | |
| | 9 | 5 | 3 | |
| 20 | 15·933 | 51·099 | 57·487 | 0·148 85 |
| | $(0·5, 0·3)$ | $(0·0, 0·5)$ | $(0·0, 0·5)$ | |
| | 9 | 5 | 3 | |
| 30 | 16·210 | 76·235 | 85·765 | 0·146 79 |
| | $(0·5, 0·3)$ | $(0·0, 0·5)$ | $(0·0, 0·5)$ | |
| | 9 | 5 | 3 | |
| 40 | 16·310 | 101·44 | 114·12 | 0·146 06 |
| | $(0·5, 0·3)$ | $(0·0, 0·5)$ | $(0·0, 0·5)$ | |
| | 9 | 5 | 3 | |
| 50 | 16·358 | 126·68 | 142·52 | 0·145 72 |
| | $(0·5, 0·3)$ | $(0·0, 0·5)$ | $(0·0, 0·5)$ | |
| | 9 | 5 | 3 | |
| 100 | 16·421 | 253·02 | 284·65 | 0·145 27 |
| | $(0·5, 0·3)$ | $(0·0, 0·5)$ | $(0·0, 0·5)$ | |
| | 9 | 5 | 3 | |

$^a$ Coordinates $(\bar{x}, \bar{y})$ of the failure initiation point.
$^b$ Failure plane number measured from upper surface of plate. Each laminate is assumed to have three potential failure planes (corresponding to its upper, mid and lower surfaces). Thus, in a three-layer plate there are nine potential failure planes.

TABLE 2—*contd*

(*b*) $N_L = 5$

| $\lambda$ | $\bar{q}_{T-H}$ | $\bar{q}_{ILS}^{L}$ | $\bar{q}_{ILS}^{T}$ | $\bar{w}_c^{T-H}$ ($\times 10^{-1}$) |
|---|---|---|---|---|
| 1 | 3·674 0 | 4·938 1 | 4·740 5 | 0·522 25 |
| | (0·5, 0·2) | (0·0, 0·5) | (0·0, 0·5) | |
| | 15 | 6 | 8 | |
| 2 | 8·106 9 | 8·089 5 | 7·765 9 | 0·369 07 |
| | (0·5, 0·5) | (0·0, 0·5) | (0·0, 0·5) | |
| | 15 | 6 | 8 | |
| 5 | 12·593 | 16·785 | 16·114 | 0·191 26 |
| | (0·5, 0·5) | (0·0, 0·5) | (0·0, 0·5) | |
| | 15 | 6 | 8 | |
| 7·5 | 13·780 | 23·961 | 23·002 | 0·162 39 |
| | (0·5, 0·5) | (0·0, 0·5) | (0·0, 0·5) | |
| | 15 | 6 | 8 | |
| 10 | 14·319 | 31·244 | 29·995 | 0·151 37 |
| | (0·5, 0·5) | (0·0, 0·5) | (0·0, 0·5) | |
| | 15 | 6 | 8 | |
| 15 | 14·765 | 46·023 | 44·182 | 0·143 16 |
| | (0·5, 0·5) | (0·0, 0·5) | (0·0, 0·5) | |
| | 15 | 6 | 8 | |
| 20 | 14·936 | 60·942 | 58·504 | 0·140 20 |
| | (0·5, 0·5) | (0·0, 0·5) | (0·0, 0·5) | |
| | 15 | 6 | 8 | |
| 30 | 15·065 | 90·946 | 87·308 | 0·138 08 |
| | (0·5, 0·5) | (0·0, 0·5) | (0·0, 0·5) | |
| | 15 | 6 | 8 | |
| 40 | 15·112 | 121·04 | 116·20 | 0·137 33 |
| | (0·5, 0·5) | (0·0, 0·5) | (0·0, 0·5) | |
| | 15 | 6 | 8 | |
| 50 | 15·134 | 151·17 | 145·13 | 0·136 99 |
| | (0·5, 0·5) | (0·0, 0·5) | (0·0, 0·5) | |
| | 15 | 6 | 8 | |
| 100 | 15·163 | 302·00 | 289·92 | 0·136 53 |
| | (0·5, 0·5) | (0·0, 0·5) | (0·0, 0·5) | |
| | 15 | 6 | 8 | |

(*continued*)

TABLE 2—*contd*

(c) $N_L = 7$

| $\lambda$ | $\bar{q}_{\text{T-H}}$ | $\bar{q}_{\text{ILS}}^{\text{L}}$ | $\bar{q}_{\text{ILS}}^{\text{T}}$ | $\bar{w}_{\text{c}}^{\text{T-H}}$ ($\times 10^{-1}$) |
|---|---|---|---|---|
| 1 | 4·121 1 | 5·065 0 | 5·011 1 | 0·561 42 |
| | (0·5, 0·2) | (0·0, 0·5) | (0·5, 0·0) | |
| | 21 | 11 | 11 | |
| 2 | 8·875 4 | 8·252 4 | 8·424 3 | 0·391 35 |
| | (0·5, 0·5) | (0·0, 0·5) | (0·0, 0·5) | |
| | 21 | 11 | 9 | |
| 5 | 13·025 | 17·221 | 17·579 | 0·193 53 |
| | (0·5, 0·5) | (0·0, 0·5) | (0·0, 0·5) | |
| | 21 | 11 | 9 | |
| 7·5 | 13·925 | 24·591 | 25·103 | 0·161 98 |
| | (0·5, 0·5) | (0·0, 0·5) | (0·0, 0·5) | |
| | 21 | 11 | 9 | |
| 10 | 14·308 | 32·050 | 32·717 | 0·150 12 |
| | (0·5, 0·5) | (0·0, 0·5) | (0·0, 0·5) | |
| | 21 | 11 | 9 | |
| 15 | 14·612 | 47·164 | 48·146 | 0·141 32 |
| | (0·5, 0·5) | (0·0, 0·5) | (0·0, 0·5) | |
| | 21 | 11 | 9 | |
| 20 | 14·726 | 62·420 | 63·720 | 0·138 19 |
| | (0·5, 0·5) | (0·0, 0·5) | (0·0, 0·5) | |
| | 21 | 11 | 9 | |
| 30 | 14·811 | 93·108 | 95·048 | 0·135 94 |
| | (0·5, 0·5) | (0·0, 0·5) | (0·0, 0·5) | |
| | 21 | 11 | 9 | |
| 40 | 14·842 | 123·89 | 126·48 | 0·135 15 |
| | (0·5, 0·5) | (0·0, 0·5) | (0·0, 0·5) | |
| | 21 | 11 | 9 | |
| 50 | 14·856 | 154·72 | 157·95 | 0·134 78 |
| | (0·5, 0·5) | (0·0, 0·5) | (0·0, 0·5) | |
| | 21 | 11 | 9 | |
| 100 | 14·875 | 309·06 | 315·50 | 0·134 29 |
| | (0·5, 0·5) | (0·0, 0·5) | (0·0, 0·5) | |
| | 21 | 11 | 9 | |

TABLE 2—*contd*

(d) $N_L = 9$

| $\lambda$ | $\bar{q}_{T-H}$ | $\bar{q}_{ILS}^L$ | $\bar{q}_{ILS}^T$ | $\bar{w}_c^{T-H} \ (\times 10^{-1})$ |
|---|---|---|---|---|
| 1 | 4·4944<br>(0·5, 0·2)<br>27 | 5·0934<br>(0·5, 0·0)<br>14 | 5·1570<br>(0·5, 0·0)<br>12 | 0·601 66 |
| 2 | 9·2122<br>(0·5, 0·5)<br>27 | 8·5839<br>(0·0, 0·5)<br>12 | 8·4779<br>(0·0, 0·5)<br>14 | 0·401 10 |
| 5 | 13·222<br>(0·5, 0·5)<br>27 | 18·031<br>(0·0, 0·5)<br>12 | 17·809<br>(0·0, 0·5)<br>14 | 0·194 77 |
| 7·5 | 13·998<br>(0·5, 0·5)<br>27 | 25·793<br>(0·0, 0·5)<br>12 | 25·475<br>(0·0, 0·5)<br>14 | 0·162 00 |
| 10 | 14·314<br>(0·5, 0·5)<br>27 | 33·634<br>(0·0, 0·5)<br>12 | 33·219<br>(0·0, 0·5)<br>14 | 0·149 73 |
| 15 | 14·560<br>(0·5, 0·5)<br>27 | 49·506<br>(0·0, 0·5)<br>12 | 48·895<br>(0·0, 0·5)<br>14 | 0·140 69 |
| 20 | 14·650<br>(0·5, 0·5)<br>27 | 65·520<br>(0·0, 0·5)<br>12 | 64·711<br>(0·0, 0·5)<br>14 | 0·137 45 |
| 30 | 14·717<br>(0·5, 0·5)<br>27 | 97·730<br>(0·0, 0·5)<br>12 | 96·523<br>(0·0, 0·5)<br>14 | 0·135 14 |
| 40 | 14·741<br>(0·5, 0·5)<br>27 | 130·04<br>(0·0, 0·5)<br>12 | 128·44<br>(0·0, 0·5)<br>14 | 0·134 32 |
| 50 | 14·752<br>(0·5, 0·5)<br>27 | 162·40<br>(0·0, 0·5)<br>12 | 160·39<br>(0·0, 0·5)<br>14 | 0·133 94 |
| 100 | 14·767<br>(0·5, 0·5)<br>27 | 324·39<br>(0·0, 0·5)<br>12 | 320·38<br>(0·0, 0·5)<br>14 | 0·133 44 |

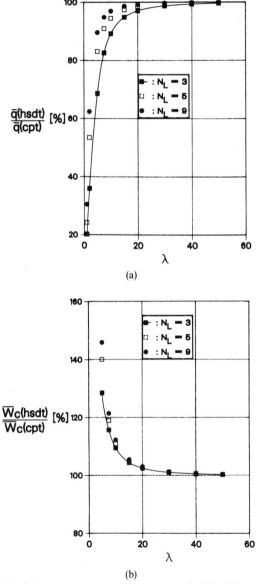

(a)

(b)

Fig. 2. (a) HSDT failure pressure as a percentage of CPT failure pressure versus width–thickness ratio for a square simply supported CFRP cross-ply plate. (b) HSDT associated centre deflection as a percentage of CPT deflection versus width–thickness ratio for a square simply supported CFRP cross-ply plate.

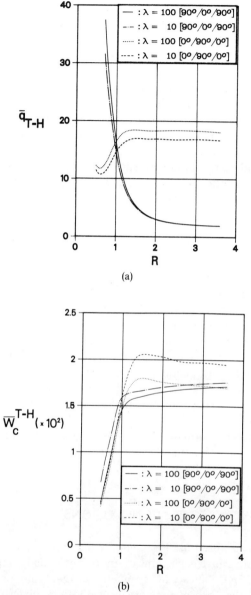

FIG. 3.   (a) Initial failure pressure versus aspect ratio for rectangular simply supported CFRP cross-ply plates. (b) Associated plate centre deflection versus aspect ratio for rectangular simply supported CFRP cross-ply plates.

($N_L = 3$). Thus, in the second study, which was intended to shed light on the effects of aspect ratio and stacking sequence on the computed initial failure pressure, it was decided to undertake computations with $N_L = 3$ and $\lambda = 10$ and 100. The plate aspect ratio, $R$, was allowed to vary between 0·5 and 3·6, and the two stacking sequences, $(0°/90°/0°)$ and $(90°/0°/90°)$, were examined. In every case, it was observed that lamina rather than interlaminar shear failure dominated. The computed initial failure pressures for each stacking sequence and each $\lambda$-value are plotted against the plate aspect ratio in Fig. 3a. Clearly, at high aspect ratios $(R > 2)$ the initial failure pressure tends to a constant value. However, for aspect ratios between 0·5 and 1·5 the failure pressure increases or decreases rapidly with increasing aspect ratio. Whether it decreases or increases is determined by the fibre direction of the outer laminae (see Fig. 3a). It is also apparent from Fig. 3a that for the $(0°/90°/0°)$ lay-up decreasing $\lambda$ produces broadly the same reduction in the failure pressure over the whole range of aspect ratios considered, whereas for the $(90°/0°/90°)$ lay-up the pressure reduction is smaller and only significant for aspect ratios up to about 1·2. Clearly though, Fig. 3a suggests that lay-up has the greatest influence on the initial failure pressure.

The associated plate centre deflections are plotted against plate aspect ratio in Fig. 3b. The general trends of the deflection curves are similar for both stacking sequences and width–thickness ratios, i.e. the deflections increase linearly and rapidly from $R = 0·5$ to $R = 1 \rightarrow 1·4$ and thereafter become sensibly independent of aspect ratio. The effect of the width–thickness ratio, $\lambda$, appears to be more significant, i.e. produces a greater increase in deflection compared with that predicted in accordance with CPT ($\lambda = 100$), for the $(0°/90°/0°)$ lay-up. However, in view of the very considerable differences in the initial failure pressures for these two lay-ups, it is somewhat surprising that the differences in their associated plate centre deflections are so small when $R > 1·5$.

## 5. CONCLUSIONS

HSDT theory has been used to compute the initial failure pressures and associated plate centre deflections of uniformly loaded, simply supported, rectangular CFRP plates with symmetric cross-ply lay-ups. It has been demonstrated that, except for unrealistically thick plates, lamina failure rather than interlaminar shear failure is the governing mode. Moreover, it has been shown that for a given width–thickness ratio lower initial failure pressures and larger deflections are predicted with HSDT than with CPT. The difference between the predicted failure pressures increases as the width–thickness ratio decreases and the number of laminae decreases. Typically, for $N_L = 3$ and $\lambda = 10$, the initial failure pressure is overestimated by about 10% using CPT. Therefore, CPT needs to be used with caution.

## ACKNOWLEDGEMENTS

The author wishes to record his appreciation for the help and guidance with the computer preparation of the figures provided by Mrs H. Drinali, and to the Department of Engineering for supporting and encouraging this research.

## REFERENCES

1. WHITNEY, J. M. and PAGANO, N. J., Shear deformation in heterogeneous anisotropic plates, *Trans. ASME, J. appl. Mech.*, **37** (1970), 1031–1036.
2. WHITNEY, J. M., The effect of transverse shear deformation on the bending of laminated plates, *J. comp. Mater.*, **3** (1969), 534–537.
3. TURVEY, G. J., Biaxial buckling of moderately thick laminated plates, *J. Strain Anal. Engng Design*, **12** (1977), 89–96.
4. REDDY, J. N. and CHAO, W. C., A comparison of closed-form and finite element solutions of thick laminated, anisotropic rectangular plates, *Nucl. Engng Design*, **64** (1981), 153–167.
5. ADALI, S., Design of shear-deformable antisymmetric angle-ply laminates to maximize the fundamental frequency and frequency separation, *Composite Structures*, **2** (1984), 349–369.
6. REISSNER, E., The effect of transverse shear deformation on the bending of elastic plates, *Trans. ASME, J. appl. Mech.*, **12** (1945), A69–A77.
7. MINDLIN, R. D., Influence of rotatory inertia and shear on flexural motions of isotropic elastic plates, *Trans. ASME, J. appl. Mech.*, **18** (1951), 31–38.
8. COWPER, G. R., The shear coefficient in Timoshenko's beam theory, *Trans. ASME, J. appl. Mech.*, **33** (1966), 335–340.
9. STEPHEN, N. G., Timoshenko's shear coefficient from a beam subjected to gravity loading, *Trans. ASME, J. appl. Mech.*, **47** (1980), 121–127.
10. WHITNEY, J. M., Shear correction factors for orthotropic laminates under static load, *Trans. ASME, J. appl. Mech.*, **40** (1973), 302–304.
11. BERT, C. W., Simplified analysis of static shear factors for beams of non-homogeneous cross-section, *J. comp. Mater.*, **7** (1973), 525–529.
12. ASHTON, J. E. and WHITNEY, J. M., *Theory of Laminated Plates*, Stamford, Conn., Technomic Publishing Co. Inc., 1970.

13. LO, K. H., CHRISTENSEN, R. M. and WU, E. M., A higher-order theory of plate deformation. Part 2: Laminated plates, *Trans. ASME, J. appl. Mech.*, **44** (1977), 669–676.
14. MURTHY, M. V. V., An improved transverse shear deformation theory for laminated anisotropic plates, NASA TP-1903, November 1981.
15. REDDY, J. N., A simple higher-order theory for laminated composite plates, *Trans. ASME, J. appl. Mech.*, **51** (1984), 745–752.
16. REDDY, J. N., A refined nonlinear theory of plates with transverse shear deformation, *Int. J. Solids Struct.*, **20** (1984), 881–896.
17. BHIMARADDI, A. A. and STEPHENS, L. K., A higher-order theory for free vibrations of orthotropic, homogeneous and laminated rectangular plates, *Trans. ASME, J. appl. Mech.*, **51** (1984), 195–198.
18. TURVEY, G. J., An initial flexural failure analysis of symmetrically laminated cross-ply rectangular plates, *Int. J. Solids Struct.*, **16** (1980), 451–463.
19. TURVEY, G. J., A study of the onset of flexural failure in cross-ply laminated strips, *Fibre Sci. Technol.*, **13** (1980), 325–336.
20. TURVEY, G. J., Flexural failure analysis of angle-ply laminates of GFRP and CFRP, *J. Strain Anal. Engng Design*, **15** (1980), 43–49.
21. TURVEY, G. J., Initial flexural failure of square, simply supported, angle-ply plates, *Fibre Sci. Technol.*, **15** (1981), 47–63.
22. TURVEY, G. J., Uniformly loaded, antisymmetric cross-ply laminated, rectangular plates: an initial flexural failure analysis, *Fibre Sci. Technol.*, **16** (1982), 1–10.
23. TURVEY, G. J., Uniformly loaded, clamped, cross-ply laminated, elliptic plates— an initial failure study, *Int. J. Mech. Sci.*, **22** (1980), 551–562.
24. TURVEY, G. J., Flexural failure analysis of anisotropic elliptic plates, *Computers and Structures*, **14** (1981), 463–468.
25. ADALI, S. and NISSEN, H., Micromechanical initial failure analysis of symmetrically laminated cross-ply plates, *Int. J. Mech. Sci.* (in press).
26. TURVEY, G. J., Effect of shear deformation on the yielding of circular plates, *Computers and Structures*, **18** (1984), 307–310.

# APPENDIX

The dimensionless coefficients, $a_{ij}$ $(i, j = 1 \rightarrow 5)$, in eqn. (13) are as follows:

$$a_{11} = -(A_{11}\alpha^2 + A_{66}\beta^2)$$

$$a_{12} = -(A_{12} + A_{66})\alpha\beta$$

$$a_{13} = (4/3)\{E_{11}\alpha^2 + (E_{12} + 2E_{66})\beta^2\}\alpha$$

$$a_{14} = -(B_{11}\alpha^2 + B_{66}\beta^2) + (4/3)(E_{11}\alpha^2 + E_{66}\beta^2)$$

$$a_{15} = -\{(B_{12} + B_{66}) - (4/3)(E_{12} + E_{66})\}\alpha\beta$$

$$a_{22} = -(A_{66}\alpha^2 + A_{22}\beta^2)$$

$$a_{23} = (4/3)\{E_{22}\beta^2 + (E_{12} + 2E_{66})\alpha^2\}\beta$$

$a_{24} = a_{15}$

$a_{25} = -(B_{22}\beta^2 + B_{66}\alpha^2) + (4/3)(E_{22}\beta^2 + E_{66}\alpha^2)$

$a_{33} = -[\lambda^2\{\alpha^2(A_{55} - 8D_{55} + 16F_{55}) + \beta^2(A_{44} - 8D_{44} + 16F_{44})\}$
$\qquad + (16/9)\{H_{11}\alpha^4 + 2(H_{12} + 2H_{66})\alpha^2\beta^2 + H_{22}\beta^4\}]$

$a_{34} = -\alpha\{\lambda^2(A_{55} - 8D_{55} + 16F_{55}) - (4/3)$
$\qquad \times [\{F_{11}\alpha^2 + (F_{12} + 2F_{66})\beta^2\} - (4/3)\{H_{11}\alpha^2 + (H_{12} + 2H_{66})\beta^2\}]\}$

$a_{35} = -\beta\{\lambda^2(A_{44} - 8D_{44} + 16F_{44}) - (4/3)$
$\qquad \times [\{(F_{12} + 2F_{66})\alpha^2 + F_{22}\beta^2\} - (4/3)\{(H_{12} + 2H_{66})\alpha^2 + H_{22}\beta^2\}]\}$

$a_{44} = -(D_{11}\alpha^2 + D_{66}\beta^2) - \lambda^2(A_{55} - 8D_{55} + 16F_{55})$
$\qquad + (8/3)\{(F_{11}\alpha^2 + F_{66}\beta^2) - (2/3)(H_{11}\alpha^2 + H_{66}\beta^2)\}$

$a_{45} = -\alpha\beta\{(D_{12} + D_{66}) - (8/3)(F_{12} + F_{66}) + (16/9)(H_{12} + H_{66})\}$

$a_{55} = -(D_{22}\beta^2 + D_{66}\alpha^2) - \lambda^2(A_{44} - 8D_{44} + 16F_{44})$
$\qquad + (8/3)\{(F_{22}\beta^2 + F_{66}\alpha^2) - (2/3)(H_{22}\beta^2 + H_{66}\alpha^2)\}$

# 12

# Micromechanical Failure Analysis of Elliptic, Cross-ply Laminates under Flexural Loads

S. ADALI and G. S. JHETAM

*Department of Mechanical Engineering, University of Natal,
Durban 4001, Republic of South Africa*

## ABSTRACT

*A first-ply failure analysis is given for a symmetric, cross-ply laminate of elliptic shape subject to flexural loads. In particular, clamped elliptic laminates under a uniformly distributed load are treated. The effects of fibre type and content, voids and hygrothermal conditions are investigated on the failure load and its location by means of micromechanical relations from which elastic and strength properties of the composite are computed. Numerical results are given in graphical form with a view towards the design of elliptic laminates by means of parametric studies.*

## 1. INTRODUCTION

Determination of the failure characteristics of laminated fibre composites is complicated by the fact that several failure modes may be active simultaneously in the failure phenomenon. As a result, increasing the fibre content of the composite does not necessarily improve the load-carrying capacity of the structure as the failure may occur in the transverse fibre direction. In fact, the strength mostly depends on the matrix strength in the transverse direction and an excessive amount of fibre may lead to a weakening of the composite by causing stress concentrations which reduce the transverse material strength. Consequently, an important design consideration for fibre composites is the investigation of the effect of fibre content on the failure load.

Other considerations involve the amount of void in the resin and the hygrothermal effects. The presence of voids in the material may come about as a result of manufacturing imperfections, and adverse environmental conditions such as high temperature and humidity induce the hygrothermal effects. Such negative factors cause deterioration in the elastic and strength properties and affect the performance of the structure.

In the present study, a micromechanical approach is given to the initial failure analysis of elliptic cross-ply plates under flexural loads in order to investigate the effects of these parameters on the failure load. The dependence of the elastic and strength properties on the fibre content, void volume and hygrothermal conditions is expressed in terms of micromechanical relations. First-ply failure loads are computed using Tsai–Hill failure criterion.

The initial failure analyses of laminated, elliptic plates subject to flexural loads were given by Turvey[1,2] using the same failure criterion. Corresponding results for symmetric, rectangular cross-ply laminates were obtained in Refs 3 and 4. In Refs 1–4, fixed elastic and strength parameters were used, i.e. the fibre contents of the materials were taken as constant. The effect of fibre content as well as those of voids and hygrothermal conditions on the failure load were studied by means of micromechanical relations by Adali and Nissen[5] for symmetric, rectangular cross-ply laminates.

Numerical results investigating the effect of various problem parameters such as fibre type, fibre content, the amount of voids, number of layers, aspect ratio and hot-wet conditions are given in graphical form. It is observed that the failure load reaches a maximum value with respect to the fibre content, indicating the possibility of an optimal design. It is found that the sensitivity of fibres to changes in various problem parameters differs considerably with respect to one another.

## 2.  PROBLEM FORMULATION

A symmetric, cross-ply laminate of elliptical shape is composed of an even number of orthotropic layers, each of thickness $h$, with the total thickness denoted by $H$. The semi-axis along the $x$-direction is denoted by $a$ and the semi-axis along the $y$-direction by $b$. The plate is clamped around the boundary and its deflection is denoted by $W(x, y)$. The equation governing the bending of the laminate subject to a uniformly distributed load $q$ is given by

$$D_{11} W_{xxxx} + 2(D_{12} + 2D_{66}) W_{xxyy} + D_{22} W_{yyyy} = q \tag{1}$$

where $D_{ij}$ are the laminate stiffnesses given by

$$D_{ij} = \frac{1}{3} \sum_{k=1}^{N} Q_{ij}^{(k)}(h_k^3 - h_{k-1}^3) \tag{2}$$

In eqn. (2), $h_k$ is the distance from the mid-surface to the surface between the $(k+1)$th and $k$th lamina, and is negative for $k < N/2$ and postive for $k \geq N/2$, where $N$ denotes the number of layers. The plane stress reduced stiffness components $Q_{ij}^{(k)}$ are given by

$$
\begin{array}{lll}
Q_{11}^{(k)} = E_L/D_v & Q_{22}^{(k)} = E_T D_v & \text{for } \theta = 0 \\
Q_{11}^{(k)} = E_T/D_v & Q_{22}^{(k)} = E_L/D_v & \text{for } \theta = \pi/2 \\
Q_{12}^{(k)} = \nu_{LT} E_T/D_v & Q_{66}^{(k)} = G_{LT} & \text{for all layers}
\end{array}
$$

where $D_v = 1 - \nu_{LT}\nu_{TL}$ and $\theta$ is the fibre orientation. $E_L$ and $E_T$ denote the Young moduli in the longitudinal and transverse directions, $\nu_{LT}$ is the Poisson ratio, $\nu_{TL} = \nu_{LT}E_T/E_L$, and $G_{LT}$ is the shear modulus in the LT plane.

The boundary conditions for a clamped plate of elliptical shape are

$$W = 0 \qquad \partial W/\partial \mathbf{n} = 0 \quad \text{for } x, y \in \partial D \tag{3}$$

where $\mathbf{n}$ is the normal vector to the boundary which is denoted by $\partial D$.

## 3.   METHOD OF SOLUTION

The failure load $q$ is evaluated by determining the stress field for each layer and by substituting stress components into the Tsai–Hill failure criterion to compute $q$.

The solution of the boundary value problem (1), (3) is given by

$$W = qw \tag{4}$$

where

$$w(x, y) = (a^4/8D_0)(1 - x^2/a^2 - y^2/b^2)^2 \tag{5}$$

$$D_0 = 3D_{11} + 2(D_{12} + 2D_{66})/R^2 + 3D_{22}/R^4 \tag{6}$$

where $R = b/a$. In the $xy$ coordinate system, the strain field for the $k$th layer is determined from the expressions

$$\varepsilon_{xk} = -zqw_{xx} \qquad \varepsilon_{yk} = -zqw_{yy} \qquad \gamma_{xyk} = -2zqw_{xy} \tag{7}$$

where $z$ indicates the distance from the mid-surface. The strain field in the material coordinate system is computed by a coordinate transformation from the $xy$-system to the LT-system, which yields

$$\varepsilon_{Lk} = \varepsilon_{xk} \qquad \varepsilon_{Tk} = \varepsilon_{yk} \qquad \gamma_{LTk} = \gamma_{xyk} \qquad \text{for } \theta = 0 \tag{8a}$$

$$\varepsilon_{Lk} = \varepsilon_{yk} \qquad \varepsilon_{Tk} = \varepsilon_{xk} \qquad \gamma_{LTk} = -\gamma_{xyk} \qquad \text{for } \theta = \pi/2 \tag{8b}$$

The stresses for the $k$th layer in the LT coordinate system are obtained from the stress–strain relations for an orthotropic lamina, viz.

$$\bar{\sigma}_{Lk} = q\sigma_{Lk} \qquad \bar{\sigma}_{Tk} = q\sigma_{Tk} \qquad \bar{\tau}_{LTk} = q\tau_{LTk} \tag{9}$$

where

$$\sigma_{Lk} = (E_L\varepsilon_{Lk} + \nu_{LT}E_T\varepsilon_{Tk})/qD_\nu \tag{10a}$$

$$\sigma_{Tk} = (\nu_{LT}E_T\varepsilon_{Lk} + E_T\varepsilon_{Tk})/qD_\nu \tag{10b}$$

$$\tau_{LTk} = G_{LT}\gamma_{LTk}/q \tag{10c}$$

We note that the expressions for $\sigma_{Lk}$, $\sigma_{Tk}$ and $\tau_{LTk}$ are independent of $q$, as can be deduced from (7), (8) and (10).

The failure load $q$ is computed by using the Tsai–Hill criterion of failure, viz.

$$(\bar{\sigma}_{Lk}/X_{t,c})^2 - (\bar{\sigma}_{Lk}\bar{\sigma}_{Tk}/X_{t,c}^2) + (\bar{\sigma}_{Tk}/Y_{t,c})^2 + (\bar{\tau}_{LTk}/S)^2 > 1 \tag{11}$$

where $X_{t,c}$ and $Y_{t,c}$ are the tensile and compressive strengths of the composite material in the fibre and transverse directions, respectively. $S$ denotes the shear strength of the material. Inserting (9) into (11), rearranging the resulting expression and noting that the highest failure load is given when (11) is an equality, we obtain

$$q = \mp [(\sigma_{Lk}/X_{t,c})^2 - (\sigma_{Lk}\sigma_{Tk}/X_{t,c}^2) + (\sigma_{Tk}/Y_{t,c})^2 + (\tau_{LTk}/S)^2]^{-1/2} \tag{12}$$

where the failure load $q$ can be positive or negative depending on the direction of the transverse load, but due to symmetric lamination its magnitude remains the same.

The failure load $q$ is computed from (12) by determining the location of the failure in the $xy$-coordinate system and the lamina where the first-ply failure occurs such that (12) yields the smallest magnitude for $q$. For symmetric cross-ply laminates, the initial failure occurs either in the outermost lamina or in the lamina next to the outermost one.


## 4. MICROMECHANICAL RELATIONS

Material and strength properties of fibre composites depend on the fibre content as well as on the void ratio and hygrothermal conditions. In computing these properties, the above parameters are taken into account by means of micromechanical relations which employ certain degradation coefficients to incorporate the void and hygrothermal effects.

The material constants are computed using the relations given in Refs 6–9, viz.

$$E_L = V_f E_f + C_{ht} V_m E_m \tag{13}$$

$$E_T = C_{ht} E_m (V_f + \eta_T V_m)/(\eta_T V_m + C_{ht} V_f E_m/E_{ft}) \tag{14}$$

$$G_{LT} = C_{ht} G_m (V_f + \eta_g V_m)/(\eta_g V_m + C_{ht} V_f G_m/G_f) \tag{15}$$

$$v_{LT} = V_f v_f + V_m v_m \tag{16}$$

where the stress partitioning parameters $\eta_T$ and $\eta_g$ are given by

$$\eta_T = 0.5 \qquad \eta_g = (1 + G_m/G_f)/2$$

In eqns (13)–(16) the subscripts f and m refer to fibre and matrix related properties, respectively, with $V_m = 1 - V_f - V_v$, where $V_v$ is the void volume ratio. $E_{ft}$ in (14) denotes the Young modulus of the fibre in the transverse direction in the case of anisotropic fibres. $C_{ht}$ denotes the hygrothermal degradation coefficient and can be computed using the empirical relations given by Chamis,[6,8,9] viz.

$$C_{ht} = (T_{GW} - T)^{1/2}/(T_{GD} - T_0)^{1/2} \tag{17}$$

where

$$T_{GW} = 0.005 M_l^2 - 0.1 M_l + 1)T_{GD} \tag{18}$$

$$M_l = M_\infty R_{hr}(3\beta_m V_m + V_v)\rho_m/\rho_l \tag{19}$$

In eqn. (17), $T_{GW}$ and $T_{GD}$ denote the glass transition temperature of wet and dry resin, respectively, $T$ is the given composite temperature and $T_0$ is the reference temperature usually taken as the room temperature (23°C). In eqns (18) and (19), $M_l$ denotes the moisture pick-up of the composite by weight, $M_\infty$ the matrix saturation moisture at 100% relative humidity and room temperature, $R_{hr}$ relative humidity ratio and $\beta_m$ the matrix moisture expansion coefficient. The specific density of the composite is computed from

$$\rho_l = V_f \rho_f + V_m \rho_m \tag{20}$$

The strength parameters $X_{t,c}$, $Y_{t,c}$ and $S$ are computed using the following relationships:

$$X_t = X_{ft}(V_f + C_{ht}V_m E_m/E_f) \tag{21}$$

$$X_c = \min(X_{c1}, X_{c2}) \tag{22}$$

$$X_{c1} = F(V_v)C_{ht}G_m/(1 - V_f + V_f C_{ht}G_m/G_f)$$

$$X_{c2} = X_{fc}(V_f + V_m C_{ht}E_m/E_f$$

$$F(V_v) = (1 - V_v/(1 - V_f))^2/(1 - V_v/(1 - V_v))$$

$$Y_t = C_v C_{ht}Y_{mt}/SCF \tag{23}$$

$$SCF = (1 - V_f(1 - C_{ht}E_m/E_f))/(1 - 2(V_f/\pi)^{1/2}(1 - C_{ht}E_m/E_{ft}))$$

$$Y_c = C_v C_{ht}Y_{mc} \tag{24}$$

$$S = C_v C_{ht}S_m \tag{25}$$

where the void degradation factor $C_v$ is given by

$$C_v = 1 - 2(V_v/(1 - V_f)\pi)^{1/2} \tag{26}$$

In eqn. (23), SCF stands for Stress Concentration Factor. The micromechanics relations for strength properties were taken from different sources. $X_t$ in (21) is derived in Ref. 7; $X_c$ in (22) is given by Chamis[6,8] as an empirical relation. $Y_t$ in (23) can be found in Ref. 10, and the $V_f$-independent values for $Y_c$ and $S$ are recommended in Refs 11 and 12. The effect of hygrothermal conditions is incorporated into the strength relations by means of the degradation factor $C_{ht}$ following the work of Chamis.[6,8,9] We note that the relations (21)–(25) produce values for strength parameters which are in close agreement with the experimental results.

## 5. NUMERICAL RESULTS

Numerical values of the failure load $q$ are given for graphite, boron and E-glass reinforced epoxy materials, for which the following material and strength data are used:

| | | | |
|---|---|---|---|
| Graphite: | $E_f = 230\,GPa$ | $E_{ft} = 16\cdot6\,GPa$ | $G_f = 8\cdot27\,GPa$ |
| | $v_f = 0\cdot20$ | $X_{ft} = 2\cdot8\,GPa$ | $X_{fc} = 2\cdot4\,GPa$ |
| | $\rho_f = 1\cdot75$ | | |
| Boron: | $E_f = 410\,GPa$ | $E_{ft} = E_f$ | $v_f = 0\cdot20$ |
| | $X_{ft} = 3\cdot45\,GPa$ | $X_{fc} = 4\cdot0\,GPa$ | $\rho_f = 2\cdot6$ |

E-glass:      $E_f = 72\cdot4\,\text{GPa}$      $E_{ft} = E_f$      $v_f = 0\cdot22$

               $X_{ft} = 1\cdot7\,\text{GPa}$      $X_{fc} = 1\cdot7\,\text{GPa}$      $v_f = 2\cdot6$

Epoxy:      $E_m = 3\cdot45\,\text{GPa}$      $v_m = 0\cdot35$      $Y_{mt} = 0\cdot055\,\text{GPa}$

             $Y_{mc} = 0\cdot110\,\text{GPa}$      $S_m = 0\cdot067\,\text{GPa}$      $\beta_m = 0\cdot33$

             $T_{GD} = 216\,°\text{C}$      $\rho_m = 1\cdot2$

The above values of the material and strength parameters were gathered from several sources.[6,7,11,13] Shear moduli $G_f$ and $G_m$ are computed from $G = E/2(1 + v)$ in all cases except for graphite which is an anisotropic fibre.

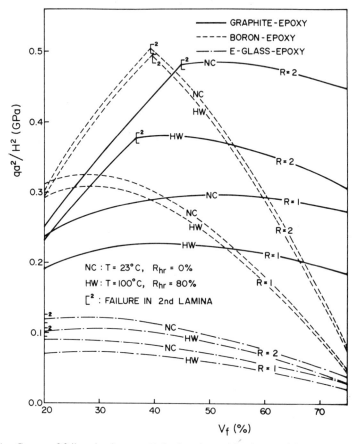

FIG. 1.   Curves of failure load versus $V_f$ for four-layered laminates with $V_v = 0\%$. Failure occurs in the second lamina for values of $V_f$ higher than the one marked by $[^2$.

The reference temperature is taken as $T_0 = 23°C$ and the matrix saturation moisture as $M = 7\%$. The unit of the failure load $q$ is GPa as all the problem parameters are given in GPa. The first-ply failure may occur in the first or the second lamina. The location of the failure with respect to layers is indicated by the sign [$^2$, to the right of which failure occurs in the second lamina. The elliptic plate is found to fail first at the ends of the minor or major axes depending on the aspect ratio $b/a$.

Figure 1 shows the curves of $qa^2/H^2$ plotted against $V_f$ for four-layered laminates with $V_v = 0\%$. It is observed that boron–epoxy laminates are not as sensitive to hot-wet conditions as graphite or E-glass–epoxy laminates, although they are most affected by changes in the fibre content.

Figure 2 shows the curves of $qa^2/H^2$ plotted against $V_f$ for six- and eight-layered laminates with $b/a = 2$ and $V_v = 0\%$. As the number of layers increases, the location of the failure shifts towards the second lamina as expected, for which $\theta = \pi/2$.

The effect of the voids in the matrix on the failure load is investigated in

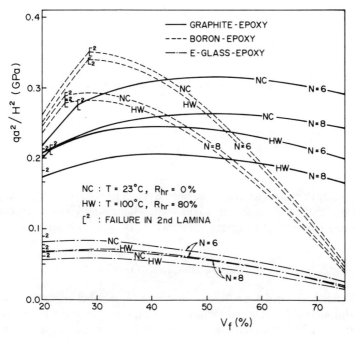

FIG. 2. Curves of failure load versus $V_f$ for laminates with $b/a = 2$ and $V_v = 0\%$.

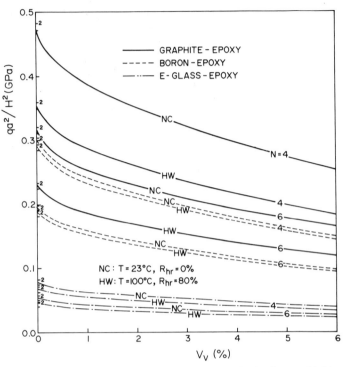

FIG. 3.    Curves of failure load versus void volume ratio $V_v$ for laminates with $b/a = 2$ and $V_f = 60\%$.

Fig. 3 for plates with $b/a = 2$ and $V_f = 60\%$. It is observed that voids affect the failure load at varying degrees depending on the fibre type.

Figure 4 shows the curves of $qa^2/H^2$ plotted against the aspect ratio $b/a$ for six-layered plates with $V_f = 60\%$ and $V_v = 0\%$. Failure curves follow essentially the same pattern as the ones given in Refs 4 and 5 for rectangular plates with the failure shifting to the second lamina and the curves levelling off once $b/a$ exceeds a certain value, which is approximately 1·5.

## 6.    CONCLUSIONS

The effects of fibre type, fibre volume ratio, number of layers, hot-wet conditions, void volume ratio and aspect ratio on the initial failure load of symmetrically laminated cross-ply plates of elliptical shape are investigated. The elastic constants and strength parameters are computed using

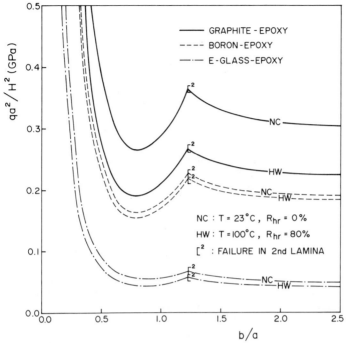

FIG. 4. Curves of failure load versus the aspect ratio $b/a$ for six-layered laminates with $V_f = 60\%$ and $V_v = 0\%$.

micromechanical relations which take the above-mentioned effects into account. Numerical results indicate the existence of an optimal fibre volume ratio, for which the first-ply failure load is the highest.

It is observed that the first-ply failure may occur in the first or second lamina depending on the fibre content, hygrothermal conditions and the void ratio. Thus, manufacturing imperfections in the form of voids in the resin or adverse environmental conditions may change the location of the failure anticipated at the design stage. A comparison with the numerical results of Ref. 5 indicates that clamping in general leads to lower failure loads as compared to simple supports as well as a change of the location of failure from the middle of the plate to its boundary.

## ACKNOWLEDGEMENT

This work was supported by a grant from SA Foundation for Research Development.

# REFERENCES

1. TURVEY, G. J., Uniformly loaded, clamped, cross-ply laminated elliptic plates—an initial failure study, *Int. J. Mech. Sci.*, **22** (1980), 551–562.
2. TURVEY, G. J., Flexural failure analysis of anisotropic elliptic plates, *Computers and Structures*, **13** (1981), 463–468.
3. TURVEY, G. J., A study of the onset of flexural failure in cross-ply laminated strips, *Fibre Sci. Technol.*, **15** (1980), 325–336.
4. TURVEY, G. J., An initial flexural failure analysis of symmetrically laminated cross-ply rectangular plates, *Int. J. Solids Struct.*, **16** (1980), 451–463.
5. ADALI, S. and NISSEN, H., Micromechanical initial failure analysis of symmetrically laminated cross-ply plates, *Int. J. Mech. Sci.*, **38** (1986).
6. CHAMIS, C. C., Simplified composite micromechanics equations for strength, fracture toughness and environmental effects, *SAMPE Quarterly*, **15** (July 1984), 41–55.
7. TSAI, S. W. and HAHN, H. T., *Introduction to Composite Materials*, Westport, Connecticut, Technomic Publishing Company, 1980.
8. CHAMIS, C. C. and SMITH, G. T., Resin selection criteria for tough composite structures, *AIAA J.*, **23** (1985), 902–911.
9. CHAMIS, C. C., Simplified composite micromechanics equations for hygral, thermal and mechanical properties, *SAMPE Quarterly*, **15** (April 1984), 14–23.
10. AGARWAL, B. D. and BROUTMAN, L. J., *Analysis and Performance of Fibre Composites*, New York, John Wiley, 1980, p. 41.
11. HOLMES, M. and JUST, D. J., *GRP in Structural Engineering*, London, Elsevier Applied Science Publishers, 1983, p. 178.
12. HULL, D., *An Introduction to Composite Materials*, Cambridge, Cambridge University Press, 1981, p. 164.
13. SCHWARTZ, M. M., *Composite Materials Handbook*, New York, McGraw-Hill, 1984.

# 13

# Failure Modes for Compression Loaded Angle-ply Plates with Holes

S. W. Burns, C. T. Herakovich

*Composite Mechanics Group, Dept of Engineering Science and Mechanics,*
*Virginia Polytechnic Institute and State University,*
*Blacksburg, Virginia 24061, USA*

and

J. G. Williams

*Production Research and Development, Conoco, Inc.,*
*P.O. Box 1267, Ponca City, Oklahoma 74603, USA*

## ABSTRACT

*A combined theoretical–experimental investigation of failure in notched, graphite–epoxy, angle-ply laminates subjected to far-field compression loading indicates that failure generally initiates on the hole boundary and propagates along a line parallel to the fiber orientation of the laminate. The strength of notched laminates with specimen width-to-hole diameter ratios of 5 and 10 are compared to the strength of unnotched laminates. The experimental results are complemented by a three-dimensional finite element stress analysis that includes interlaminar stresses around holes in $[\pm\theta]_s$ laminates. The finite element predictions indicate that failure is initiated by shear stresses at the hole boundary.*

## INTRODUCTION

The class of composite laminates known as angle-ply laminates illustrates a basic advantage of composite materials, the ability to tailor laminate stiffness and strength by varying the orientation of the fibers in the individual layers. Use of this property in the design of notched laminates requires a thorough knowledge and understanding of the strength properties as a function of fiber orientation, stacking sequence, notch geometry, and loading configuration.

Failure of notched laminates has been considered previously by a number of investigators. The previous works have included micromechanics, plane stress of homogeneous, anisotropic plates, and 3-D stress analysis including interlaminar stresses. Various failure theories have also been proposed. A review of the pertinent literature can be found in Ref. 1. Previous works on compression loaded notched plates includes those by Rhodes *et al.*[2] and Shuart and Williams.[3]

In the present study we consider the case of far-field compression loading of an angle-ply laminate with a circular hole. Compression loading was chosen because it is, so often, the controlling factor in the design of aerospace structures which experience both tension and compression. Compression is usually the controlling design consideration because composites generally exhibit significantly lower strength in compression than in tension. Angle-ply laminates were chosen to isolate the influence of fiber orientation. The stacking sequence was held constant throughout the study.

This chapter presents the results of an experimental investigation on four angle-ply laminates: $[(\pm 10)_{12}]_s$, $[(\pm 20)_{12}]_s$, $[(\pm 30)_{12}]_s$, and $[(\pm 45)_{12}]_s$. The experimental results are discussed in light of a three-dimensional finite element stress and failure analysis using the tensor polynomial failure criterion. Due to computer storage limitations, the analysis considered only four-ply $[\pm \theta]_s$ laminates.

## EXPERIMENTAL PROGRAM

The experimental program included three basic types of tests: (1) compression and shear testing of unidirectional laminae to determine material mechanical and strength properties; (2) compression testing of unnotched angle-ply laminates to determine laminate mechanical response and strength properties; (3) compression testing of notched angle-ply laminates to determine the effects of circular holes on the strength and mode of failure.

All compression specimens were 48-ply thick AS4/3502 graphite–epoxy. The shear properties were determined from 12-ply thick unidirectional off-axis 45° and 10° tensile coupons with a specimen aspect ratio of 20 to minimize shear coupling effects induced by the rigid end grips. The nominal fiber volume fraction of all specimens was 62%. All laminates were cured at 350°F and ultrasonically scanned to detect flaws prior to testing. The ends of compression specimens were ground flat to ensure uniform compression loading. Holes were made with a diamond impregnated core drill.

Specimen widths were sized using an orthotropic plate buckling criterion[4] to ensure that compressive failure preceded buckling. The basic test configuration for compression tests is shown in Fig. 1, and the dimensions of individual specimens are presented in Table 1. As indicated in Table 1, all specimens were 10 in in length and approximately 0·25 in thick; specimen width to hole diameter ratios of 5 and 10 were investigated. The maximum number of specimens tested for any configuration was three. During compression, the specimens were held in a rigid frame which provided clamped conditions at the loaded ends and simply-supported conditions along the sides (Fig. 2). Notched compression specimens were strain gaged near the hole, on the hole edge (for larger diameter holes), and at far field (Fig. 2). Direct current displacement transducers (DCDT's) measured out-of-plane displacements near the hole and axial head displacements.

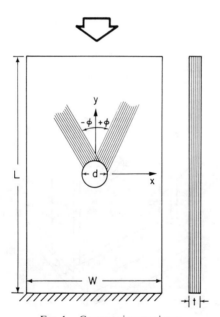

FIG. 1.    Compression specimen.

TABLE 1
*Specimen dimensions*

| Laminate | No. specimens | Width (in) | Hole dia. (in) | $W/d$ |
|---|---|---|---|---|
| $[(\pm 10)_{12}]_s$ | 2 | 2·5 | 0·25 | 10 |
| $[(\pm 20)_{12}]_s$ | 2 | 2·5 | 0·25 | 10 |
| $[(\pm 10)_{12}]_s$ | 2 | 2·5 | 0·5 | 5 |
| $[(\pm 20)_{12}]_s$ | 2 | 2·5 | 0·5 | 5 |
| $[(\pm 30)_{12}]_s$ | 3 | 5 | 0·5 | 10 |
| $[(\pm 45)_{12}]_s$ | 3 | 5 | 0·5 | 10 |

$L = 10$ in and $t = 0·25$ in for all laminates.

All ultimate stress values given in this chapter for notched specimens are based upon the unnotched far-field cross-sectional area of the specimens. Thus the average stress in the net section is higher than these quoted values because of the presence of the hole. The net section average stress is as much as 25% larger than the far-field stress with this maximum being the case for the specimens with $W/d$ ratios of 5.

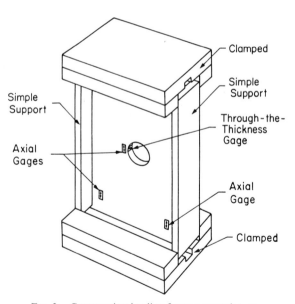

FIG. 2.  Compression loading frame nomenclature.

## EXPERIMENTAL RESULTS

**Material Properties**

Details of the experimental results for basic material properties and response of unnotched laminates will not be presented here due to limitations on space. These results can be found in Ref. 1. The material elastic and strength properties are summarized in Table 2. It should be noted that the ultimate values given in Table 2 for the $[(\pm 45)_{12}]_s$ laminate are the maximum values obtained during the test. They do not correspond to failure since these specimens did not fail, but rather exhibited large deformations which exceeded the travel limit of the load frame prior to failure.

**Laminates with $\frac{1}{4}$ in Diameter Holes**

*General considerations*

As indicated in Table 1, two laminate configurations with $\frac{1}{4}$ in holes were tested: $[(\pm 10)_{12}]_s$, and $[(\pm 20)_{12}]_s$. Typical failed specimens from these tests are shown in Figs 3 and 4. The white color in the photographs is paint that was used for Moiré interferometry studies of out-of-plane deformations. Failure is generally denoted by the dark lines (regions) emanating from the hole. Other dark regions are artifacts of strain gages which were taken off the specimens prior to C-scan and X-ray examination.

TABLE 2
*Properties of AS4/3502 graphite–epoxy*
*Unidirectional properties*

| | |
|---|---|
| $E_1 = 19.85$ msi | $X_c = -122.4$ ksi |
| $E_2 = 1.53$ msi | $Y_c = Z_c = -34.6$ ksi |
| $v_{12} = 0.32$ | $S_{12} = S_{13} = 9.4$ ksi |
| $v_{23} = 0.40$ | $S_{23} = 8.0$ ksi |
| $G_{12} = 0.85$ msi | |
| $G_{23} = 0.60$ msi | |

*Laminates*

| Laminate | Compressive ultimate $\bar{\sigma}$ |
|---|---|
| $[(\pm 10)_{12}]_s$ | $-97.0$ ksi |
| $[(\pm 20)_{12}]_s$ | $-42.5$ ksi |
| $[(\pm 30)_{12}]_s$ | $-34.3$ ksi |
| $[(\pm 45)_{12}]_s$ | $-28.6$ ksi[a] |

[a] Not tested to failure.

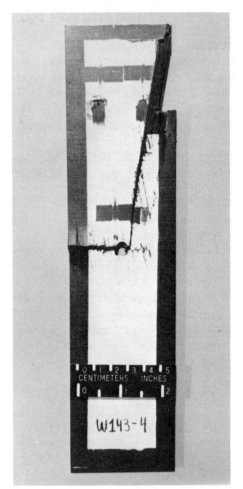

FIG. 3.    Failed $[(\pm 10)_{12}]_s$ specimen with $\frac{1}{4}$ in dia. hole.

$[(\pm 10)_{12}]_s$ *laminate*

A failed $[(\pm 10)_{12}]_s$ laminate is shown in Fig. 3. The mode of failure is compressive failure across the net section combined with a crack parallel to the fiber direction emanating from the hole edge to a corner of the specimen. Since these two failures occurred at essentially the same time, it was not possible to determine which mode preceded the other. This failure is different from the unnotched $[(\pm 10)_{12}]_s$ laminate which exhibited only

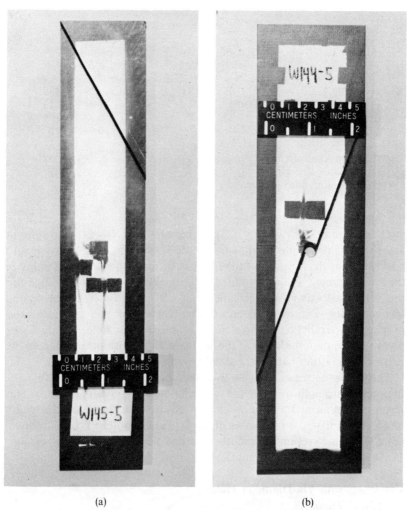

(a)                                    (b)

FIG. 4.    Failed $[(\pm 20)_{12}]_s$ specimens with $\frac{1}{4}$ in dia. hole.

one mode of failure, a fracture parallel to the fiber direction. Thus, the introduction of a hole has induced a second mode of failure for this laminate. Strain gage readings indicated significant nonlinearity near the hole prior to failure. The two specimens failed at ultimate loads of $-76$ ksi and $-77$ ksi (Table 3) which corresponds to 79% of the unnotched strength $(-97$ ksi).

TABLE 3
*Average notched compressive strength ( AS4/3502)*

| Laminate | $W/d = 5$ | | | $W/d = 10$ | | |
|---|---|---|---|---|---|---|
| | $\sigma^u$ (ksi) | $\sigma^u/\bar{\sigma}$ | $\sigma^u/\sigma^*$ | $\sigma^u$ (ksi) | $\sigma^u/\bar{\sigma}$ | $\sigma^u/\sigma^*$ |
| $[(\pm 10)_{12}]_s$ | $-62$ | 0·65 | 0·81 | $-76$ | 0·79 | 0·88 |
| $[(\pm 20)_{12}]_s$ | $-36$ | 0·87 | 1·1 | $-45$ | 1·07 | 1·19 |
| $[(\pm 30)_{12}]_s$ | — | — | — | $-26$ | 0·77 | 0·86 |
| $[(\pm 45)_{12}]_s$ | — | — | — | $-25$ | 0·94 | 1·0 |

$\sigma^u$ (ksi), notched strength.
$\bar{\sigma}$, unnotched strength.
$\sigma^*$, net section notched strength.

### $[(\pm 20)_{12}]_s$ *laminate*

The failure of the two $[(\pm 20)_{12}]_s$ laminates (Fig. 4) was interesting in that the specimens failed in the same mode at essentially the same strength ($-45$ ksi), but the location of the fracture surface differed considerably in the two specimens. Failure initiated at the intersection of the hole edge and the horizontal axis in one specimen, but not in the other. In the second specimen, the failure initiated at one corner of the specimen and propagated parallel to the fiber direction missing the hole entirely. Fiber breakage in half of the layers of the laminate, along with shear failure of the matrix in the other half, is the dominant mode of failure.

The ultimate stresses for these two laminates are actually higher than the unnotched strength of a $[(\pm 20)_{12}]_s$ laminate ($-42·5$ ksi) (Table 3). This result is even more dramatic when it is recalled that the notched strength has been calculated on the bases of the far-field unnotched cross-sectional area. If the net section is used, the notched strength is $-50$ ksi, which is 118% of the unnotched strength.

### Laminates with $\frac{1}{2}$ in Diameter Holes

### $[(\pm 10)_{12}]_s$ *laminate*

The mode of failure in the $[(\pm 10)_{12}]_s$ laminate with a $\frac{1}{2}$ in diameter hole was the same as that with a $\frac{1}{4}$ in hole. Only one $[(\pm 10)_{12}]_s$ specimen was tested to failure. The ultimate stress for that $\frac{1}{2}$ in diameter specimen ($W/d = 5$) was $-62$ ksi which is 81% of the ultimate stress in specimen with the $\frac{1}{4}$ in hole ($W/d = 10$) and 64% of the unnotched strength. It is noted that the $\frac{1}{4}$ in hole specimen corresponds to a $W/d$ ratio of 10 whereas the $\frac{1}{2}$ in hole specimen corresponds to a $W/d$ of 5. Thus there is a significant hole size effect for this laminate.

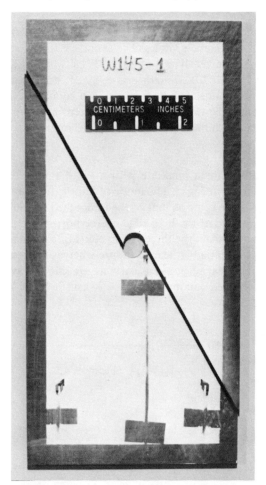

FIG. 5.   Failed $[(\pm 30)_{12}]_s$ specimen with $\frac{1}{2}$ in dia. hole.

Through-the-thickness strain gages mounted on the $\frac{1}{2}$ in diameter hole indicated that delamination occurred at a far-field stress of approximately $-30$ ksi. However, this could not be verified with X-rays when one specimen was removed from the testing machine after the indication of delamination by the strain gages. The inability to verify delamination could be due to the fact that the specimen was not under load while the X-ray was taken.

$[(\pm 20)_{12}]_s$ *laminate*

The mode of failure in these laminates was again the same as in the specimens with $\frac{1}{4}$ in diameter holes. The average ultimate stress of the $\frac{1}{2}$ in hole diameter specimens ($W/d = 5$) was $-37$ ksi which is 82% of the $\frac{1}{4}$ in hole diameter ($W/d = 10$) strength and 87% of the unnotched strength. Again, there is a significant hole size effect. In contrast to the $[(\pm 10)_{12}]_s$ laminates, through-the-thickness strain measurements from the $[(\pm 20)_{12}]_s$ laminates did not exhibit any indication of delamination.

$[(\pm 30)_{12}]_s$ *laminate*

A typical failure for a laminate with a $\frac{1}{2}$ in diameter hole ($W/d = 10$) $[(\pm 30)_{12}]_s$ is shown in Fig. 5. The mode of failure (fracture parallel to the fiber direction) is the same as that of the unnotched $[(\pm 30)_{12}]_s$ specimen. The fracture initiated at the hole edge on the horizontal axis of the hole. Stress–strain curves from the three specimens (Fig. 6) indicate that the far-field response is essentially linear to failure whereas the response near the hole exhibits nonlinear response as early as $-8$ ksi. The average ultimate stress of the notched laminates was $-26$ ksi, 76% of the unnotched strength.

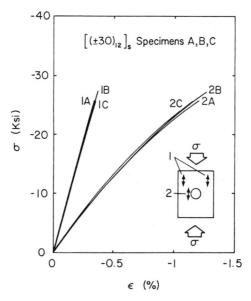

FIG. 6.    Axial strain for $[(\pm 30)_{12}]_s$ specimens with $\frac{1}{2}$ in dia. hole.

The through-the-thickness strains for the $[(\pm30)_{12}]_s$ laminate were extremely small and, therefore, inclusive as to delamination. The wide range of through-the-thickness strains as the fiber orientation is changed in these angle-ply laminates is consistent with theoretical considerations for the through-the-thickness Poisson's ratio.

### $[(\pm45)_{12}]_s$ *laminate*

None of the three 45° laminates exhibited failure. In all cases, the axial deformation prior to failure exceeded the travel limit on the load frame. One of the three specimens did exhibit damage zones emanating from the hole edge and propagating parallel to the fiber direction as shown by an ultrasonic C-scan (Fig. 7). Shuart and Williams[3] have shown previously that these bands initiate at the hole edge and increase in length with increasing load until they reach the outer edges of the specimen. C-scans at

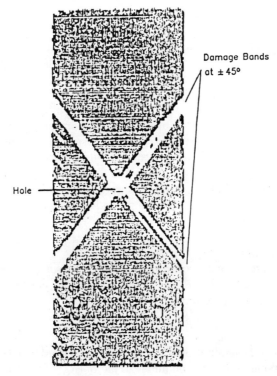

Damage Bands
at ± 45°

Hole

FIG. 7.   Ultrasonic C-scan of $[(\pm45)_{12}]_s$ specimen A.

successively higher strain levels indicated that the damage develops at far-field strain levels in excess of $-1.1\%$. The strain response for these $[(\pm45)_{12}]_s$ laminates indicated nonlinear behavior both at far-field and near the hole with significantly larger strains near the hole. The maximum stresses attained with the notched laminates were typically about 90% of those attained from the unnotched laminates.

## FAILURE ANALYSIS

### The Finite Element Model

The predictions of failure presented in this section are based upon a 3-D, linear elastic finite element stress analysis which is fully described in Ref. 1. Because of computer storage limitations, it was not possible to model a full 48-ply symmetric laminate. The model included four plies and by invoking midplane symmetry the number of plies modelled was reduced to two. An additional approximation was made to reduce the size of the problem to acceptable limits. The $y$ axis (Fig. 1) was assumed to be an axis of symmetry. A thorough study involving models in which $y$-axis symmetry was not invoked indicated this approximation did not adversely affect the prediction of failure in the vicinity of the horizontal $x$-axis where failure was observed experimentally to initiate.[1]

Each ply was subdivided into three layers of 16 node elements. Stresses were determined at the center of each layer, giving six locations through the half thickness of the laminate at which stresses were predicted. Stresses and strains were calculated at the element Gauss points and at points on the element boundaries, making it possible to obtain the distribution of stress directly on the hole boundary. Stresses were calculated on the hole edge at $2.5°$ increments; obviously, this is a limiting factor on the prediction of the exact failure location.

### Failure Predictions

Failures were predicted on the bases of the tensor polynomial failure criterion.[5] A load factor $R$ was defined to be the positive root of the quadratic equation describing the failure criterion:

$$F_i\sigma_i R + F_{ij}\sigma_i\sigma_j R^2 = 1$$

Every point in the laminate has a load factor associated with failure of that point. The point with the minimum load factor ($R_{min}$) is the point at which failure is predicted to occur first. The minimum load factor, $R_{min}$, when multiplied by the far-field strain used in the finite element analysis, ($-0.1\%$), gives the predicted far-field, first failure strain. The stresses used in the failure analysis were corrected to provide for better satisfaction of the stress-free boundary conditions on the hole edge. The rationale for the corrections is fully discussed and justified in Ref. 1.

A thorough investigation of the through-the-thickness variations indicated that failure was always predicted to initiate at the Gauss points closest to the $\pm\varphi$ interface. Figures 8 and 9 show the variation in the normalized load factor, above and below the interfaces, as a function of location about the hole edge, for all four angle-ply laminates. Also presented in each figure are the far-field strains corresponding to the predicted first failure of each laminate. Table 4 presents numerical values for the $\theta$ location of the initial failure, the corresponding far-field strain, and each term in the tensor polynomial.

The results show that there is excellent correlation between the far-field failure initiation strain as predicted above and below the $\pm\varphi$ interface.

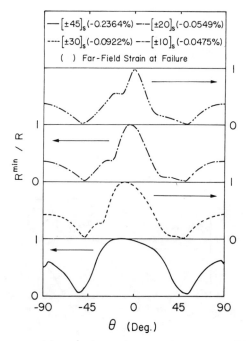

FIG. 8.    Load factor ratio $R_{min}/R$ versus $\theta$ at $r = a$, $z = 5H/12$.

This is particularly significant because, as indicated in the figures and the table, the location relative to the horizontal axis ($\theta = 0$) is a function of both the fiber orientation and which side of the interface is considered. For the 10° laminate failure is predicted at $\theta = 0$ on both sides of the interface; however, as the fiber angle is increased, the location of initial failure moves away from the horizontal axis with a distorted nearly symmetric shape about the axis in the + and $-\varphi$ layers.

Close examination of the individual terms of the tensor polynomial in Table 4 indicates that the $\tau_{12}$ term is the largest contributor to the polynomial in all laminates. For the $\pm 45$ laminate, this term is essentially equal to or greater than 1·0 on both sides of the interface. For all other laminates, failure is predicted to be primarily a combination of both $\tau_{12}$ and $\tau_{13}$. For the cases of 10° the $\sigma_1$ term is also significant which may explain the two failure modes in Fig. 3.

A disappointing aspect of the failure analysis is the fact that the predicted far-field failure strains are very conservative compared to the strains

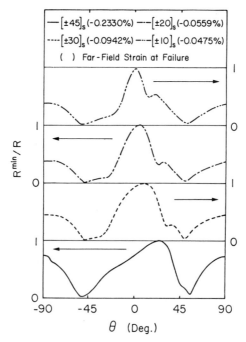

FIG. 9.   Load factor ratio $R_{\min}/R$ versus $\theta$ at $r = a$, $z = 7H/12$.

TABLE 4

*Terms in corrected-stress tensor polynomial at failure in notched angle-ply laminates $(r = a)$*

| Laminate | $[\pm 10]_s$ | $[\pm 20]_s$ | $[\pm 30]_s$ | $[\pm 45]_s$ |
|---|---|---|---|---|
| $z$ | 5H/12 | 5H/12 | 5H/12 | 5H/12 |
|  | 7H/12 | 7H/12 | 7H/12 | 7H/12 |
| $\theta_c$ | 0·0 | 2·5 | 7·5 | 22·5 |
|  | 0·0 | −5·0 | −10·0 | −22.5 |
| $\varepsilon^u$ (%) | −0·047 5 | −0·055 9 | −0·094 2 | −0·233 0 |
|  | −0·047 5 | −0·054 9 | −0·092 2 | −0·236 4 |
| $F_1 \sigma_1$ | 0·131 5 | 0·070 7 | 0·060 3 | 0·071 8 |
|  | 0·131 3 | 0·080 5 | 0·062 2 | 0·021 4 |
| $F_{11} \sigma_1^2$ | 0·069 4 | 0·020 0 | 0·014 6 | 0·020 7 |
|  | 0·069 2 | 0·026 0 | 0·017 6 | 0·021 4 |
| $F_2 \sigma_2$ | −0·146 2 | −0·251 1 | −0·369 8 | −0·440 3 |
|  | −0·146 2 | −0·206 6 | −0·313 6 | −0·448 4 |
| $F_{22} \sigma_2^2$ | 0·006 7 | 0·019 7 | 0·042 7 | 0·060 6 |
|  | 0·006 7 | 0·013 3 | 0·030 7 | 0·062 8 |
| $F_3 \sigma_3$ | −0·004 4 | 0·014 6 | 0·005 4 | 0·050 9 |
|  | −0·003 1 | 0·036 6 | 0·062 4 | 0·015 8 |
| $F_{33} \sigma_3^2$ | 0·000 0 | 0·000 1 | 0·000 0 | 0·000 8 |
|  | 0·000 0 | 0·000 4 | 0·001 2 | 0·000 1 |
| $F_{44} \tau_{23}^2$ | 0·015 4 | 0·089 9 | 0·161 5 | 0·045 2 |
|  | 0·015 1 | 0·067 0 | 0·116 1 | 0·042 6 |
| $F_{55} \tau_{13}^2$ | 0·332 0 | 0·486 4 | 0·394 7 | 0·211 4 |
|  | 0·332 9 | 0·468 0 | 0·376 2 | 0·217 1 |
| $F_{66} \tau_{12}^2$ | 0·595 6 | 0·549 8 | 0·690 6 | 0·978 9 |
|  | 0·593 9 | 0·514 9 | 0·643 1 | 1·015 4 |

determined experimentally. This may be due to a number of factors including: (1) low input values for the shear strength; (2) nonlinear material behavior near the hole edge; (3) the idealization of a distinct interface which gives rise to a mathematical singularity and theoretical stresses which are much higher than those present in a real laminate.[6] Future work will address these questions.

## CONCLUSIONS

Experimental results for failure of notched angle-ply laminates under compression loading and a failure analysis based upon the tensor polynomial failure criterion using stresses obtained from a 3-D finite element stress analysis have been presented. The following conclusions can be drawn from the theoretical/experimental study:

- $10°$, $20°$, and $30°$ laminates fail by fracture parallel to fiber direction breaking fibers in half the layers along with shear failure of the matrix in the other layers.
- The $10°$ laminate also exhibits compressive failure across the horizontal axis.
- $45°$ laminates exhibit matrix damage in bands emanating from the hole and propagating in the $\pm$ fiber directions.
- The $20°$ laminate exhibits notch insensitive behavior for small holes.
- $10°$ and $20°$ laminates exhibit notch sensitive behavior as $W/d$ is varied from 5 to 10.
- The tensor polynomial predicts that failure initiates at the ply interface, on the hole boundary.
- Failure is primarily due to the combined influence of the inplane shear stress $\tau_{12}$ and the out-of-plane shear stress $\tau_{13}$.
- The predicted location and mode of failure are consistent with experimental results.
- The predicted failure loads are very conservative.

## ACKNOWLEDGEMENT

The authors gratefully acknowledge the support of the NASA Virginia Tech Composites Program under NASA Grant NAG 1-343. J. G. Williams was associated with the Structural Mechanics Branch, NASA Langley Research Center when this research was performed.

## REFERENCES

1. BURNS, S. W., HERAKOVICH, C. T. and WILLIAMS, J. G., Compressive failure of notched angle-ply composite laminates: Three-dimensional finite element analysis and experiment, CCMS-85-11, VPI-E-85-22, Virginia Polytechnic Institute and State University, Blacksburg, VA, Nov. 1985.
2. RHODES, M. D., MIKULAS, M. M., JR. and McGOWAN, P. E., Effects of orthotropy and width on the compression strength of graphite–epoxy panels with holes, *AIAA Journal*, **22** (1984), 1283–1291.

3. SHUART, M. J. and WILLIAMS, J. G., Investigating compression failure mechanisms in composite laminates with a transparent fiberglass–epoxy birefringent materials. *Experimental Techniques*, **8** (1984), 24–25.
4. TIMOSHENKO, S. P., *Theory of Elastic Stability*, New York, McGraw-Hill, 1936.
5. TSAI, S. W. and WU, E. M., A general theory of strength for anisotropic materials, *J. Comp. Mater.*, **5** (1971), 58.
6. HERAKOVICH, C. T., POST, D., BUCZEK, M. B. and CZARNEK, R., Free edge strain concentrations in real composite laminates: Experimental-theoretical correlation, *J. appl. Mech.*, **52** (1985), 787–793.

# 14

# Fabrication and Mechanical Properties of Hybrid Composites with Braiding Construction

ZENICHIRO MAEKAWA, HIROYUKI HAMADA and TSUNEO HORINO

*Faculty of Textile Science, Kyoto Institute of Technology,*
*Matsugasaki, Sakyo-ku, Kyoto 606, Japan*

ATSUSHI YOKOYAMA

*Osaka Prefectural Industrial Research Institute,*
*Enokojima, Nishi-ku, Osaka 550, Japan*

and

YASUHIKO IWASAKI

*Graduate Student, Kyoto Institute of Technology,*
*Matsugasaki, Sakyo-ku, Kyoto 606, Japan*

## ABSTRACT

*The composite pipe with braided fabric is produced from various types of fibres and combinations of these fibres under various braiding angles. The through-the thickness braided pipe is fabricated. The behaviour of braided pipe under lateral compressive load is investigated in an experimental and theoretical way.*

*It is made clear that the hybrid construction and through-the-thickness braiding is effective in strengthening the braided pipes. The braiding angle effect is also discussed in this chapter.*

## 1. INTRODUCTION

The three-dimensional reinforced construction has attracted great interest in the improvement of the performance of fibre-reinforced composites.[1-3] The structure of the three-dimensional fabric is, however, complex as compared to that of the two-dimensional one. Therefore the weaving machine for the three-dimensional fabric becomes considerably expensive. This is a reason why the three-dimensional fabric does not extend over such a wide range of applications. Braiding one of the three-dimensional fabrications has an advantage that it can easily make various shaped constructions. Therefore the composite with integrally-formed braiding may be a promising candidate for materials with a bright future.

This study deals with the braided composite pipes. The braided composites are fabricated from glass, carbon and aramid fibres, and the combination of these fibres under various braiding angles. The through-the-thickness braided pipes, which are bound to each lamina, are also fabricated. The lateral compressive tests are carried out in order to specify the mechanical properties of braided composites. The theoretical prediction of the braided composite pipes is performed by using the curved beam theorem and a computer simulation of the mechanism of the braiding process is briefly stated.

## 2. EXPERIMENTAL PROCEDURE

### 2.1. Fabrication

The manufacturing procedure in this study can be classified into four processes, as shown in Fig. 1. In braiding process, the fibres are braided around the mandrel with a mould releasing agent according to draw up the mandrel, as shown in Fig. 2. In this figure, the fibres are supplied from the spindle which moves on the orbit, then winds around the mandrel. The braided fabric is woven on the mandrel according to the intercross of the moving spindle. It is possible to wave the various braiding types by

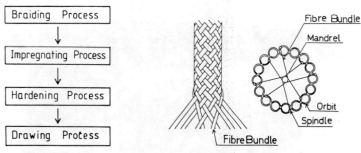

FIG. 1. Fabrication procedure of FRP pipes with braiding construction.

FIG. 2. Schematic diagram of braiding process.

the number, positioning or moving direction of spindle or shape of orbit. Furthermore, the braided pipe having a non-circular section can be woven by a suitable arrangement of the above conditions. In this study, the plane weave is adopted as the braiding type so that plane weave is the most basic woven type. Figure 3 shows the shape of the braider. Dimensions of the mandrel which is the core of the pipe are 17·5 mm diameter and 400 mm length respectively. The braider can change the braiding angle of the pipe by varying the ratio of the gears.

In the resin impregnating process, the mandrel with braiding fibres and the resin mixed with the hardener are set in the resin impregnating

FIG. 3.   Outside view of braider.

apparatus. This apparatus is kept under vacuum for 10 min at 50°C to impregnate the resin into the fibres.

In the hardening process, the pipe is maintained under a dry atmosphere for 12 h at 50°C. Then the polyester film is coiled around the impregnated specimen to form a pipe. The holder is made of glass cloth tape which is tightly gripped in the drawing process.

Lastly, the pipe is drawn out from the mandrel after polymerization, then specimens are cut out of the pipe.

## 2.2. Specimen Preparation

Three kinds of fibre are prepared; they are micro-glass roving (RER-231) consisting of 1600 filaments, carbon fibre (T300, Toray Industry Co. Ltd) and aramid fibre (Kevlar 29). The matrix is epoxy resin (EPOMIK R-140, Mitsui Petro-Chemical Industry Co.). All specimens have double layers. The method of the through-the-thickness braiding is that the binding fibre is woven simultaneously with braiding fibres, as shown in Fig. 4. The binding fibre fastens the inner and outer layers of the specimen. The through-the-thickness braided pipes are prepared in three types, which are radial type, longitudinal type and spiral type, as shown in Fig. 5. The binding fibres are woven by 3, 4 and 8 bundles in the longitudinal-type specimen. In the radial-type specimen, the binding fibre is arranged in every 22·5 mm length. In this figure, carbon fibres are used as the binding fibre in order to be clearly seen. However, in the experiment the binding fibres are made of glass fibres, which are also used as the braiding fibre.

Table 1 shows the list of the specimens. The braiding angle represents the angle of fibre to the longitudinal direction of the pipe. The values of the braiding angle are calculated by a computer simulation which is developed for this study, as shown in Fig. 6. This simulation can estimate the

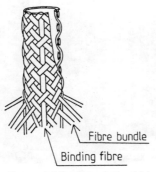

Fibre bundle

Binding fibre

Fig. 4.   Schematic diagram of the through-the-thickness braiding pipe.

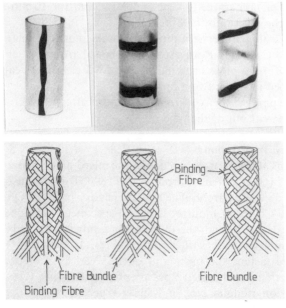

FIG. 5.   Shape of the through-the-thickness braiding pipe.

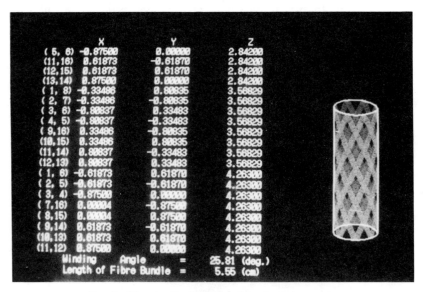

FIG. 6.   Photograph of display of the simulation result.

TABLE 1
*List of specimens*

| Sign | | $G_LG_L$ | $G_MG_M$ | $G_HG_H$ | $G_SG_S$ | $G_LG_S$ | $G_MG_S$ | $G_HG_S$ | $G_SG_L$ | $G_SG_M$ | $G_SG_H$ |
|---|---|---|---|---|---|---|---|---|---|---|---|
| Material of fibre | Inner | Glass | Glass | Glass | Glass | Glass | Glass | Glass | Glass | Glass | Glass |
| | Outer | Glass | Glass | Glass | Glass | Glass | Glass | Glass | Glass | Glass | Glass |
| Braiding angle (°) | Inner | 25·8 | 32·6 | 42·5 | 54·65 | 25·8 | 32·6 | 42·5 | 54·65 | 54·65 | 54·65 |
| | Outer | 25·8 | 32·6 | 42·5 | 54·65 | 54·65 | 54·65 | 54·65 | 25·8 | 32·6 | 42·5 |
| Number of ply | | 2 | 2 | 2 | 2 | 2 | 2 | 2 | 2 | 2 | 2 |

| Sign | | $C_HG_H$ | $G_HC_H$ | $C_HC_H$ | $A_HG_H$ | $G_SG_S$ | $G_LG_S$ | $G_HA_H$ | $A_HA_H$ |
|---|---|---|---|---|---|---|---|---|---|
| Material of fibre | Inner | Carbon | Glass | Carbon | Aramid | Aramid | Glass | Glass | Aramid |
| | Outer | Glass | Carbon | Carbon | Glass | Glass | Aramid | Aramid | Aramid |
| Braiding angle (°) | Inner | 42·5 | 42·5 | 42·5 | 42·5 | 42·5 | 42·5 | 42·5 | 42·5 |
| | Outer | 42·5 | 42·5 | 42·5 | 42·5 | 42·5 | 42·5 | 42·5 | 42·5 |
| Number of ply | | 2 | 2 | 2 | 2 | 2 | 2 | 2 | 2 |

| Sign | | LL' | HH' | LL'4 | HH'3 | HH'4 | HH'8 | LL'R | HH'R | LL'S | HH'S |
|---|---|---|---|---|---|---|---|---|---|---|---|
| Braiding angle (°) | Inner | 25·8 | 42·5 | 25·8 | 42·5 | 42·5 | 42·5 | 25·8 | 42·5 | 25·8 | 42·5 |
| | Outer | 25·8 | 42·5 | 25·8 | 42·5 | 42·5 | 42·5 | 25·8 | 42·5 | 25·8 | 42·5 |
| Number of ply | | 2 | 2 | 4 | 2 | 2 | 2 | 2 | 2 | 2 | 2 |
| Type of binding fibre | | — | — | Longi-tudinal | Longi-tudinal | Longi-tudinal | Longi-tudinal | Radial | Radial | Spiral | Spiral |
| Number of binding fibre | | — | — | 4 | 3 | 4 | 8 | — | — | — | — |

intercross point of the fibres, the braiding angle and the length of fibre. The specimens with one braiding angle are of 12 types. The specimens with different braiding angles are of six types, and the hybrid specimens which have two layers of different materials of fibre are of four types. The through-the-thickness braided pipes are of 10 types. To express the structure of the braided pipes, each specimen is indicated by two characters and two suffixes. The first character indicates the material of the fibre of the inner layer, and the suffix indicates the braiding angle of the inner layer. The second character and its suffix represent those of the outer layer. The characters G, C and A mean glass, carbon and aramid fibre, and the suffixes L, M, H and S indicate $25.8°$, $32.6°$, $42.5°$ and $54.65°$ ($58.77°$, $58.52°$) respectively. The apostrophe indicates the through-the-thickness braided pipe. The third character indicates the type of binding fibre and the third numeral indicates the number of binding fibres. In non-binding pipes, there are 16 fibre bundles in each layer. The number of fibres in the through-the-thickness type is 8 bundles in the inner layer and 16 bundles in the outer layer.

## 2.3. Testing

Figure 7 shows the shape of the specimen. The diameter of the specimen is $17.5$ mm and the length is 45 mm. The specimen is placed between the upper plate and the lower plate and is deformed by pushing down the upper plate, as shown in Fig. 8. The deformation speed is 2 mm/min.

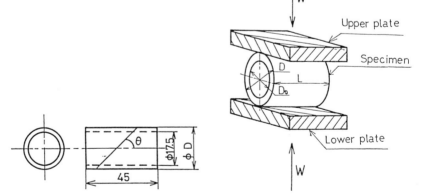

FIG. 7.　Dimension of specimen.　　　　　FIG. 8.　Lateral compressive test.

## 3. THEORETICAL APPROACH

To estimate the experimental results, the lateral compressive stress is calculated by the following equations. In this approach, each layer of the braided pipe is regarded as having angle-ply structure.

### 3.1. Elastic Constant [4]

We shall calculate the elastic modulus in the radial direction. The elastic modulus in the fibre direction is given by

$$E_L = E_f V_f + E_m(1 - V_f)$$

where $E_f$ is the elastic modulus of the fibre, $E_m$ is the elastic modulus of the matrix and $V_f$ is the volume fraction of the fibre.

The elastic modulus $E_T$ in the direction normal to the fibre, the shearing modulus $G_{LT}$, Poisson's ratio $v_L$ in the fibre direction and $v_T$ in the normal direction are calculated by

$$\frac{1}{E_T} = \frac{1\cdot36(K_f - K_m)}{(K_f - K_m)^2 - (v_f K_f - v_m K_m)^2} + \frac{1 - 1\cdot05\sqrt{V_f}}{E_m}$$

$$\frac{1}{G_{LT}} = \frac{1\cdot36}{G_f - G_m} + \frac{1 - 1\cdot05\sqrt{V_f}}{G_m}$$

$$v_L = \frac{1\cdot05\sqrt{V_f}(v_f - v_m)}{K_f - K_m} + v_m$$

$$v_T = \left(\frac{E_T}{E_L}\right)v_L$$

where

$$K_f = \frac{E_f}{1 - v_f^2} \qquad K_m = \frac{E_m}{1 - v_m^2} \qquad G_f = \frac{E_f}{2(1 + v_f)} \qquad G_m = \frac{E_m}{2(1 + v_m)}$$

$v_f$ is Poisson's ratio of the fibre and $v_m$ is Poisson's ratio of matrix.

The elastic modulus in the radial direction $E_\theta$ is given by

$$\frac{1}{E_\theta} = \frac{1}{E_{\theta_0}} - \psi_0^2 G_{xy_0}$$

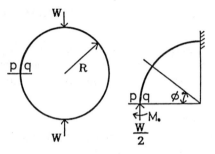

FIG. 9.  Cross-section of the pipe under lateral compressive load.

where

$$\frac{1}{E_{\theta_0}} = \frac{\sin^4 \alpha}{E_L} + \frac{\cos^4 \alpha}{E_T} + \left(\frac{1}{G_{LT}} - \frac{2v_L}{E_L}\right) \sin^2 \alpha \cos^2 \alpha$$

$$\psi_0 = \left[\frac{\cos^2 \alpha}{E_T} - \frac{\sin^2 \alpha}{E_L} - \frac{1}{2}\left(\frac{1}{G_{LT}} - \frac{2v_L}{E_L}\right) \cos 2\alpha\right] \sin 2\alpha$$

$$\frac{1}{G_{xy_0}} = \left(\frac{1+v_L}{E_L} + \frac{1+v_T}{E_T}\right) \sin^2 2\alpha + \frac{1}{G_{LT}} \cos^2 2\alpha$$

where $\alpha$ is the winding angle.

### 3.2. Lateral Compressive Stress

The analytical region is limited to a quarter of the pipe, as shown in Fig. 9. The bending moment at angle $\phi$ is given by

$$M = M_0 + \frac{WR}{2}(1 - \cos \phi) \qquad M_0 = -\frac{WR}{2}\left(1 - \frac{2}{\pi}\right)$$

where $W$ and $R$ are the lateral compressive load and the radius of the pipe, respectively. Therefore, the maximum bending moment arises at the loading point:

$$M_{max} = 0\cdot318WR$$

The bending stress of curved beam (Fig. 10) caused by the above moment is calculated by

$$\sigma_\theta = \frac{-E_\theta(y)(y - e)}{(R - y)} \frac{M}{\int E_\theta(\delta)(\delta - e)\,\mathrm{d}A}$$

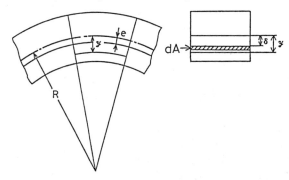

FIG. 10. Cross-section of the pipe (regard as curved beam).

where $E_\theta(y)$ is the elastic modulus at distance $y$ from the middle line and $e$ is the distance from the middle line to the neutral line, which is calculated from the following equation:

$$e = \frac{\int [E_\theta(\delta)/(R-\delta)]\delta \, dA}{\int [E_\theta(\delta)/(R-\delta)] \, dA}$$

### 3.3. Deformation under Lateral Compressive Load

When the compressive load is $W$, as shown in Fig. 9, the elastic strain energy of a quarter of the pipe is given by

$$U = \frac{1}{2} \frac{1}{[\int E_\theta(\delta)(\delta-e) \, dA]^2} \int \frac{E_\theta(\delta)(\delta-e)^2}{(R-\delta)^2} \, dA \frac{W^2 R^3}{4} \left\{ \frac{\pi}{4} - \frac{2}{\pi} \right\}$$

From Castigliano's theorem, the deflection $\eta$ of the pipe under lateral compressive load is given by

$$\eta = 2\eta' = 2\frac{\partial U}{\partial(W/2)} = \frac{WR^3}{[\int E_\theta(\delta)(\delta-e) \, dA]^2} \int \frac{E(\delta)(\delta-e)^2}{(R-\delta)^2} \, dA \left\{ \frac{\pi}{4} - \frac{2}{\pi} \right\}$$

The equivalent strain and the equivalent elastic modulus of the lateral compressive test are defined as follows:

$$\varepsilon_{eq} = \frac{\delta}{D} \qquad E_{eq} = \frac{\sigma_\theta}{\varepsilon_{eq}}$$

where $D$ is the outer diameter.

## 4.   RESULTS AND DISCUSSION

### 4.1.  Influence of Braiding Angle

The relation between lateral compressive stress and displacement is shown in Fig. 11. In this figure the pipes have the same braiding angle in both layers. Sudden reductions in stress are observed in both cases of the type $G_H G_H$ and $G_S G_S$. On the other hand, in the types $G_L G_L$ and $G_M G_M$ the drop of stress is not observed. This difference is because the interlaminar delamination, as shown in Fig. 12, occurs in $G_H G_H$ and $G_S G_S$. Figure 13 shows the relation between the equivalent elastic modulus and the braiding angle. In this figure, the broken lines indicate the theoretical predictions described in Section 3. The equivalent elastic modulus increases as the braiding angle increases. The pipe with the glass fibre has the maximum equivalent elastic modulus. This is because the thickness of the glass fibre pipe is large as compared to other fibre pipes. The theoretical prediction is in good agreement with experimental results at low braiding angle. However, at high braiding angle, the values of the theoretical prediction are larger than the values in the experimental result. This difference is caused by the fact that the theory used here assumes inplane anisotropy, though the FRP has three-dimensional anisotropic properties. Therefore, the difference between the theoretical result and the experimental result increases as the specimen becomes thick. Figure 14 shows the relation

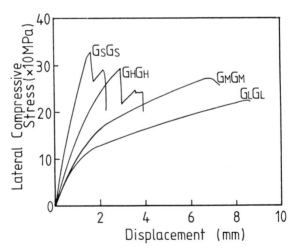

Fɪɢ. 11.    Relation between lateral compressive stress and displacement in the case of pipes with same kind of fibre and same braiding angle.

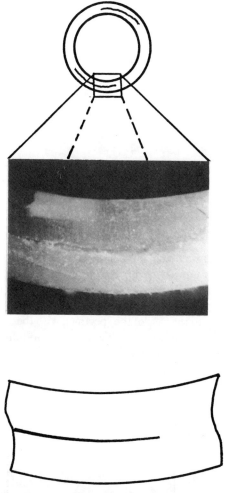

FIG. 12. Aspect of interlaminar delamination.

between the maximum stress and the braiding angle in the case of the pipe with the same kind of fibres and the same braiding angle in both layers. The maximum stresses of the carbon fibre pipe and the aramid fibre pipe increase with braiding angle similar to the equivalent elastic modulus. Nevertheless, the maximum stress of the glass fibre pipe approaches a constant level as a result of the occurrence of interlaminar delamination (Fig. 12).

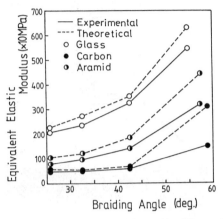

FIG. 13.   Relation between equivalent elastic modulus and braiding angle in the case of pipes with same kind of fibre and same braiding angle.

The relation between the equivalent elastic modulus and the braiding angle in the case of pipes with different braiding angles is shown in Fig. 15. In this figure, the basic braiding angle is fixed at 54·65°, the other braiding angle is varied. The equivalent elastic modulus decreases as braiding angle decreases similarly to the case of the specimen with the same braiding angle. However, the tendency of the equivalent elastic modulus to decrease is not so severe as compared to the specimen with one braiding angle. The

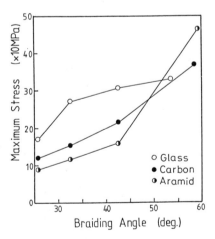

FIG. 14.   Relation between maximum stress and braiding angle in the case of pipes with same kind of fibres and same braiding angle.

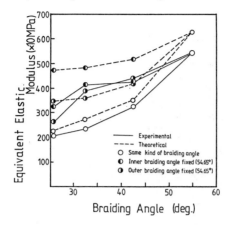

FIG. 15. Relation between equivalent elastic modulus and braiding angle in the case of pipes with different braiding angle.

specimens with high braiding angles in the inner layer show high equivalent elastic modulus. Therefore, it is clear that the combination of winding angles affects the performance of the braided pipe. The theoretical results are in good agreement with the experimental results. Figure 16 shows the relation between the maximum stress and the braiding angle. The maximum stresses of all cases reach constant level at high braiding angle caused by interlaminar delamination.

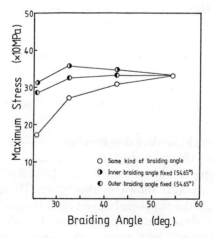

FIG. 16. Relation between maximum stress and braiding angle in the case of pipes with different braiding angle.

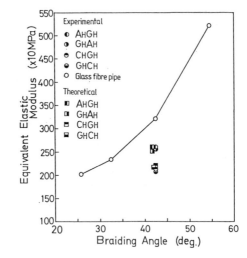

FIG. 17.    Equivalent elastic modulus of hybrid pipes.

## 4.2.  The Effect of Hybrid Pipe

Figure 17 shows the equivalent elastic modulus of the hybrid pipes obtained by experimental and theoretical analyses. The braiding angle is fixed at 42·5°. For the purpose of comparison, the results of glass fibre pipes are included in this figure. The equivalent elastic modulus of hybrid pipes is lower than that of glass fibre pipes. This result indicates that the hybrid pipe has flexibility. The theoretical results are in good agreement with the experimental results. Figure 18 shows the maximum stress of the hybrid pipes. The maximum stress of the type $G_H A_H$ is at the highest level in the hybrid pipes. Furthermore, the maximum stress of the type $G_H A_H$ is larger than that of glass fibre pipe $(G_S G_S)$. This is because interlaminar delamination does not occur in the hybrid pipes owing to its flexibility.

## 4.3.  Effect of Through-the-thickness Braiding

Figure 19 shows the equivalent elastic modulus of the through-the-thickness braided pipes. The braiding angle is fixed at 42·5°. The results of glass fibre pipes are also shown in this figure. The equivalent elastic modulus of through-the-thickness braided pipes is nearly a fixed quantity. The results of the glass fibre pipes are larger than that of through-the-thickness braided pipes. This is because the construction of the through-the-thickness braided pipe is 8 bundles in the inner layer and 16 bundles in

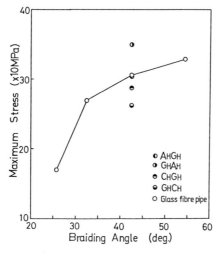

Fig. 18.   Maximum stress of hybrid pipes.

the outer layer. Figure 20 shows the maximum stress of through-the-thickness braided pipes. The maximum stress of the radial binding type LL'R, HH'R and the spiral binding type LL'S, HH'S are at a maximum in through-the-thickness braided pipes. The results from the radial and spiral binding types are larger than the results from the non-binded pipe. It is clear

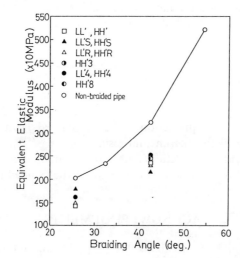

Fig. 19.   Equivalent elastic modulus of the through-the-thickness braided pipes.

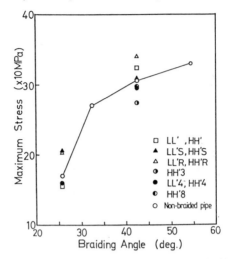

FIG. 20.   Maximum stress of the through-the-thickness braided pipes.

that through-the-thickness braiding is effective for the increase of strength of the braided pipes.

## 5.   CONCLUSION

This chapter dealt with the fabrication and mechanical properties of the braided construction. The following interesting results were obtained.

(1)   The braided construction can be fabricated by using the simple mechanism machine. The through-the-thickness braided pipe can be fabricated by application of braiding techniques.

(2)   The braiding angle affects the flexibility under lateral compressive load.

(3)   The theoretical prediction using the curved beam theory is in agreement with experimental results.

(4)   The hybrid construction and through-the-thickness braiding is effective for the increase of strength of the braided pipes.

## ACKNOWLEDGEMENT

The authors would like to thank Dr Sinya Motogi (Osaka Municipal Technical Research Institute) for helpful discussion.

# REFERENCES

1. BROWN, R. T., Through-the-thickness braiding technology, 30th National SAMPE Symposium, March 1985.
2. FLORENTINE, R. A., CHOU, TSU-WEI and KO, F., Magnetically woven composite I-beam—a structural alternate—Phase I final report, 40th Annual Conf. Reinforced Plastic/Composites Institute, The Society of the Plastic Industry, Jan. 1985.
3. KO, F., BRUMER, J., PASTORE, A. and SCARDINO, F., Development of muti-bar weft-insert warp knit glass fabrics for industrial applications, *Journal of Textile Engineering for Industry*, **102** (1980).
4. YAMAWAKI, K. and UEMURA, M., The fracture strength of helical-wound composite cylinders. 1. Tensile strength, *Zairyo*, **19** (1970), 957.

# 15

# Reprocessing of Carbon Fibre/PEEK Laminates

J. F. Liceaga and J. J. Imaz San Miguel

*INASMET, Barrio de Igara s/n, 20.009 San Sebastian, Spain*

## ABSTRACT

*In this work, the reprocessing possibilities of fibre reinforced thermoplastics are analysed. The material utilized is a carbon fibre reinforced polyether-ether-ketone (PEEK) unidirectional preimpregnated system aimed at high performance applications.*

*The preimpregnate is pressed in a heated flat plates press in order to obtain a flat laminate. After this process, the laminate is heated and moulded in a cool curved mould. This process is repeated several times in order to measure loss of properties in the material. After the reprocessing, the mechanical properties and damage caused are studied. The study shows that the properties of carbon/PEEK laminates after the reprocessing are good, although slight degradation is observed. The influences of temperature and mould curvature angle are also measured. A microscopic study was realized by SEM.*

## INTRODUCTION

Carbon fibre reinforced composite materials are extensively utilized in many applications. The current matrices used in carbon fibre reinforced plastics for high performance structural applications are usually thermosetting matrices, mainly epoxy resins. Thus, carbon fibre composite materials are extensively used in very different industrial fields: aeronautics industry, military industry, sporting equipment, etc.

But over the last few years there has been a major effort in the development of new matrix materials to improve performance.

Among these matrices, thermoplastic resins have a short moulding cycle, which is very advantageous from an economic viewpoint, because this contributes to increased production speed.

Another advantageous property of preimpregnated thermoplastic resins is the indefinite shelf life which permits us to stock this prepreg for a long time. Thermoplastic resins also resolve several bonding problems and so cohesive bonding offers an easy way to assemble parts. Moreover, thermoplastic resins are reusable to a certain extent. When the technical or economical life of a thermoplastic composite is finished it can be processed again, as an additive to an injection moulding compound by grinding it for example. Also thermoplastic products can be repaired simply, by any heat source, which can be used to correct the dimensions of a part of the piece, or to apply additional material.

On the other hand, thermoplastic resins have advantageous properties since they contribute to improve toughness in composite materials.[1]

Long fibre reinforced thermoplastics can be obtained in laminate form by hot press moulding. But another interesting aspect of these materials is that after flat pressing, curved parts can be obtained by moulding process which show resemblance to the deformation process of sheet metal[2] and this property of reprocessing capability should give us an advantage for the processing, because sometimes it could be interesting to obtain the product by indirect means, that is to say, moulding of curved parts of laminates previously processed by flat pressing.

In order to simplify the piece production, the laminate which has been obtained from the first process in flat pressing can be characterized and homologized. This standardization guarantees choice of materials, because we now have an initial laminate with known mechanical properties that has been processed at recommended conditions (pressure, temperature, moulding time, etc.).[3]

After the flat laminate fabrication, we can heat the laminate again for reprocessing in a curved mould. In order to carry out consecutive cycles of curved moulding it is necessary to know the fibre/matrix interface behaviour during this process, as well as the thermoplastic matrix flow towards new geometrical form. It is also possible that after the curved moulding, the structural component, carbon fibre, would not have the same quality in the same direction.

Therefore, when curved moulding is finished, it is necessary to study fibre capacity in the presence of these deformations when the carbon fibre is inserted in the resin matrix. For this reason and because successive processing can give rise to carbon fibre/PEEK system degradation it is interesting to study the mechanical properties of the laminate after reprocessing.

The fibre deformation and resin degradation cause a loss of the mechanical properties in the composite laminate which could perhaps make the reprocessing disadvantageous.

## MATERIALS AND TESTING METHODS

The relatively recently developed carbon fibre/polyether-ether-ketone (PEEK) preimpregnated system, aimed at high performance applications is employed as the research material. This system is commercially called APC-2.

The resin component of APC-2, polyether-ether-ketone (PEEK) commercially called Victrex, is a aromatic semicrystalline polymer, which has a glass transition temperature of 143°C and a melting point of 343°C. PEEK resin has superior solvent resistance and is unaffected by almost all the aggressive solvents encountered in an aerospace environment. For this reason, polyether-ether-ketone and the composites where PEEK is the matrix are suited to the environmental needs of airframe designers.[4]

Mechanical performance of APC-2 in the fibre direction is principally controlled by the properties of the fibre, which is a polyacrylonitrile based carbon fibre. The fibres are unidirectionally aligned and they are well dispersed and thoroughly wetted in the PEEK matrix to give a fibre content of 61% by volume and 68% by weight. The unidirectional single ply prepregs have a nominal thickness of 0·125 mm and a nominal density of 213 g/m$^2$.

Twenty reinforcement layers, unidirectionally disposed, were respectively stacked in order to prepare the carbon fibre reinforced composite laminate. Therefore, all the laminates together had a nominal thickness of 2·5 mm. Sheet production from the preimpregnate was realized by placing these stacked layers between press plates, heating to 380–400°C under low pressure, and cooling rapidly. This ensured that the best properties were obtained. In order to optimize the performance of our APC-2 laminates and flats by means of adequate crystallinity and morphology control of PEEK polymer, the cooldown rate was approximately 100°C/min to a temperature of 100°C.

Thus, PEEK/carbon laminate was obtained for posterior curving. After this process, the flat laminates were heated to a temperature of 380°C in order to reach an adequate resin fluidity for pressing the flat laminate in a curved mould. The mould was a plywood mould. In order to maintain the resin fluidity, the transport time between the heating stage and the mould was as short as 5 s.

After this process, the curved laminate was placed between the flat press plates, heating to 380°C to obtain a flat laminate again. This flat laminate was curved again, and this process was repeated the times required in each case (Fig. 1). In this curving process, different curvature angles were utilized in order to measure the influence of the angle on the damage caused.

After these fabrication and reprocessing cycles, each one of the specimens was submitted to a fractographic study showing possible damage. This study was realized by scanning electron microscopy (SEM).

As well as the fractographic study, the material mechanical properties were studied in order to detect any decrease in characteristic values. For this detection, a flexural static test was selected, according to ASTM Standard, D-790 M, because this is probably the most sensitive mechanical test for evaluating any damage to the carbon fibre reinforced unidirectional composite, as well as the bending between layers and the properties lost in the laminate when it was reprocessed. In this test we calculated Young's modulus, material deformation and flexural strength.

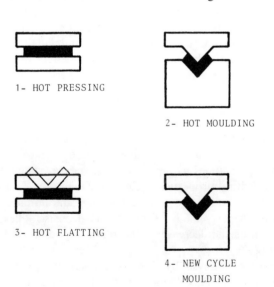

1- HOT PRESSING

2- HOT MOULDING

3- HOT FLATTING

4- NEW CYCLE
MOULDING

Fig. 1.   Process—cycle diagram.

## RESULTS AND DISCUSSION

In the study of the processing and the subsequent carbon fibre/PEEK laminate curving, is important to analyse the variables which influence this processing. We have studied these four: angle and radius of curvature, number of processes and process temperature.

### Influence of the Curvature Angle and Radius

For curved part moulding we utilized a plywood mould designed with a variable angle. The inside angles we used to damage the material were 120° and 90°. The specimens reprocessed by this plywood were subsequently studied to measure damage caused. We have observed that when the reprocessing mould angle is reduced, the damage to and loss of mechanical properties of the laminate increases. This loss of mechanical properties is manifested by means of flexural strength decrease, as well as by a decrease of the rigidity module.

A resin displacement and a very light fibre pull-out were observed in the curved zone by SEM. Fibre pull-out was greater when the mould inside angle was 90°.

On the other hand, when the radius of curvature of the mould where the laminate was curved was reduced, an increase in structural damage was expected. In this step of the study, the curvature radius utilized was constant, therefore, its influence was not measured.

At present, we are continuing to work utilizing wide range angles and radius for curving the laminate, in order to deduce some rule which describes the influence of both moulding parameters on damage caused in the material.

### Influence of Number of Processes

The curved laminate was reprocessed in a flat plate press after the curving in order to curve it again. Therefore, each specimen was reprocessed several times (complete process: curve + flat plate pressing). The study of flexural properties in each specimen showed a reduction of these properties. Property loss increases when the number of processes increases. This property loss is reflected by the decrease of flexural strength (Fig. 2) as well as the decrease of the rigidity module, and the fracture deflection (Fig. 3).

With respect to analysis by SEM, slight damage to the fibre can be observed (Fig. 4a, b). This damage is identified as fibre pull-out, and it increases when reprocessing times are also increased.

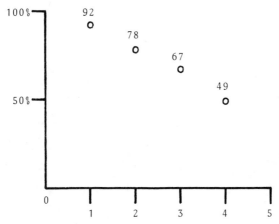

FIG. 2. Flexural strength variation in relation to number of processes (angle 90°).

This property loss can be explained by the effort that the fibres have to make in order to adapt successively to the mould form. We think that a slight resin degradation by maintenance at high temperatures can also contribute to property loss.

## Influence of Moulding Temperature

Similarly, we utilize different temperatures between 340 and 400°C for moulding laminates. By means of these results, it can be observed that lower

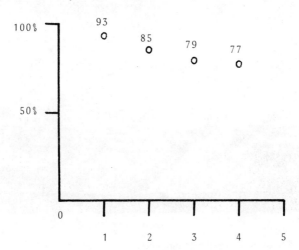

FIG. 3. Fracture deflection variation in relation to number of processes (angle 90°).

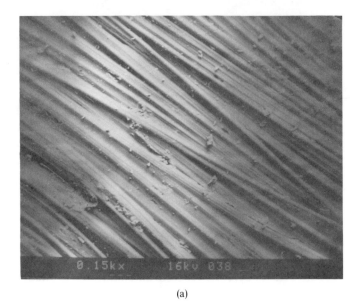

(a)

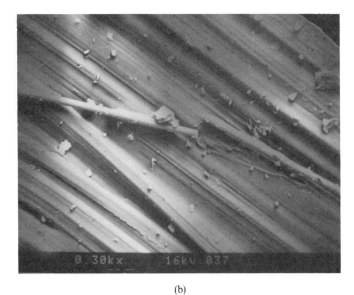

(b)

FIG. 4.    (a) SEM of the material before reprocessing. (b) SEM showing fibre pull-out after being reprocessed four times for the 90° curved sample.

temperatures complicate the adoption from flat laminates to curved parts because resin fluidity is excessively low, while higher temperatures give rise to slight degradation of the resin, which produces a loss of mechanical properties in the laminate.

## CONCLUSIONS

Carbon fibre/PEEK prepreg layers were moulded in a hot press containing the laminates. After this first process, we have a laminate whose mechanical properties can be homologized.

When the first process is finished, we pressed the laminate again in a curved mould obtaining a piece in a curved form.

This moulding process of continuous fibre reinforced laminates shows much resemblance to the deformation process of sheet metals.

When the laminate carbon fibre/PEEK which has been previously pressed is moulded, a slight loss of mechanical properties can be detected. This loss increases when the inside angle of curvature decreases.

Successive processing of the laminate promotes degradation in the material, produced by the continuous adaptation of the carbon fibres in the mould direction and by the continuous maintenance at high temperatures to raise the fluidity. This degradation is observed in the decrease of mechanical properties such as flexural rigidity modulus and flexural strength, and by observation of fibre pull-out in SEM.

These results have been obtained in a wider program on reprocessing and it is connected with the mechanical properties of the thermoplastic matrix composites. At present we are studying the influence of some design or geometric variables on the structural damage and residual mechanical properties. This research program has the financial support of Diputación Foral de Guipuzcoa and Gobierno Vasco.

## REFERENCES

1. BISHOP, S., The mechanical and impact performance of advanced carbon fibre reinforced plastics, ECCM, Bordeaux, Sept-1985, pp. 222–227.
2. POST, L. and VAN DREUMEL, W. H. M., Advances in composites technology, Reinforced Thermoplastic PEI for use in aircraft interiors, JEC, Paris, March 1986.
3. Aromatic Polymer Composites—APC 2, Fiberite Corp., 1986.
4. SWERDLOW, M. S. AND COGSWELL, F. N., Development of thermoplastic composites for airplane construction, Aeroplas, 1986. Use of plastics in civil aircraft construction and furnishing, May 1986.

# 16

# The Manufacture of Composite Aerofoil Section Models for Dynamic Wind Tunnel Tests

P. J. Mobbs

*Royal Aircraft Establishment, Farnborough, Hants GU14 6TD, UK*

ABSTRACT

*The increasing importance of the study of time-dependent aerodynamics has given rise to a requirement for wind tunnel models capable of withstanding the dynamic loads involved. The good specific properties of carbon fibre composites make these materials ideal for the manufacture of such models. A model aerofoil has been fabricated using conventional design and construction concepts. A further model was fabricated using newly developed co-curing techniques in an attempt to reduce manufacturing costs. Both models met the design requirements but the latter demonstrated both the potential for substantial costs savings and the ability to make design modifications without incurring significant re-cooling costs.*

## 1. INTRODUCTION

A fundamental aspect of aircraft design is the optimisation of the aerodynamic properties of the airframe. To validate the design it is often required to measure the behaviour of a suitably scaled model in an airstream—usually a wind tunnel. The main requirements of a model for static tests involving steady flow are that it should be strong enough to sustain the induced loads, for example those of lift or drag. It must also have adequate stiffness to restrict distortion to an acceptable level. These design requirements are not particularly difficult to meet and static models are made from a variety of materials including steel, wood and fibre-reinforced plastic.

However, it is increasingly necessary to study time-dependent aerodynamics in which the time dependence is caused by movement of the aerodynamic surface.[1] Typical examples of this branch of research are the pitching of a canard or tailplane with possible application to active control technology (ACT). Design of a model for this type of testing is influenced by many of the parameters affecting aircraft construction, and specific strength, stiffness and fatigue life become dominant factors. In addition, since the monitoring of transient parameters requires the installation of a large number of internally mounted transducers, the model must have a payload carrying capability.

One material that has attractive specific strength and stiffness characteristics is carbon fibre composite (CFC). This material also has good fatigue performance and offers potential for reduced fabrication costs.

Models have been constructed from CFC by the Aerodynamics Department, RAE, Bedford, using carbon fibres in a cold-setting resin matrix.[2] However, one such model disintegrated during the test programme while subjected to loads well below its design capability. A fractographic investigation into the cause of failure indicated that the quality of the composite was inferior to that which could be produced using a hot-setting pre-preg system and autoclave processing.[3] The failure of a model under test imposes an extremely high cost penalty. An estimate for this particular model is £50K, comprising £30K for the model plus £20K for the damaged transducers. In order to minimise the risk of model failure it was decided that improved fabrication techniques should be used in the construction of future high performance wind tunnel models.

This paper outlines a collaborative manufacturing programme involving the RAE Farnborough composites fabrication facilities and the RAE Bedford Model Shop to construct such a model. A description is included of the manufacturing routes followed to fabricate wind tunnel models using (a) the fabrication of pre-cured components, referred to as model type A, and (b) specialised co-curing techniques, referred to as model type B.

## 2.  MODEL—TYPE A

### 2.1.  General Design Considerations

The model, as designed by RAE Bedford, was a tailplane of aerofoil section, having a low aspect ratio which is similar to the control surfaces on many military aircraft.

The projected test programme required that the model should be capable of being force driven about its pitching axis with an amplitude of 2°, at a frequency of 100 Hz in an airstream moving at a velocity of Mach 1·8. Additionally, it was required that model deflection under these conditions should be negligible.

To monitor the behaviour of the model, provision had to be made for the installation of approximately 100 internally mounted pressure transducers and accelerometers. This precluded the use of honeycomb or foam core material and placed restrictions on the location of substructural members.

The design concept was similar to that of conventional metallic aircraft, being fabricated from a large number of separate parts. As seen in Fig. 1, the substructure consists of a leading edge spar, centre spar and ribs. An analysis by BAe Warton using finite element modelling techniques identified areas of the aerofoil skins that would be prone to local buckling. These areas were stiffened by the addition of L-shaped stringers.

## 2.2. Materials Selection

The Ciba-Geigy XAS/914 pre-preg system was chosen for the manufacture of all CFC components. It has tolerant processing characteristics and considerable experience in using this material had been gained at RAE. In addition, a large data bank of material properties existed.

To prevent damage to the installed instrumentation and to prevent possible thermal distortion of the model it was essential that all bonding operations should be carried out at room temperature. A number of adhesives were subjected to comparative trials to assess their lap shear performance. The Ciba-Geigy Redux 410 adhesive was selected as it gave the most consistent results at an acceptable level.

## 3.   MODEL MANUFACTURE—TYPE A

### 3.1. Tooling

A total of nine moulds were required, comprising two spar moulds, three stringer moulds, two surface skin moulds and two assembly moulds. Due to the small number of components to be manufactured in each, nominally three, tooling costs had to be minimised. For this reason it was decided to use resin-based tooling and the Ciba-Geigy tooling resin LY 568 was selected. Since this system cures at room temperature it was possible to cast the moulds directly from hardwood patterns. It was anticipated that location of individual components would be a critical aspect of this

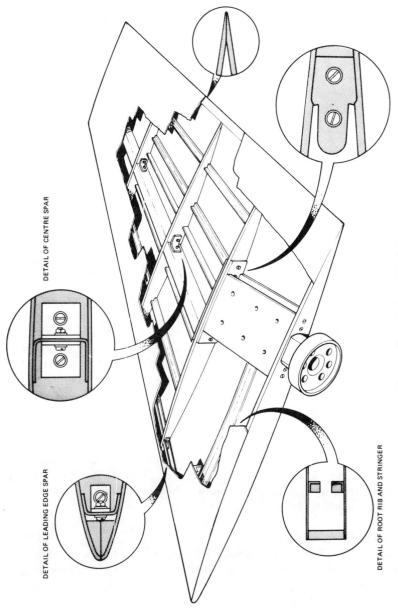

DETAIL OF CENTRE SPAR

DETAIL OF LEADING EDGE SPAR

DETAIL OF ROOT RIB AND STRINGER

FIG. 1. General view of Model A.

manufacturing process. Therefore trimming lines were engraved on the surface of the hardwood patterns and raised datum points provided to locate the spars. These trimming lines and datum points were subsequently reproduced on the mould surface and thence on the surface of the CFC components. Previous experience of moulding CFC channel sections had demonstrated that spring-back would occur when moulding the spar members. An experimentally derived spring-back allowance of 1·5° was provided at each corner of the wooden spar patterns and hence accurately reproduced in the resin tooling.

It was essential that the moulds should have a hard non-porous working surface, and be thermally stable to prevent unacceptable distortion during the CFC curing process. These conditions were met by using a sand-filled resin mix with a gel coat on all working surfaces.

After removal from the wooden patterns all moulds were heat treated at 170°C prior to their use to cure CFC components.

### 3.2. Fabrication

All CFC components were layed up and cured in an autoclave following the manufacturer's recommended cure cycle.

Assembly of the models was carried out using jigs having datum surfaces and holes to allow precise location of the model skins and substructure. In order to achieve optimum bond line clearance for the spars and ribs, repeated checks were made closing the two model halves in the assembly jigs using engineer's witness wax to check the clearance between substructure and skin. Apart from the stringers which were bonded to both skins, all substructural components were assembled and bonded to the right-hand skin (see Fig. 2).

Three type A models were produced. The first was to prove the manufacturing process and it therefore contained no instrumentation. For models 2 and 3 it was necessary to install the instrumentation before finally bonding the two model halves together. The transducers and accelerometers were located in mounting blocks which were bonded into position using Redux 410 adhesive.

### 3.3. Quality Assurance

All carbon fibre components were subjected to ultrasonic inspection prior to assembly. They were immersion-scanned using a 10-MHz focused probe in the C-scan mode. The inspection gave an indicated void content of less than 1% for all components.

After assembly the completed models were subjected to further

FIG. 2. Model A including substructure prior to instrumentation.

ultrasonic inspection using a 10-MHz contact probe in order to detect possible flaws in the bonded joints. No significant flaws were found.

Finally, the models were mounted on an hydraulic actuator and driven in still air through their operating frequency and amplitude range. Not only were the models able to withstand these loading conditions but distortion was of an acceptable magnitude.[4]

Models 2 and 3 of design A were subjected to wind tunnel tests in order to measure the aerodynamic characteristics of the aerofoil section. Monitored output from on-board accelerometers and strain gauges indicated that the structural behaviour of the model was conforming to predictions. In addition, useful data relating to the time-dependent aerodynamic behaviour of the model was obtained.[5] However, manufacturing costs were high due to the requirement to hand-fit all mating components to precise tolerances. It was felt that an alternative design philosophy employing co-curing techniques could reduce the number of component parts that would require fitting operations, thus allowing a significant reduction in manufacturing costs. A new manufacturing technique was examined, as described in the following section.

## 4.   MODEL—TYPE B

### 4.1.  General Design Considerations

Since this second design was intended to demonstrate a fabrication concept, model geometry was not critical. In order to ease the development of the process, an aerofoil shape of constant chordal section was chosen. A CFC wind tunnel model of this shape had already been constructed by BAe Woodford employing conventional manufacturing methods, as shown in Fig. 3. Thus, direct comparison between identically shaped models built using differing fabrication methods was possible.

The design requirements for the model were a capability of being force-driven about its pinching axis at an amplitude of 2°, in a transonic air flow at a frequency of 80 Hz. Due to financial constraints it was not planned to fit instrumentation to this model. However, the same design constraints were assumed in that maximum internal volume was designed into the model to allow for possible installation of instrumentation.

An important feature of the design of the substructure was that it allowed each half of the model to be fabricated in a single cure cycle using elastomeric tooling. Each half-model section was subdivided into 18 rectangular bays with integral T-section spars and ribs co-cured to the

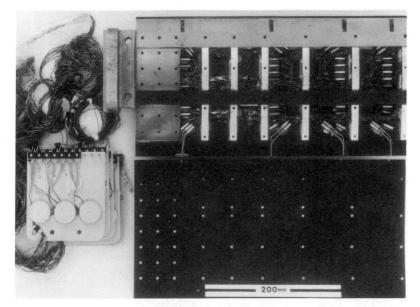

FIG. 3.   Model type B conventional construction including installed transducers.

inside skin surface (see Fig. 4). This use of elastomeric tooling allowed extra ply-reinforcement to be incorporated into highly stressed regions of the skins without the requirement to mould precisely located joggles into substructural components. The concept for this type of model construction was developed at RAE Farnborough.[6] However, manufacture of model type B was undertaken by BAe Chadderton under a research contract funded by RAE.

### 4.2.  Material Selection

During early development trials several resin-based curing moulds of the type used successfully to fabricate model type A fractured during the autoclave curing cycle. It was felt that this failure could be attributed to excessively high pressures being generated by thermal expansion of the elastomeric tooling at the cure temperature of 175°C. Therefore, in order to minimise the risk of tool failure, it was decided to use a controlled flow matrix resin having a cure temperature of 120°C for model B. The selected system was a Ciba-Geigy XAS/920 fibre/matrix pre-preg. As before, Redux 410 cold-set adhesive was selected for all bonding operations.

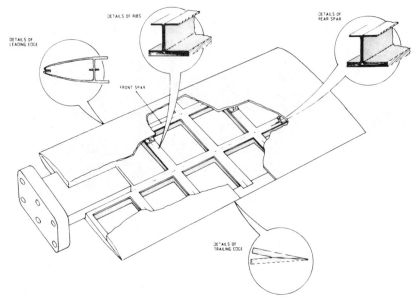

FIG. 4.   General view of Model B.

## 5.   MODEL MANUFACTURE—TYPE B

### 5.1.  Tooling

Since it was of constant chordal section only one mould was required to manufacture the model, it being possible to produce both halves in the same mould. This was constructed using the techniques developed for model A.

The box sections forming the substructure bays were formed independently. This was achieved by vacuum forming the pre-preg over epoxy resin form blocks which had been cast to the same profile as each rectangular bay.

To hold the pre-preg in shape and to consolidate it during cure, a silicon rubber block incorporating a removable aluminium alloy core plug was cast exactly the same size as each model bay. The function of the plug was to provide support for the sides of the elastomer blocks. In addition, when curing of the model was completed removal of the core plugs enabled the rubber blocks to be easily removed from the model.

### 5.2.  Fabrication

The model was fabricated by laying up the outer skin surface in the mould. A cruciform-shaped section was layed up for each bay and vacuum formed to shape on the specially prepared form blocks. Each pre-formed cruciform was then wrapped around the appropriate silicon rubber block and placed in the tool. A flat caul plate was placed on the assembly to ensure that all the spar and rib caps were formed at a level plane which would subsequently become the model centre line.

Compaction of the spars and ribs was effected to a large extent by expansion of the rubber tooling while being heated during the autoclave cure cycle. After being stripped from the tooling, each model half was virtually ready for final assembly (see Fig. 5). A small amount of hand finishing was required to remove resin flash. The mating surfaces of the two model halves were prepared for bonding by hand abrading with a fine grade silicon carbide paper. Finally, the two model halves and the root block were bonded together in one process using a vacuum bagging technique to apply a uniform clamping pressure.

FIG. 5.   Model B prior to assembly.

## 5.3. Quality Assurance Model B

Both model halves were subjected to ultrasonic inspection prior to assembly using a 10-MHz contact probe. The inspection indicated an overall void content of less than 1%. After assembly the model was subjected to radiographic inspection. A dispersion of bubble-shaped cavities in the bond line was detected. This would not be acceptable in a model intended for wind tunnel testing. However, it is considered that it would be relatively easy to avoid this defect by eliminating exposure of the model to a negative pressure in the vacuum bag during the bonding operation. To achieve this the clamping force would have to be provided by a positive external pressure using an autoclave.

A series of tests were then conducted to evaluate the frequency response functions of the models. Using an impulse excitation technique it was possible to evaluate the resonant frequency, damping and generalised mass factors, and model stiffness ($n$). The tests demonstrated that the modal frequency was higher than for the control model (mentioned in Section 4.1) which had been manufactured by BAe Woodford, using conventional techniques. However, when due account was taken of the effect of mounting 100 transducers with a mass of 1·5 g/unit, both models appeared to have a similar frequency response.


## 6. CONCLUDING REMARKS

Comparative trials have been made using alternative design and manufacturing techniques to provide high performance wind tunnel models for the evaluation of time-dependent aerodynamic characteristics. It was found that the form of construction employing elastomeric rubber co-curing techniques substantially reduced the number of separate parts (from 19 down to 2). This had a significant effect on the cost of deployment of autoclave facilities and it also eliminated the expensive requirement for extensive precision hand fitting.

The dynamic response of the co-cured model type B compared favourably with that of the corresponding model made following a conventional manufacturing route. An additional attractive feature of the co-curing process is that it allows scope for refinement of the dynamic performance of the model; the number of plies in the skins or substructure and even the substructure geometry can very easily be changed with minimal re-tooling or fitting costs.

In view of the success of this project it is likely that the co-curing technique developed here will be used for the manufacture of future models for evaluation of time-dependent aerodynamics.

## REFERENCES

1. MABEY, D. G., A review of experimental research on time-dependent aerodynamics, RAE Technical Memo Aero 2002, 1984.
2. LLEWELYN DAVIS, D. I. T. P., The use of CFRP in the construction of models for testing in wind tunnels, RAE Technical Report 75017, 1975.
3. PURSLOW, D., The application of composites fractography for failure of a wind tunnel model, RAE Technical Report 83020, 1983.
4. MABEY, D. G., Private communication, RAE.
5. MABEY, D. G., WELSH, B. L. and CRIPPS, B. E., Measurement of steady and oscillatory pressures on a low aspect ratio model at subsonic and supersonic speeds, RAE Technical Report 84095, 1984.
6. MOBBS, P. J., Fibre reinforced plastics forming methods, UK Patent GB2141660B, RAE Farnborough.
7. GABRI, B. S., Dynamic response characteristics of two RAE carbon fibre models, technical evaluation report, Cranfield Data Systems Limited, March 1986.

# 17

# The Influence of Prepreg Ageing on the Chemorheology of Carbon Fibre Laminate Processing

A. APICELLA, J. KENNY, L. NICOLAIS

*Department of Materials and Production Engineering,*
*University of Naples, Naples, Italy*

and

M. IANNONE

*Aeritalia S.A.I.P.A., Pomigliano d'Arco, Naples, Italy*

## ABSTRACT

*The ordinary working procedure of laying-up of composite laminates involves the exposure of the prepregging materials for several days in an ambient held at controlled temperature and humidity. The ageing of the material influences the reactivity of the epoxy prepolymer. The laminates are then subjected to high temperatures in order to provide the heat required for initiating and maintaining the chemical reactions needed for the epoxy polymerization and cross-linking. The changes of the molecular structure and weight distribution induced by the chemical reactions modify the rheological properties of the system. The cure cycle is generally optimized for assigned viscosity modifications, sorbed moisture devolatilization rate, resin flux, and laminate thickness; however, the hygrothermal ageing of the prepreg alters the chemorheological behaviour of the epoxy system leading to undesired local modifications of the actual cure process. First, an integrated analysis procedure providing the base for the optimization between resin chemistry, factory operations and curing is developed; and the calorimetric and rheological behaviours of aged prepreg are experimentally and theoretically illustrated.*

## INTRODUCTION

Composite structures based on high performance polymeric materials are commonly prepared by autoclave operations. The autoclave temperature and pressure application program is determined by the rheological characteristics of the fluid thermosetting resin in order to ensure a proper out-gassing, cure and compactness of the composite.[1,2] The cure of a low molecular weight prepolymer is associated with a significant increase in the material viscosity and glass transition temperature and accompanied by an intense heat generation due to the exothermic nature of the thermosetting reactions. The relative rates of heat generation and transfer may strongly affect the actual temperature and viscosity profiles through the thickness of the part.

The understanding of the chemorheological relationships results is very important in the fabrication of advanced polymeric composite parts where precise resin cure control is required.[2,3] Microscopic defects in the network structure of the matrix phase and macroscopic defects such as voids, bubbles, debonded and broken fibres may derive from an uncontrolled polymerization process. The cure of a polymeric composite based on thermosetting matrices needs therefore optimization of the cure cycle parameters as well as the choice of an adequate formulation of the reacting system as a function of the geometry of the part. The problem becomes more complex when the initial reactivity of the system is not constant but depends on the previous history of the material. Prepregging materials used for laying-up of composite laminates are stored for several days in an ambient held at controlled temperature and humidity. A description of the changes in the reactivity and in the chemorheological behaviour of the epoxy prepolymer produced by the hygrothermal ageing of prepregs is presented in this work.

Finally, a theoretical model relating the thermokinetic parameters, the imposed external temperature, the reagent characteristics, the geometrical factors with the actual viscosity and temperature profiles achieved in the laminates is developed in order to provide a base for the optimization between resin chemistry, factory operations and curing.

## CHEMICAL BEHAVIOUR

In the synthesis of thermosetting polymers a series of independent reactions involving monomers and molecules composed of a sequence of two (dimers), three (trimers) and more structural units occur. The molecular constitution of the products of the reaction is described by the complexity distribution functions given in Flory[4] and Stockmayer.[5]

An outline of the mechanisms of reaction leading to the cross-linking of epoxy prepolymers may be useful to enable appropriate modelling of the network formation. Chain extension and cross-linking of epoxide resins depend upon the reaction of epoxy groups with themselves and with the hydrogen of donor compounds. The occurrence in these systems of differently activated and kinetically distinct reactions suggests further consideration of the chemorheology of the cure, gelation and vitrification processes with respect to the prepolymer composition and temperature cure path. In fact, the gelation and vitrification phenomena, which usually occur in two phases of the material processing such as the initial polymerization and postcure, may strongly influence the thermoset molecular morphology and material properties.

Previous investigations[1,2,6] agreed with the conclusion that three reactions dominate the cure behaviour of the TGDDM–DDS (tetraglycidyl diaminodiphenyl methane and the aromatic amine diaminodiphenylsulfone) systems, namely, epoxide–primary amine addition, epoxide etherification with the hydroxyls and the epoxide–secondary amine addition. The kinetic constant of the former reaction in the range of temperature between 177 and 200°C has been reported to be at least one order of magnitude higher than the other two values.[7] DSC has been applied to particular conditions of temperature and composition in order to isolate the single reactions.[1,2]

It has been observed that in the cure with the aromatic amines, the faster and less activated reaction regime is the addition of the primary amine to the epoxy, and the slower and more highly activated reactions are the homopolymerization of epoxies, secondary amine addition, and hydroxyl–epoxide reactions. Due to its high rate constant, primary amine addition will principally occur in TGDDM–DDS mixtures containing more than an equimolar content of DDS primary amines and epoxy groups. The reaction of etherification, conversely, characterizes the thermal polymerization of a TGDDM sample containing some initial concentration of hydroxyls (i.e. isopropyl alcohol). The kinetic analysis of the dynamic DSC scans has been carried out according to the method proposed by Prime. The activation energy, $E_a$, the frequency factor, $K_0$, and the overall heat of reaction for the two limiting systems were reported in Refs 1 and 2. The overall heat of reaction of the primary amine addition is significantly higher than that of the etherification, i.e. 255 and 170 cal/g of

TGDDM, respectively. The values found for the activation energy of the primary amine addition and etherification were, respectively, 16·6 and 41·0 kcal/mole. These two reactions are principally governing the cure behaviour of the TGDDM–DDS systems. FT–IR analysis has shown that the hydroxyl concentration is almost constant once all the primary amines are exhausted. These facts suggest that the secondary amine addition, at least at temperatures below 200°C,[2] does not play a relevant role in the cure reactions. The heat developed in the early stages of the cure or scan, where only the primary amine addition is assumed to occur, is hence proportional to the heat of reaction of the epoxide with the primary amines. The values of the overall heats of reaction for the TGDDM–DDS mixtures of different compositions calculated according to the previous assumption, have been found to favourably compare with those experimentally determined in the DSC scans.[2]

Therefore, the relationship between the heat developed during cure and the corresponding value of the epoxy conversion for TGDDM–DDS mixtures of different composition may be derived by considering that the DDS molecule is essentially acting as a bifunctional curing agent while the TGDDM is tetrafunctional. On the other hand the use of $n$th-order kinetics with a temperature dependent rate constant given by an Arrhenius expression is commonly adopted, although it could result in an over-simplified expression of the entire cure reactions of most of the commercial systems:

$$dP_e/dt = K_0 \exp(-E/RT)(1 - P_e)^n \qquad (1)$$

where $P_e$ is the extent of reaction related to the epoxy resin, $t$ is the time variable, $K_0$ is the kinetic constant, $E$ is the activation energy, $R$ is the universal gas constant, $T$ is the temperature and $n$ is the reaction order.

The values of the parameters obtained from the calorimetric analysis of a commercial prepreg containing TGDDM–DDS systems are given in Table 1.

TABLE 1

*Thermokinetical analysis of a TGDDM–DDS prepreg system at different storage times*

| | | |
|---|---|---|
| Storage time (days) | 0 | 22 |
| Heat of reaction, $H$ (J/g) | 420 | 420 |
| Activation energy, $E$ (J/mol) | 67 | 107 |
| Rate constant, ln $K$ (ln 1/s) | 12·6 | 23·3 |
| Reaction order, $n$ | 1·93 | 3·10 |
| Glass transition temperature: $T_{g0}$ (K) | 280 | 288 |

The analysis was carried out at different times during the storage of the prepregs in the 'clean room' at controlled temperature and humidity (25°C, 50%). Experimental results show how the reactivity changes with the exposure time; while the overall heat of reaction is nearly constant the kinetic parameters change considerably. These results may be interpreted by a modification in the activity of the catalyser during the prepreg ageing.

## CHEMORHEOLOGY

The rheological behaviour of a reacting system is governed by two effects; the first is related to the molecular structural changes induced by the cure reactions and the second is associated with the variation of the segmental mobility determined by temperature variations. A relationship has been proposed by Valles and Macosko[7] for a non-linear copolymerization such as that considered here

$$\eta = K(gMW)^{3.4} \tag{2}$$

where $\eta$ is the viscosity, $K$ is a constant, $MW$ is the weight-average molecular weight of the polymer and $g$ is the ratio of the radii of giration of a branched chain to a linear chain of the same molecular weight. The Stockmayer approach for three-dimensional condensation polymers may be applied for the computation of the $MW$ value;[5] while the value of $g$ is related to the reagent functionality and branching coefficient.[8]

On the other hand, the influence of the temperature on the viscosity in a reacting system is considered according to the William, Landel and Ferry equation where the glass transition temperature $(T_g)$ of the system is a function of the reaction extension:[1,6]

$$\frac{\eta(T, P_e)}{\eta(T_r)} = (gMW)^{3.4} \frac{\exp(C_1(T_r - T_{g0})/(C_2 + T_r - T_{g0}))}{\exp(C_1(T - T_g(P_e))/(C_2 + T - T_g(P_e)))} \tag{3}$$

In eqn. (3) $(T_r)$ is the viscosity of the unreacted system at the reference temperature $T_r$ and $C_1$, $C_2$ are the constants of the W–L–F equation.

The results given in Table 1 show the variation in $T_{g0}$ during storage. The system becomes more rigid producing an increase in $T_{g0}$ and then in the initial viscosity, thus losing the capacity to be modelated.

## MATHEMATICAL MODELLING

Equation (3) leads to the computation of viscosity profiles during laminate curing once temperature and degree of cure of the resin are known. For thin laminates the process may be considered as isothermal and the degree of cure can be computed solving the appropriate kinetic equation. However, as was discussed earlier the cure of an epoxy resin is always coupled with a sensible development of heat. The reacting system must be viewed as a non-isothermal bulk reactor with volumetric heat generation and aerial heat transfer for the initial heating and for the dissipation of the heat of reaction. Then, profiles of temperature and degree of cure inside the composite must be computed, taking into account the system geometry, the thermal diffusivity of the composite and the resin reaction rate. This can be done by solving the energy balance together with an appropriate expression for the cure kinetics as reported in the literature.[9,10] In the case of the cure of epoxy based laminates the energy equation becomes:

$$\rho C \frac{\partial T}{\partial t} = k_x \frac{\partial T}{\partial x} + \rho H \frac{\partial P_e}{\partial t} \tag{4}$$

where $\rho$ is the composite density, $C$ is the composite heat capacity, $k_x$ is the thermal conductivity coefficient in the direction transverse to the fibres and $x$ is the thickness coordinate. Equation (4) is coupled with eqn. (3) and the entire system must be solved numerically.

## RESULTS AND DISCUSSION

The proposed model was applied to the evaluation of the processing of laminates based on TGDDM–DDS mixtures. The values of the heats of reaction and kinetic constants obtained by DSC characterization and presented in Table 1 were used as input data of the model.

Generally, the laminate is placed in the autoclave at room temperature and the system is heated at a controlled rate. The numerical results for the temperature on the skin and at the centre of the laminate made from prepreg aged at different time intervals are shown in Fig. 1. While the temperature on the skin follows the imposed autoclave temperature, the imbalance between the rate of heat generation and the thermal diffusivity of the composite produces an increment of temperature in the body of the laminate. When the balance between these two quantities is reached the temperature profile is a maximum. The differences between the kinetical behaviour owing to the ageing of the prepregs are also evident from Fig. 1. The unaged formulation shows a moderate temperature peak at the centre of the laminate, while the aged material behaves in a more drastic manner showing a sharp temperature peak.

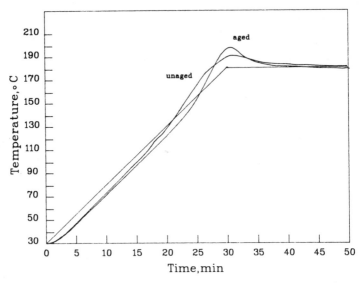

FIG. 1.   Temperature versus time for the aged and the unaged formulations at the centre of the 10 mm half-thickness laminate. The right hand axis corresponds to the skin temperature.

The values of temperature and degree of cure have been used to compute the viscosity as a function of cure time by means of the eqn. (3). Numerical results corresponding to the viscosity of the unaged and aged formulations on the skin of the composite are shown in Fig. 2. Due to the gradual increase in the temperature, the viscosity of the unreacted resin decreases from a relatively high value at room temperature until it reaches its minimum value. Then, the reaction starts and the viscosity begins to increase to infinity when the material reaches the gel point. The initial viscosity of the aged resin is higher than the value of the unaged material according to the highest value of $T_{g0}$. The form of the viscosity curve is significantly different showing a retarded start to the chemical reaction with a shifted position of the minimum of the viscosity. In this case, due to the lower values of the resin viscosity, starvation problems may occur. Pressure is one of the most critical variables in the cure of thermosetting composites. The application of pressure must be programmed as a function of the resin viscosity. In the first part of the autoclave process the pressure must be small enough in order to avoid the degasification of volatile substances. When the viscosity is increasing, after the minimum, the pressure must be increased to provide the energy necessary for the compaction of the prepregs. If the viscosity is too small at the moment of application of pressure, an excess of resin flow should be produced with a loss of the laminate resin content and with the corresponding loss of mechanical properties of the composite part. Then, the difference observed between the behaviour of both formulations is very important in order to determine the best process conditions.

The numerical results obtained, assuming that the prepreg has been placed into an autoclave and has been exposed suddenly to the cure temperature, are discussed in this section. First, the viscosity development

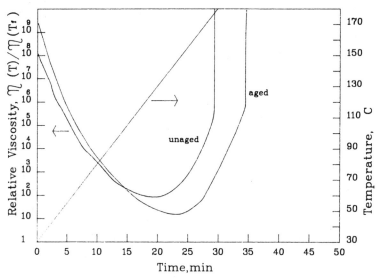

Fig. 2. Relative viscosity versus time for the aged and the unaged formulations at the centre of the 5 mm half-thickness laminate. The right hand axis corresponds to the skin temperature.

as a function of time for the unaged resin is discussed on the basis of the numerical results shown in Fig. 3. Due to the contribution of the thermal conductivity of the fibres the temperature at the centre of the laminate reaches the external imposed temperature and increases as a consequence of the heat developed by the chemical reaction. The viscosity curves relative to the skin ($x = 1$) and to the core ($x = 0$) of the laminate indicate that the heat accumulated in the centre leads locally to a higher reaction rate with the consequence of a faster increase of the viscosity. Therefore, the gelation phenomenon moves from the core to the skin.

Figure 3 also shows the effect of the thickness of the laminate on cure, variables. With increasing thickness, the thermal conductivity of the composite decreases and more time is needed to reach the cure temperature at the centre. However, once the reaction starts the relative heat accumulation inside the laminate becomes higher. Then, the thicker laminate approaches adiabatic-like conditions and the reaction practically explodes at the centre suddenly reaching the gel point. The coexistence of gelled and ungelled regions during the process and the high temperature

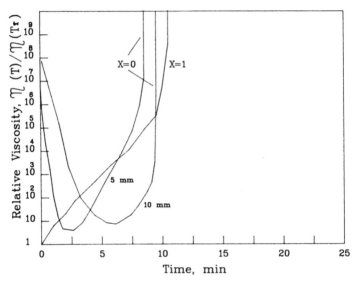

FIG. 3.   Relative viscosity versus time for the unaged formulation and for different laminate thicknesses during the isothermal process.

differences across the laminate must be avoided in order to obtain a homogeneous structure.

The results obtained applying the same analysis for the aged resin are represented in Fig. 4. Due to the different reactivities, the aged resin shows different viscosity curves between the centre and the skin of the laminate, with respect to the behaviour of the unaged resin shown in Fig. 3. The smaller value of the reaction rate constant is evident from the slower increase in the viscosity on the skin of the composite, while the higher value of the activation energy is manifested in the more relevant effect of the increase in the thickness of the laminate.

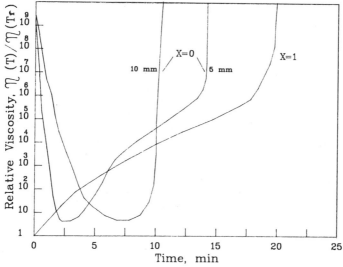

FIG. 4.   Relative viscosity versus time for the aged formulation and for different laminate thickness during isothermal process.

## CONCLUSIONS

A general model for the prediction of the rheological behaviour of epoxy resins during the processing of epoxy based laminates has been developed. The model is able to predict the temperature and the extent of reaction across the laminate thickness during processing. The model leads to the computation of the viscosity inside the composite where experimental measurements are not possible. Thermal characterization of the reacting systems gives the input data necessary for the mathematical modelling.

The numerical simulation has shown how the previous history of prepregs may affect the process behaviour of epoxy laminates and has confirmed the necessity for a controlled temperature ramp in order to obtain a more homogeneous laminated structure. Moreover, the numerical results indicated that the process variables may present anomalous behaviour as a consequence of the ageing process, which may affect the physical properties and durability of the final product.

## REFERENCES

1. APICELLA, A., Effect of chemorheology on epoxy resin properties, in: *Developments in Reinforced Plastics—5*, Chap. 5 (G. Pritchard ed.), London, Elsevier Applied Science, 1986.

2. NICOLAIS, L. and APICELLA, A., Processing of composite structures, *Pure appl. Chem.*, **57** (1985), 1701.
3. HALPIN, J. C., KARDOS, J. L. and DUDUKOVIC, M. P., Processing science: an approach for prepreg composite systems, *Pure appl. Chem.*, **55** (1983), 893.
4. FLORY, P. J., Condensation polymerization and constitution of condensation polymers, in: *High Molecular Weight Organic Compounds* (R. E. Burk and O. Grumitt eds), Chap. 5, New York, Interscience, 1949.
5. STOCKMAYER, W. H., Molecular distribution in condensation polymers, *J. Polym. Sci.*, **9** (1952), 69.
6. APICELLA, A., NICOLAIS, L., IANNONE, M. and PASSERINI, P., Thermokinetics and chemorheology of the cure reactions of the TGDDM–DDS epoxy systems, *J. appl. Polym. Sci.*, **29** (1984), 2083.
7. VALLES, E. and MACOSKO, C., Structure and viscosity of Poly(dimethyl-siloxanes) with random branches, *Macromolecules*, **22** (1979), 521.
8. ZIMM, B. H. and STOCKMAYER, W. H., The dimensions of chain molecules containing branches and rings, *J. Chem. Phys.*, **17** (1949), 1301.
9. LOOS, A. C. and SPRINGER, G. S., Curing of epoxy matrix composites, *J. Comp. Mater.*, **17** (1983), 135.
10. APICELLA, A., D'AMORE, A., KENNY, J. and NICOLAIS, L., True viscoelastic properties of high performance composite materials, in: *Proceedings of the 44th ANTEC*, SPE Publ., Boston, 1986, p. 557.

# 18

# An Energy Release Rate Approach for Free-edge Delamination Problem in Composite Laminates

KYOHEI KONDO and TAKAHIRA AOKI

*Department of Aeronautics, University of Tokyo,
7-3-1 Hongo, Bunkyo-ku, Tokyo 113, Japan*

## ABSTRACT

*Location where delamination occurs in a balanced, symmetric laminate with free edge under uniaxial mechanical loading and/or hygrothermal expansion is predicted by comparing the energy release rates due to delamination growth from the free edge on various interfaces. The change of release rate with the delamination crack length is obtained by utilizing the conventional finite element method. And a new method is proposed to evaluate a saturated value of energy release rate in the delamination crack extension. The saturated value of energy release rate is used as a measure to predict the occurrence of delamination which has yielded good agreement with observations on tension–tension and compression–compression fatigue tests conducted on carbon–epoxy composite laminates.*

## INTRODUCTION

It is well recognized that the strengths of straight-sided composite laminate coupons under uniaxial loading are lower than those predicted by the first ply failure criterion based on the two-dimensional stress field obtained through the classical lamination theory. The discrepancy may be attributed to the free-edge delamination caused by the interlaminar stresses which are three-dimensional in nature. Therefore it is of great importance to study the influence of the free-edge interlaminar stresses on the initiation and growth of a delamination crack.

Following the initial work of Pipes and Pagano,[1] numerous investigators have studied the problem of free-edge stresses in laminated composites. Wang and Choi[2] revealed that the interlaminar stresses are singular at the free edge as well as at the crack tip of delaminated coupons. And the experimental observations[3] have shown that the thicker the laminae, the smaller the stress required for the onset of free-edge delamination in the laminates with the same stacking sequence. To account for the effect of thickness of each lamina, Kim and Soni[4] predicted the onset of delamination based on the average value of normal stress over a characteristic distance from the free edge with the results which are in good agreement with the experimental results. However, it is more reasonable to predict the onset and propagation of delamination based on the fracture mechanics. Wang[5] analytically obtained the stress intensity factors and the energy release rates at the delamination crack tip. Kim and Hong[6] obtained the strain energy release rates by the crack closure integral utilizing the conventional finite element method.

In this paper the energy release rate due to propagation of a delamination crack is obtained by various methods such as the stiffness method, the crack closure integral method and the J-integral method. To obtain the stress field, we utilize the Pipes and Pagano[1] quasi-three-dimensional model and the transformed quasi-three-dimensional model which is similar to that of the superposition method proposed by Whitcomb and Raju.[7] And a new method is proposed to obtain the saturated value of the energy release rate during the interlaminar crack growth in a semi-infinite laminate. Then the energy release rate due to delamination crack growth at each interlaminar boundary in carbon–epoxy composite laminates is numerically calculated, and the theoretical predictions for the location of delamination based on the energy release rate are compared with the observations on tension–tension and compression–compression fatigue tests.

## THE CONVENTIONAL QUASI-THREE-DIMENSIONAL PROBLEM

We consider a balanced, symmetric semi-infinite laminate with a straight edge under uniaxial mechanical loading parallel to its edge and/or uniform hygrothermal expansion, as shown in Fig. 1. The laminate is modeled as an assembly of unidirectional fiber-reinforced plies with delamination cracks at interlaminar boundaries.

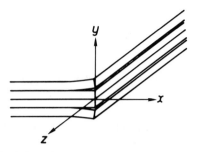

Fig. 1. Semi-infinite, balanced, symmetric laminate with free-edge delaminations.

For the Pipes–Pagano quasi-three-dimensional model,[1] the displacements can be written as

$$\omega \equiv \begin{bmatrix} u \\ v \\ w \end{bmatrix} = \begin{bmatrix} 0 \\ 0 \\ \hat{\varepsilon}_z z \end{bmatrix} + \begin{bmatrix} U(x, y) \\ V(x, y) \\ W(x, y) \end{bmatrix} \tag{1}$$

where $\hat{\varepsilon}_z$ is a prescribed uniform axial strain. The stress–strain relation is given by

$$E - E_0 = S\Sigma \tag{2}$$

where $E^T \equiv [\varepsilon_x \, \varepsilon_y \, \varepsilon_z \, \gamma_{yz} \, \gamma_{zx} \, \gamma_{xy}]$ are strains, $E_0^T \equiv [\varepsilon_{x0} \, \varepsilon_{y0} \, \varepsilon_{z0} \, \gamma_{yz0} \, \gamma_{zx0} \, \gamma_{xy0}]$ are initial strains due to hygrothermal loading, $S \equiv [S_{ij}]$ is the compliance tensor of the material and $\Sigma^T \equiv [\sigma_x \, \sigma_y \, \sigma_z \, \tau_{yz} \, \tau_{zx} \, \tau_{xy}]$ are stresses. The initial strains can be derived from the equation

$$E_0 = \alpha \, \Delta T + \beta \, \Delta M \tag{3}$$

where $\Delta T$, $\Delta M$ are changes of temperature and moisture content, and $\alpha$ and $\beta$ are thermal and hygroscopic expansion coefficients, respectively.

Considering that $\varepsilon_z = \hat{\varepsilon}_z$ and rearranging terms, eqn. (2) may be partitioned as

$$\begin{bmatrix} \bar{E} - \bar{E}_0 \\ \hat{\varepsilon}_z - \varepsilon_{z0} \end{bmatrix} = \begin{bmatrix} \bar{S} & \bar{S}_{i3} \\ \bar{S}_{i3}^T & S_{33} \end{bmatrix} \begin{bmatrix} \Sigma \\ \sigma_z \end{bmatrix} \tag{4}$$

where the overbar denotes the reduced form. It follows from eqn. (4) that

$$\sigma_z = (\hat{\varepsilon}_z - \varepsilon_{z0} - \bar{S}_{i3}^T \Sigma)/S_{33} \tag{5}$$

Introduction of this equation into eqn. (4) yields

$$\bar{E} - \bar{E}_0 = (\bar{S} - \bar{S}_{i3}\bar{S}_{i3}^T/S_{33})\bar{\Sigma} + (\bar{S}_{i3}/S_{33})(\hat{\varepsilon}_z - \varepsilon_{z0}) \tag{6}$$

The potential energy function $\Pi$ for this problem is written as follows:

$$\Pi = \frac{1}{2} \int_V [(\bar{E} - \bar{E}_0)^T, \hat{\varepsilon}_z - \varepsilon_{z_0}] \begin{bmatrix} \bar{\Sigma} \\ \sigma_z \end{bmatrix} dV - \int_{S_\sigma} \omega^T f \, ds \tag{7}$$

where $f$ is the prescribed traction on the boundary $S_\sigma$.

## TRANSFORMED QUASI-THREE-DIMENSIONAL PROBLEM

Whitcomb and Raju[7] proposed the superposition method for analysis of free-edge stresses of straight-sided coupons. Similarly, we transform the free-edge problem of semi-infinite laminates under uniaxial mechanical loading and/or uniform hygrothermal loading to a problem where a system of self-equilibrating loads is applied to the straight side.

At first, an infinite plate is assumed to be subjected to a uniform axial mechanical loading and/or uniform hygrothermal loading. To obtain a semi-infinite laminate, we make a cut parallel to the axial loading. The classical lamination theory is valid for the infinite plate even if interlaminar delamination cracks are present. Hence, applying tractions derived from the classical lamination theory to the cut as shown in Fig. 2a, the stress distribution in the semi-infinite laminate can be made identical to that in the original infinite laminate. Then, by applying tractions which are of the same magnitude but of opposite sign to those by the classical lamination theory to the side of the semi-infinite laminate as shown in Fig. 2b, the boundary condition that the side is free from tractions is satisfied. Therefore, superposition of the edge traction problem (Fig. 2b) to the uniform stress problem of the classical lamination theory (Fig. 2a) gives the free-edge problem of the semi-infinite laminate as shown in Fig. 2c.

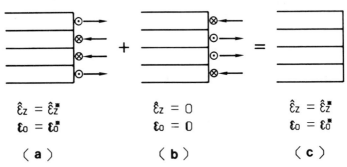

$$\hat{\varepsilon}_z = \hat{\varepsilon}_z^* \qquad\qquad \hat{\varepsilon}_z = 0 \qquad\qquad \hat{\varepsilon}_z = \hat{\varepsilon}_z^*$$
$$\varepsilon_0 = \varepsilon_0^* \qquad\qquad \varepsilon_0 = 0 \qquad\qquad \varepsilon_0 = \varepsilon_0^*$$

$$(\,\text{a}\,) \qquad\qquad\qquad (\,\text{b}\,) \qquad\qquad\qquad (\,\text{c}\,)$$

FIG. 2.   Superposition for analysis of semi-infinite laminate.

Since the problem of the classical lamination theory gives no interlaminar stresses, the edge traction problem yields the interlaminar stresses which are identical to those in the free-edge problem. Consequently, the quasi-three-dimensional problem discussed in the preceding section can be changed to the transformed quasi-three-dimensional problem represented in Fig. 2b, where $\hat{\varepsilon}_z = 0$ and $E_0 = 0$. The condition that $\hat{\varepsilon}_z = 0$ is derived from the fact that the system of tractions applied to the edge is self-equilibrating.

The self-equilibrium of the tractions also implies that the interlaminar stresses are present only near the edge by virtue of Saint Venant's principle. Therefore the stresses in a semi-infinite laminate are identical to those in a coupon with considerable width. It means that the width of the laminate is not important for the interlaminar stresses except for extremely narrow coupons. In the following, we investigate characteristics of the interlaminar stresses by analyzing either a semi-infinite laminate or a coupon with considerable width depending on simplicity of analysis.

Transformation of the free-edge problem to the edge traction problem clarifies the relationship between the mechanical and hygrothermal loadings. We consider a cross-ply laminate with $(0°/90°)_s$ fiber orientations and a ply thickness $t$. The stress–strain relations in principal material coordinates of each ply under plane stress are

$$\begin{bmatrix} \sigma_1 \\ \sigma_2 \\ \tau_{12} \end{bmatrix} = \begin{bmatrix} Q_{11} & Q_{12} & 0 \\ Q_{12} & Q_{22} & 0 \\ 0 & 0 & Q_{66} \end{bmatrix} \left( \begin{bmatrix} \varepsilon_1 \\ \varepsilon_2 \\ \gamma_{12} \end{bmatrix} - \begin{bmatrix} \varepsilon_{1_0} \\ \varepsilon_{2_0} \\ \gamma_{12_0} \end{bmatrix} \right) \tag{8}$$

where 1 and 2 denote the directions parallel and perpendicular to the fiber direction, respectively. And the initial strains may be derived from the relations

$$\begin{bmatrix} \varepsilon_{1_0} \\ \varepsilon_{2_0} \\ \gamma_{12_0} \end{bmatrix} = \begin{bmatrix} \alpha_1 \\ \alpha_2 \\ 0 \end{bmatrix} \Delta T + \begin{bmatrix} \beta_1 \\ \beta_2 \\ 0 \end{bmatrix} \Delta M \tag{9}$$

The classical lamination theory gives the tractions applied to the $0°$ and $90°$ plies at the edge in the transformed problem as follows:

$$\begin{bmatrix} T_x \\ T_z \end{bmatrix}_{0°} = \begin{bmatrix} t_x \\ 0 \end{bmatrix} \qquad \begin{bmatrix} T_x \\ T_z \end{bmatrix}_{90°} = \begin{bmatrix} -t_x \\ 0 \end{bmatrix} \tag{10}$$

where

$$t_x = -\frac{Q_{12}(Q_{11}-Q_{22})}{(Q_{11}+Q_{22})^2-4Q_{12}^2}\frac{N_z}{2t} - \frac{Q_{11}Q_{22}-Q_{12}^2}{Q_{11}+Q_{22}+2Q_{12}}$$
$$\times \{(\alpha_1-\alpha_2)\Delta T+(\beta_1-\beta_2)\Delta M\} \tag{11}$$

in which $N_z$ is the uniaxial mechanical load per width. This equation implies that for the cross-ply laminate the interlaminar stresses due to mechanical and hygrothermal loadings differ only by a scale factor. It is true even for any kind of symmetric angle-ply or cross-ply laminate.

As one of the advantages of the transformed edge problem, we can easily understand the physical mechanism of the interlaminar stresses considering the equilibrium of the edge stresses with the applied tractions instead of considering the equilibrium of the edge stresses with the far field stresses in the original free-edge problem as proposed by Pagano and Pipes.[8]

## FINITE ELEMENT CALCULATION OF ENERGY RELEASE RATE

We analyze a finite width laminated coupon which has a width of $2b$ and a thickness of $2h$, as shown in Fig. 3. Because of the symmetries in the problem, only the shaded region is considered. The displacements $U$ and $W$ are prescribed as zero on the $x = 0$ line and $V$ is prescribed as zero on the $y = 0$ line. A typical idealization of the shaded region by eight-noded isoparametric elements is shown in Fig. 4.

The finite element method is formulated for the conventional quasi-three-dimensional problem based on the potential energy given by eqn. (7) or for the transformed quasi-three-dimensional problem based on the potential energy which is derived by prescribing that $E_0 = 0$ and $\dot{\varepsilon}_z = 0$ in

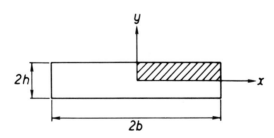

FIG. 3.   Finite-width laminate and coordinate system.

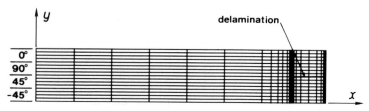

FIG. 4.   Finite element discretization for $(0°/90°/45°/-45°)_s$ laminate.

eqn. (7). Considering a unit length in the $z$ direction, the energy release rate is calculated through the stiffness method or the crack closure integral method.

### (a) The Stiffness Method

For a prescribed axial strain $\hat{\varepsilon}_z$ in the conventional quasi-three-dimensional problem, the energy release rate $G$ is given by

$$G = -\frac{\partial U}{\partial a} \tag{12}$$

where $U$ is the strain energy defined by

$$U = \tfrac{1}{2}\hat{\varepsilon}_z^2 C \tag{13}$$

In this equation, $C$ is the stiffness of the coupon as

$$C = T_z/\hat{\varepsilon}_z \tag{14}$$

where $T_z$ is the total axial load. Substituting eqn. (13) into eqn. (12), we have

$$G = -\tfrac{1}{2}\hat{\varepsilon}_z^2 \frac{dC}{da} \tag{15}$$

In the finite element analysis, the energy release rate is calculated from the change of the stiffness by the crack growth, as shown in Fig. 5.

### (b) The Crack Closure Integral Method

In the conventional or transformed quasi-three-dimensional problem, if a delamination extends by a small amount, $\Delta a$, the energy absorbed in the process, $-\Delta U$, is equal to the work required to close the crack to its original length, $\Delta W$. It follows from Clapeyron's principle that

$$-\Delta U = \Delta W = \frac{1}{2}\int_0^{\Delta a} \omega^\mathrm{T} X \, ds \tag{16}$$

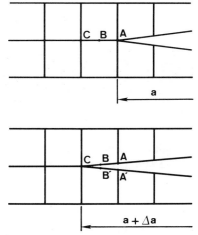

FIG. 5.   Delamination crack opening in finite element analysis.

where X is the force required to close the crack. Substitution of eqn (16) into eqn. (12) gives

$$G = \lim_{\Delta a \to 0} \frac{1}{2\Delta a} \int_0^{\Delta a} \omega^T X \, ds \tag{17}$$

In the finite element analysis, the energy release rate is calculated from the forces transmitted through the nodes in front of the crack tip and the relative displacement at those nodes after the crack growth, as shown in Fig. 5. It follows from eqn. (17) that

$$G = \frac{1}{2\Delta a} \{ (d_{A'} - d_A)^T X_A + (d_{B'} - d_B)^T X_B \} \tag{18}$$

where $d^T \equiv [U, V, W]$ is the nodal displacement vector after the crack growth and $X^T \equiv [X, Y, Z]$ is the nodal force vector before the crack growth. The total energy release rate may be divided into the three components as

$$G = G_I + G_{II} + G_{III} \tag{19}$$

where

$$G_I = \frac{1}{2\Delta a} \{ (V_{A'} - V_A)Y_A + (V_{B'} - V_B)Y_B \}$$

$$G_{II} = \frac{1}{2\Delta a} \{ (U_{A'} - U_A)X_A + (U_{B'} - U_B)X_B \} \tag{20}$$

$$G_{III} = \frac{1}{2\Delta a} \{ (W_{A'} - W_A)Z_A + (W_{B'} - W_B)Z_B \}$$

## SATURATED VALUE OF ENERGY RELEASE RATE IN DELAMINATION GROWTH

Wang[5] indicated that the energy release rate of a laminated coupon with considerable width subjected to mechanical loading remains constant as the delamination exceeds a length which is characteristic of the coupon. Here the saturated value of the energy release rate of a semi-infinite laminate subjected to mechanical and hygrothermal loading is calculated in the transformed edge stress problem by using the classical plate theory.

We consider a semi-infinite laminate with delaminations of length $a$ subjected to edge tractions which are of the same magnitude but of the opposite sign to those derived from the classical lamination theory in the original problem, as shown in Fig. 6. Note that the delaminations divide the

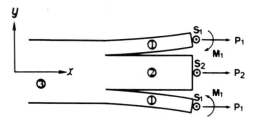

FIG. 6. Stress resultants in transformed edge-stress problem.

laminate into three regions, 1–3. From the applied tractions, the axial force, the bending moment and the shearing force per unit length at the region 1 or 2 can be calculated as

$$P_i = \int T_{x_i} \, dy = \int \sigma_{x_i} \, dy$$

$$M_i = \int T_{x_i} y \, dy = \int \sigma_{x_i} y \, dy \qquad (i = 1, 2) \qquad (21)$$

$$S_i = \int T_{z_i} \, dy = \int \tau_{xz_i} \, dy$$

It is assumed that the deformations of each region may be analyzed by the plate theory for the sufficiently long cracks. Then, because of the Kirchhoff hypothesis, the displacements are given by

$$\begin{bmatrix} U \\ V \\ W \end{bmatrix} = \begin{bmatrix} \bar{U}(x) \\ \bar{V}(x) \\ \bar{W}(x) \end{bmatrix} + y \begin{bmatrix} -\dfrac{d\bar{V}(x)}{dx} \\ 0 \\ 0 \end{bmatrix} \qquad (22)$$

and the nonzero strains are

$$\begin{bmatrix} \varepsilon_x \\ \gamma_{xz} \end{bmatrix} = \begin{bmatrix} \bar{\varepsilon}_x \\ \bar{\gamma}_{xz} \end{bmatrix} + y \begin{bmatrix} K_x \\ 0 \end{bmatrix} \tag{23}$$

where the general strains are

$$\bar{\varepsilon}_x = \frac{d\bar{U}(x)}{dx} \qquad \bar{\gamma}_{xz} = \frac{d\bar{W}(x)}{dx} \qquad \bar{K}_x = -\frac{d^2 \bar{V}(x)}{dx^2} \tag{24}$$

By using the stress–strain relations and eqns (21) and (23), the strains can be expressed in terms of the stress resultants.

The energy release rate can be directly obtained by the *J*-integral as

$$G = J \equiv \oint \left( -\sigma_{ij}\varepsilon_{ij}\,dy + \sigma_{ij}n_j \frac{\partial u_i}{\partial x}\,ds \right) \tag{25}$$

The integration path is taken as shown in Fig. 7, where the path CD is away from the delamination front. Since the paths BC, CD and DE make no

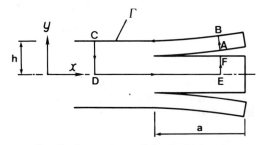

FIG. 7.   Integration path for the *J*-integral.

contribution, *J* can be obtained from the integration along the paths AB and EF. Hence, eqn. (25) can be reduced to

$$G = \int_0^h \frac{1}{2}(\sigma_x\varepsilon_x + \tau_{xz}\gamma_{xz})\,dy \tag{26}$$

For a cross-ply laminate with $(90°/0°)_s$ fiber orientation, the energy release rate is obtained as

$$\frac{G}{t} = \frac{Q_{11} + Q_{22}}{2Q_{11}Q_{22}} \left[ -\frac{Q_{12}(Q_{11} - Q_{22})}{(Q_{11} + Q_{22})^2 - 4Q_{12}^2} \frac{N_z}{2t} - \frac{Q_{11}Q_{22} - Q_{12}^2}{Q_{11} + Q_{22} + 2Q_{12}} \right.$$
$$\left. \times \{(\alpha_1 - \alpha_2)\Delta T + (\beta_1 - \beta_2)\Delta M\} \right]^2 \tag{27}$$

It is found that the $G$ is proportional to the ply thickness for the prescribed axial stress and changes of temperature and moisture content.

The new method proposed here for the evaluation of $G$ will be referred to as the delaminated plate approximation throughout.

## NUMERICAL RESULTS

The elastic, hygrothermal properties of each ply of a carbon–epoxy unidirectional composite are assumed to be transversely isotropic with

$$E_L = 130.9\,\text{GPa} \qquad E_T = 8.934\,\text{GPa}$$
$$G_{LT} = 4.648\,\text{GPa} \qquad G_{TT} = 4.648\,\text{GPa} \qquad v_{LT} = 0.3213$$
$$\alpha_L = 0.36 \times 10^{-6}/°C \quad \alpha_T = 28.8 \times 10^{-6}/°C$$
$$\beta_L = 0.0/\text{wt}\%\,H_2O \qquad \beta_T = 6.67 \times 10^{-3}/\text{wt}\%\,H_2O$$

where L and T refer to the longitudinal and transverse directions, respectively. The laminate geometric configuration for the finite element analysis is shown in Fig. 4, where the width of coupon is $2b = 25.40\,\text{mm}$. The energy release rate has been calculated for the delamination growth from the edge on each interlaminar boundary by taking the crack growth as $\Delta a = 0.05\,\text{mm}$ at each step.

The energy release rates for the cross-ply $(90°/0°)_s$ laminate with the delamination on the interface between the 90° and 0° laminae are shown as a function of the crack length in Fig. 8. Since it was found that the results based on the transformed quasi-three-dimensional model are completely identical to those based on the conventional quasi-three-dimensional model, either of the two models has been utilized for the finite element calculation. It can be seen from Fig. 8 that the stiffness method gives almost the same $G$ as the crack closure integral method. As the delamination extends from the edge, $G$ increases monotonically and attains a constant value after the delamination length becomes approximately equal to one ply thickness. The saturated values of $G$ coincide with the predictions for a semi-infinite laminate given by eqn. (27) based on the delaminated plate approximation. It is also revealed from Fig. 8 that most of the energy release rate is due to the mode II crack opening mode. It can be explained by the fact that the tractions applied on the edge are a combination of tensile and compressive forces in the transformed edge stress problem.

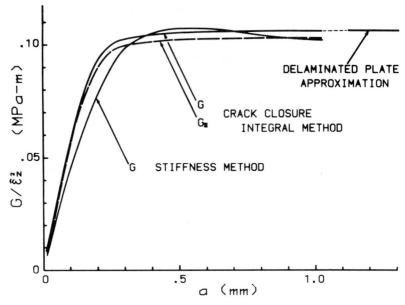

FIG. 8.     Energy release rates $G$ and $G_{II}$ versus delamination length $a$ in $(90^\circ/0^\circ)_s$ carbon–epoxy composite laminate under mechanical loading (delamination on $90^\circ/0^\circ$ interface, $t = 0.28$ mm).

## Effects of Stacking Sequence

To examine the influences of stacking sequence on $G$, the quasi-isotropic $(0^\circ/90^\circ/45^\circ/-45^\circ)_s$, $(45^\circ/-45^\circ/0^\circ/90^\circ)_s$ and $(45^\circ/-45^\circ/90^\circ/0^\circ)_s$ laminates have been analyzed. The changes of $G$ with the crack length $a$ for various delaminated interfaces are shown in Figs 9–11, where $G$ and $G_{II} + G_{III}$ are presented. It is assumed that the total energy release rate is $G$ when the normal stress $\sigma_y$ at the crack tip is tensile, and it is $G_{II} + G_{III}$ when the $\sigma_y$ is compressive. For the $(0^\circ/90^\circ/45^\circ/-45^\circ)_s$ laminate, the $\sigma_y$ at each interface is compressive when the axial tensile load is applied and vice versa. For the $(45^\circ/-45^\circ/0^\circ/90^\circ)_s$ and $(45^\circ/-45^\circ/90^\circ/0^\circ)_s$ laminates, the $\sigma_y$ at each interface is tensile when the axial tensile load is applied and vice versa.

Comparing the saturated values of the energy release rate $G$ or $G_{II} + G_{III}$ in Fig. 9, it is predicted that the delamination in the $(0^\circ/90^\circ/45^\circ/-45^\circ)_s$ laminate occurs on the interface between the $90^\circ$ and $45^\circ$ plies for the tensile axial load, and on the $45^\circ/-45^\circ$ interfaces for the compressive axial load. Figure 10 implies that in the $(45^\circ/-45^\circ/0^\circ/90^\circ)_s$ laminates the delamination occurs on either the $0^\circ/90^\circ$ or $45^\circ/-45^\circ$ interfaces for the tensile load and on the $45^\circ/-45^\circ$ interfaces for the compressive load. It is revealed from Fig.

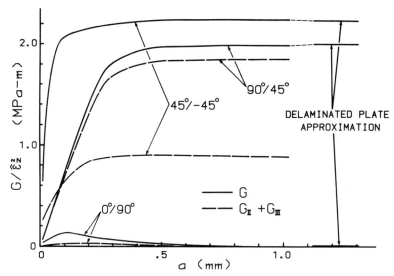

FIG. 9. Energy release rates $G$ and $G_{II} + G_{III}$ versus delamination length $a$ in $(0°/90°/45°/-45°)_s$ carbon–epoxy composite laminate under mechanical loading by the crack closure integral method for various delaminated interfaces ($t = 0.14$ mm).

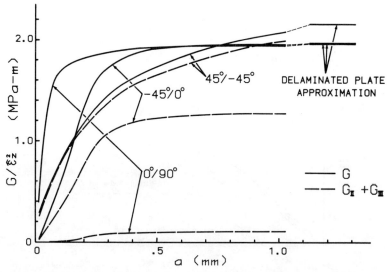

FIG. 10. Energy release rates $G$ and $G_{II} + G_{III}$ versus delamination length $a$ in $(45°/-45°/0°/90°)_s$ carbon–epoxy composite laminate under mechanical loading by the crack closure integral method for various delaminated interfaces ($t = 0.14$ mm).

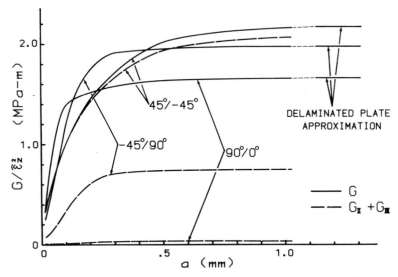

FIG. 11.   Energy release rates $G$ and $G_{II} + G_{III}$ versus delamination length $a$ in $(45°/-45°/90°/0°)_s$ carbon–epoxy composite laminate under mechanical loading by the crack closure integral method for various delaminated interfaces ($t = 0.14$ mm).

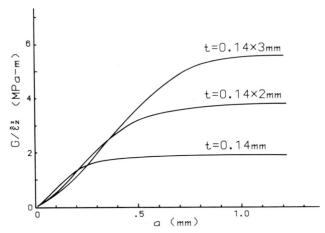

FIG. 12.   Energy release rate $G_{II} + G_{III}$ versus delamination length $a$ in $(0°/90°/45°/-45°)_s$ carbon–epoxy composite laminate under mechanical loading with various ply thicknesses (delamination on $90°/45°$ interface).

11 that the delamination in the $(45°/-45°/90°/0°)_s$ laminate occurs on the $45°/-45°$ or $-45°/90°$ interfaces for the tensile load and on the $45°/-45°$ interfaces for the compressive loads.

## Effects of Ply Thickness

The influence of the thickness of each ply on the energy release rate has been examined by analyzing the laminates with the same stack sequence but the different ply thicknesses. The $G_{II} + G_{III}$ for the $(0°/90°/45°/-45°)_s$ laminates with the delamination at the $90°/45°$ interfaces is shown in Fig. 12, which is the measure for the delamination growth under the tensile axial load as mentioned before. It is revealed that the curves are enlarged to the horizontal and vertical directions by an amount proportional to the ply thickness. It means that the $G/t$ and $a/t$ curves for different ply thicknesses coincide with each other. Therefore the saturated values of the strain energy release rates are proportional to the ply thicknesses as predicted before. This explains the well-known fact that the thicker the lamina, the smaller the load required for the delamination growth.

## COMPARISONS WITH EXPERIMENTAL RESULTS

The tension–tension and compression–compression fatigue tests have been conducted for the quasi-isotropic $(0°/90°/45°/-45°)_s$, $(45°/-45°/0°/90°)_s$ and $(45°/-45°/90°/0°)_s$ carbon–epoxy composite laminates with a width $2b = 25.40$ mm and a thickness $2h = 1.12$ mm (ply thickness $t = 0.14$ mm). For the tension–tension tests, the locations where the delamination occurred in the $(0°/90°/45°/-45°)_s$ laminate were the $90°/45°$ interfaces as predicted by the theoretical analysis. The boundaries delaminated in the $(45°/-45°/0°/90°)_s$ and $(45°/-45°/90°/0°)_s$ laminates were the $0°/90°$ and $-45°/90°$ interfaces, respectively. It is noted that in each of the latter two laminates which has different interfaces with almost the same values of the total energy release rate, the delamination occurred on the interfaces associated with the greater value of the mode I component $G_I = G - (G_{II} + G_{III})$. For the compression–compression tests, the delamination in the $(0°/90°/45°/-45°)_s$ and $(45°/-45°/0°/90°)_s$ laminates occurred on the $45°/-45°$ interfaces as theoretically predicted.

## CONCLUSIONS

The free-edge delamination of composite laminates subjected to mechanical and hygrothermal loadings has been analyzed by means of the energy release rate concept in fracture mechanics. The well-known Pipes–Pagano quasi-three-dimensional problem for the analysis of the interlaminar stresses has been transformed to the quasi-three-dimensional problem, where the edge is subjected to the self-equilibrating forces associated with stresses derived through the classical lamination theory in the original problem. The transformation of the problem has clarified the physical mechanism of interlaminar stresses and the relations between the mechanical and hygrothermal loadings. And the delaminated plate approximation in the transformed problem has been proposed to analytically obtain the energy release rate due to delamination propagation.

Numerical calculations have been performed on the carbon–epoxy composite laminates utilizing the conventional finite element method for the conventional and transformed problems, which have been observed to yield the identical results. The energy release rate has been obtained by the stiffness method or the crack closure integral method. It has been revealed that the energy release rate increases monotonically with the delamination extension from the free edge and attains the constant value which coincides with the prediction based on the delaminated plate approximation. The finite element calculations and the delaminated plate approximation have explained the well-recognized fact that the thicker the ply, the less the stress for the delamination growth. The predictions based on the saturated value of the energy release rate for the interface where delamination occurs have been in good agreement with the results of the tension–tension and compression–compression fatigue tests, which have been conducted on the carbon–epoxy composite laminates with different stacking sequences. The simple analytical solution by the delaminated plate approximation is suitable for preliminary design analysis, which requires a lot of configurations to be examined.

## REFERENCES

1. PIPES, R. B. and PAGANO, N. J., Interlaminar stresses in composite laminates under uniform axial extension, *J. Comp. Mater.*, **4** (1970), 538–548.
2. WANG, S. S. and CHOI, I., Boundary-layer effects in composite laminates. Part I: Free-edge stress singularities, *J. appl. Mech.*, **49** (1982), 541–548.
3. HERAKOVICH, C. T., Influence of layer thickness on the strength of angle-ply laminates, *J. Comp. Mater.*, **16** (1982), 216–227.
4. KIM, R. Y. and SONI, S. R., Experimental and analytical studies on the onset of delamination in laminated composites, *J. Comp. Mater.*, **18** (1984), 70–80.

5. WANG, S. S., Edge delamination in angle-ply composite laminates, *AIAA Journal*, **22** (1984), 256–264.
6. KIM, K. S. and HONG, C. S., Delamination growth in angle-ply laminated composites, *J. Comp. Mater.*, **20** (1986), 423–438.
7. WHITCOMB, J. D. and RAJU, I. S., Superposition method for analysis of free-edge stresses, *J. Comp. Mater.*, **17** (1983), 492–507.
8. PAGANO, N. J. and PIPES, R. B., The influence of stacking sequence on laminate strength, *J. Comp. Mater.*, **5** (1971), 50–57.

# 19

# Rate Effects on Delamination Fracture Toughness of Graphite/Epoxy Composites

I. M. DANIEL, G. YANIV and J. W. AUSER

*Department of Civil Engineering, Northwestern University, Evanston, Illinois 60201, USA*

## ABSTRACT

*The effects of loading rate on interlaminar fracture toughness were investigated for two graphite/epoxy materials, one with a brittle matrix and the other with a toughened one. Mode I delamination fracture toughness was studied by means of uniform-width, width-tapered, and height-tapered double cantilever beam specimens. The specimens were loaded at various crosshead rates corresponding to crack extension rates up to 26 m/s. It was found that for the brittle-matrix composite the energy release rate increases up to a certain crack velocity and thereafter it decreases. In the case of the toughened-matrix composite the energy release rate decreases monotonically with increasing crack velocity.*

## INTRODUCTION

Delamination, or interlaminar cracking, is considered one of the predominant types of damage in composite materials. It can occur under opening, shearing, tearing or a combination thereof, therefore delamination fracture toughness can be characterized by stress intensity factors or strain energy release rates in Modes I, II or III. The most commonly used specimens for Mode I characterization are the double cantilever beam (DCB) specimen, discussed by Whitney et al.,[1] and the width-tapered double cantilever beam (WTDCB) introduced by Brussat et al.[2] More recently Yaniv and Daniel[3] used a height-tapered (HTDCB) specimen.

The importance of delamination fracture toughness has stimulated a number of investigations on the effects of various parameters on this property. The effects of loading rate have been studied and discussed by Hunston and Bascom,[4] Miller *et al.*,[5] Aliyu and Daniel[6] and Daniel *et al.*[7] Some investigators found no noticeable rate effects in the composite, even when the matrix resin is rate sensitive.[4,5] Hunston and Bascom found a pronounced rate dependence in an elastomer-modified epoxy, with the fracture energy decreasing with increasing loading rate.[4] In the case of the composite, however, they found a slight increase in fracture energy with loading rate. Their results may be influenced by the fact that they used glass cloth instead of unidirectional plies. Aliyu and Daniel[6] and Auser[8] found that the strain energy release rate for AS4/3501-6 graphite/epoxy, having a brittle matrix, increases with loading rate over four decades of crack propagation velocity. Subsequently, Yaniv and Daniel[3] found that this trend is reversed at some higher value of crack propagation velocity. In the case of T300/F-185 graphite/epoxy, having a rubber-toughened matrix, Daniel *et al.*[7] and Auser[8] found a monotonic decrease of fracture toughness with crack velocity.

This chapter reviews and summarizes procedures and results obtained to date on the effects of loading rate on delamination fracture toughness for two graphite/epoxy materials.


## ANALYSIS METHODS

The various types of double cantilever beam specimens are illustrated in Fig. 1. In the beam analysis method the specimen is assumed to consist of two identical cantilever beams with build-in ends and length equal to the length of the crack. The total energy balance is expressed as

$$W = U + T + D \tag{1}$$

where  $W$ = external work
$\quad\quad U$ = elastic strain energy
$\quad\quad T$ = kinetic energy
$\quad\quad D$ = dissipative energy associated with fracture.

The elastic strain energy, including the effects of normal and shear stresses, is given by

$$U = \frac{1}{2} \int_{V(t)} (\sigma_1 \varepsilon_1 + \tau_{12}\gamma_{12}) \, dV \tag{2}$$

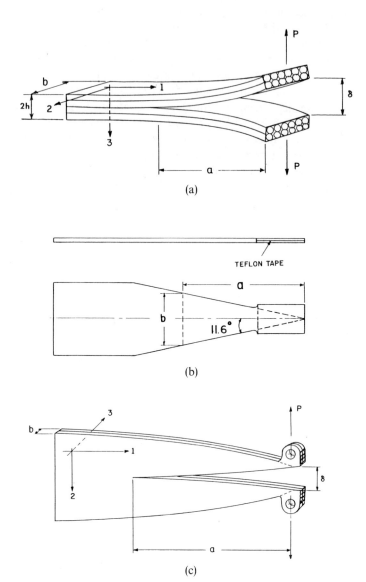

FIG. 1.   Double cantilever beam (DCB) specimens used (a) uniform, (b) width tapered and (c) height tapered.

where $\sigma_1$ and $\tau_{12}$ are the longitudinal normal and in-plane shear stresses; $\varepsilon_1$ and $\gamma_{12}$ the corresponding strains; and $V(t)$ the time dependent volume of the cantilever beams.

The kinetic energy according to the Bernoulli–Euler beam theory is given by

$$T = \frac{1}{2} \int_0^{a(t)} A(x)\rho \left[ \frac{\partial y(x, t)}{\partial t} \right]^2 \mathrm{d}x \tag{3}$$

where $A(x)$ = cross-sectional area
$\rho$ = material density
$a(t)$ = current length of the beam (crack)
$y$ = beam deflection.

The energy release rate per unit of crack extension obtained from eqn. (1) is expressed as

$$G_1 = \frac{2}{b} \frac{\partial D}{\partial a} = \frac{2}{b} \left( \frac{\partial U}{\partial a} - \frac{\partial T}{\partial a} \right) \tag{4}$$

For the simplest constant cross-section DCB specimen the expression above takes the form[6]

$$G_1 = \frac{24P^2}{E_1 b^2 h} \left\{ \frac{1}{2} \left( \frac{a}{h} \right)^2 + \frac{1}{20} \left( \frac{E_1}{G_{13}} \right) - 10 \left( \frac{\dot{a}}{c_{1L}} \right)^2 \right.$$
$$\left. \times \left[ \left( \frac{a}{h} \right)^4 + \frac{3}{20} \left( \frac{a}{h} \right)^2 \left( \frac{E_1}{G_{13}} \right) + \frac{1}{240} \left( \frac{E_1}{G_{13}} \right)^2 \right] \right\} \tag{5}$$

where $P$ = applied load
$b$ = specimen width
$h$ = cantilever beam thickness
$E_1$ = longitudinal modulus (in fiber direction)
$G_{13}$ = transverse shear modulus
$\dot{a}$ = crack propagation velocity
$c_{1L}$ = longitudinal wave propagation velocity

The effect of kinetic energy becomes significant only at very high crack propagation velocities, since the longitudinal wave velocity, $c_{1L}$, is of the order of $10^4$ m/s. For the materials considered in this investigation, the contribution of shear deformation becomes negligible for large values of $a/h$, which is the case for the first two types of specimens shown in Fig. 1.

The use of the uniform DCB (UDCB) specimen at high rates is limited because of the relatively high compliance of the specimen and the need to apply very high crosshead rates with the testing machine. This problem is somewhat alleviated with the width-tapered DCB (WTDCB) specimen, which has the property of constant rate of change of compliance with respect to crack length. Furthermore, it eliminates the need for monitoring the crack length exactly and yields a constant crack velocity for a constant opening deflection rate. The strain energy release rate for the WTDCB specimen, neglecting kinetic energy, is given by

$$G_1 = \frac{12P^2k^2}{E_1h^3}\left[1 + \frac{1}{10}\frac{E_1}{G_{13}}\left(\frac{h}{a}\right)^2\right] \tag{6}$$

where $k = a/b$
   $b =$ beam width at crack length $a$.

In order to achieve higher crack propagation velocities and to avoid the 'stick–slip' phenomenon present in thin specimens at high loading rates, the height-tapered DCB (HTDCB) specimen was introduced.[3] This specimen has a width equal to the laminate thickness and is loaded in the normal to the fiber direction in the plane of the laminate (2-direction) (Fig. 1). The contour of the specimen was designed so that the compliance decreases with increasing crack length, so that a smooth and stable crack propagation is achieved at high rates. The height $h$ of the cantilever beam was chosen to satisfy the relation

$$h^3 = c^3x^2 \tag{7}$$

where $c$ is a constant. Then, the strain energy release rate was obtained as

$$G_1 = \frac{12P^2}{E_1b^2c^3}K_1\left[1 - \left(\frac{\dot{a}}{c_{1L}}\right)^2 K_1K_2\right] \tag{8}$$

where

$$K_1 = 1 + \frac{1}{10}\frac{E_1}{G_{12}}c^2a^{-2/3} \tag{9}$$

and

$$K_2^{-1} = \frac{5}{12}c^2a^{-2/3} \tag{10}$$

The strain energy release rate in eqn. (8) consists of two terms, the first one related to the potential elastic strain energy and the second one to the kinetic energy.

The expression for $K_1$ in eqn. (9) consists of two terms, the first one related to bending of the beam and the second one to shear. In the present work the latter term did not exceed 8% of the total value, however, the shear term is expected to be more pronounced for shorter beams.

## EXPERIMENTAL PROCEDURE

The materials used in this investigation were AS4/3501-6 graphite/epoxy (Hercules, Inc.) and T300/F-185 graphite/epoxy (Hexcel Corp.). The latter contains an elastomer-modified epoxy matrix. Twenty-four and forty-eight-ply unidirectional plates were fabricated. A 0·025 mm thick teflon film was inserted at the midsurface of the laminate along one edge to initiate the crack in the UDCB and WTDCB specimens.

The uniform DCB specimens were 22·9 cm long, 2·54 cm wide and 3·56 mm thick with an initial artificial crack of 3·81 cm length at one end. The WTDCB specimens were 11·68 cm long with a width tapering from 1·27 cm to 3·81 cm over a length of 6·60 cm at the center. Metallic hinges were bonded to the cracked end of these specimens to allow for load introduction without rotation. The HTDCB specimens were routed to the calculated contour with the aid of a template. A 5·08 cm long starter crack was machined at the narrow end of the specimen.

All specimens were tested in an Instron electrohydraulic testing machine at constant crosshead rates ranging from $7·5 \times 10^{-3}$ mm/s ($1·8 \times 10^{-3}$ in/min) to 460 mm/s ($1·07 \times 10^3$ in/min). Continuous records were obtained of the load, opening deflection, and crack extension. Crack extension was monitored in several different ways. At the very low loading rates with the DCB and WTDCB specimens crack extension was monitored visually. At the higher rates the crack was monitored either by means of strain gages or by means of a conductive paint circuit. The latter method was found more practical and reliable. One edge of the UDCB or WTDCB specimens or one face of the HTDCB specimen was painted with a red oxide paint to provide an insulating layer. Then, an electric circuit was painted with a conductive silver paint, and the conductors were connected to a battery (Fig. 2). As the crack propagates, it breaks the various conductors of the circuit and a stepwise variation of the voltage across the circuit is recorded.

Measurements of load, deflection and crack length were obtained with a digital processing oscilloscope (Norland 3001). The records were recorded digitally by this oscilloscope, transferred to a microcomputer (Apple IIe) for further processing, and plotted on a plotter (HP 7470A).

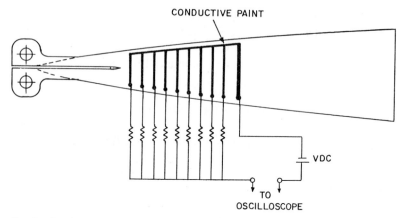

FIG. 2.   Specimen with conductive paint circuit for monitoring of crack propagation.

## RESULTS AND DISCUSSION

Typical load–time curves for UDCB specimens of T300/F-185 are shown in Figs 3 and 4. At the low crosshead rate of 8·5 mm/s (20 in/min), crack extension, which was monitored visually, seems to be stable. However, at the higher crosshead rate of 212 mm/s (500 in/min) the correlation of load and crack velocity is difficult as seen by the irregular form of the curve, which is due to unstable crack propagation.

In the case of the WTDCB specimen the load remains constant with time after crack initiation. At low loading rates the load–deflection curve is smooth; however, at higher crack propagation velocities the load shows fluctuations due to unstable crack propagation. Figure 5 shows load–time curves for AS4/3501-6 graphite/epoxy for four different deflection rates or crack propagation velocities. It is seen that crack propagation becomes unstable beyond a crack velocity of 725 mm/s (1713 in/min).

A similar phenomenon occurs in the case of T300/F-185 graphite/epoxy except that the unstable crack propagation occurs at a lower crack velocity. Figure 6 shows a load–time curve for a T300/F-185 graphite/epoxy specimen loaded at a crosshead rate of 212 mm/s (500 in/min) corresponding to a crack propagation velocity of 363 mm/s. This is an illustration of the 'stick–slip' phenomenon with alternating periods of stable and unstable crack propagation. For example, crack growth is stable between points 1 and 2 associated with a constant load. Thereafter, the crack accelerates in an unstable manner accompanied by a sharp drop in the load between

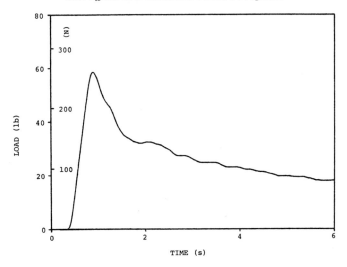

FIG. 3.   Load versus time curve for uniform DCB specimen of T300/F-185 graphite/epoxy loaded at a crosshead rate of 8·5 mm/s (20 in/min).

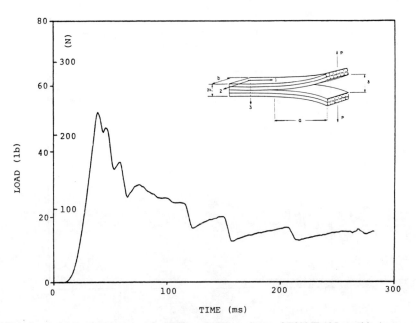

FIG. 4.   Load versus time curve for uniform DCB specimen of T300/F-185 graphite/epoxy loaded at a crosshead rate of 212 mm/s (500 in/min).

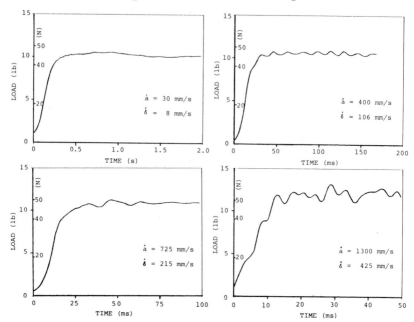

FIG. 5.   Load versus time curves for width-tapered DCB specimens of AS4/3501-6 graphite/
epoxy loaded at various crosshead rates.

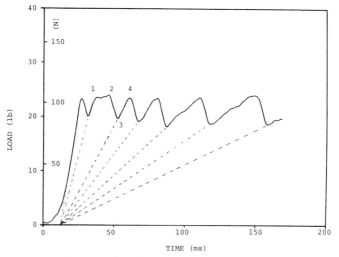

FIG. 6.   Load versus time curve for width-tapered DCB specimen of T300/F-185 graphite/
epoxy loaded at a crosshead rate of 212 mm/s (500 in/min).

points 2 and 3. At point 3, where the load is sufficiently low, the crack is arrested. The load then increases linearly without any further crack growth until it reaches the previous peak level at point 4 and causes continued crack growth.

The fracture surface of the specimens with load fluctuations show dark and light bands as shown in Fig. 7. The dark bands are caused by brittle failure of the matrix and correspond to rapid crack growth. The light bands, also referred to as 'stress whitening bands', are caused by ductile or high deformation failure of the matrix and correspond to slow crack growth. As the loading rate increases the number of dark and light bands also increases as shown in Fig. 7.

This phenomenon of crack instability is related to the rate sensitivity of the fracture toughness of polymers, since Mode I fracture toughness in a composite is a matrix dominated property. It has been observed that polymers exhibit three types of behavior under varying loading rate: rate insensitivity, positive rate sensitivity, and negative rate sensitivity.[9] It is possible for a polymer to exhibit more than one type of rate sensitivity depending on the temperature, molecular structure and range of crack velocity. As long as the fracture toughness has a positive rate sensitivity, the applied load will control the crack propagation velocity. However, at the crack velocity where the rate sensitivity changes from positive to negative

|     |     |     |
| --- | --- | --- |
| (a) | (b) | (c) |

FIG. 7.   Fracture surfaces of WTDCB specimens of T300/F-185 graphite/epoxy for various crack propagation velocities. (a) $\dot{a} = 158$ mm/s, (b) $\dot{a} = 486$ mm/s, (c) $\dot{a} = 758$ mm/s.

the crack will start accelerating. As the fracture toughness starts to decrease, the load also increases.

To overcome these difficulties the height-tapered DCB specimen was designed so that its compliance rate is a decreasing function of the crack length. Thus as the crack propagates, the balancing load increases regardless of the current type of rate sensitivity. Typical records of load, opening deflection and crack tip location as a function of time are shown in Fig. 8.

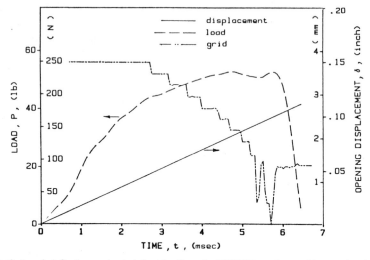

FIG. 8. Load, deflection and painted grid voltage for HTDCB specimen with a crack velocity of 26 m/s.

The energy release rate was determined by the beam analysis method using eqns (5), (6) and (8) for the three types of specimens used. Results for the two materials tested are plotted as a function of crack propagation velocity in Figs 9 and 10. In the case of the AS4/3501-6 graphite/epoxy with a brittle matrix (Fig. 9) the strain energy release rate increases with crack velocity up to a value of approximately 1 m/s and thereafter it decreases. This is in agreement with the onset of crack instability observed with the WTDCB specimens (Fig. 5). Results from the HTDCB specimens seem to be slightly lower than those obtained from the UDCB and WTDCB specimens. The difference is attributed to the 'fiber bridging' phenomenon which, although small in the case of the UDCB and WTDCB specimens, is virtually negligible in the case of the HTDCB specimen.

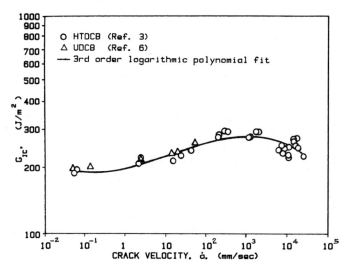

FIG. 9. Critical strain energy release rate for Mode I delamination of AS4/3501-6 graphite/epoxy as a function of crack propagation velocity.

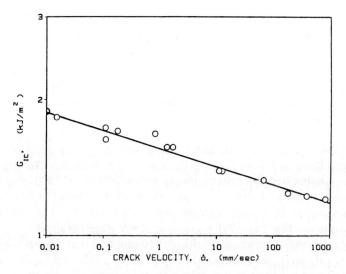

FIG. 10. Critical strain energy release rate for Mode I delamination of T300/F-185 graphite/epoxy as a function of crack propagation velocity.

An attempt was made to express the critical strain energy release rate as a logarithmic polynomial of the form

$$\log G_{\text{Ic}} = \sum_{n=0}^{N} A_n (\log \dot{a})^n \qquad (11)$$

where $A_n$ are constants. For the range of crack velocities studied for the AS4/3501-6 graphite/epoxy, extending over seven decades, it was found that $N = 3$ gives a standard deviation of less than 3% (Fig. 9).

In the case of the T300/F-185 graphite/epoxy material with the elastomer-toughened matrix (Fig. 10) the strain energy release rate decreases monotonically over the five decades of crack velocity investigated. A straight line was fitted through the data yielding a power law relation as a special case of eqn. (11) for $N = 1$,

$$G_{\text{Ic}} = A(\dot{a})^{A_1} \qquad (12)$$

Results are in qualitative agreement with those described by Hunston and Bascom for the neat resin.[4] They showed that the fracture energy of an elastomer-modified epoxy increases sharply with temperature by decreases in loading rate. In the case of the composite, as predicted by Hunston and Bascom, the dependence on loading rate is not as dramatic.

## SUMMARY AND CONCLUSIONS

The effects of loading rate on interlaminar fracture toughness were investigated for two graphite/epoxy materials, one with a brittle matrix (AS4/3501-6) and the other with an elastomer toughened matrix (T300/F-185). Mode I interlaminar fracture was studied by means of three types of specimens, uniform, width-tapered and height-tapered double cantilever beam specimens. These specimens were loaded at opening deflection rates ranging between 0·0085 mm/s (0·02 in/min) and 460 mm/s (1070 in/min) corresponding to crack extension rates of up to 26 m/s (61 500 in/min). At the lower rates crack extension was monitored visually. At higher rates it was monitored by means of strain gages or a conductive paint circuit.

Results for the AS4/3501-6 graphite/epoxy were obtained over seven decades of crack propagation velocity. It was found that the strain energy release rate increases with crack velocity up to a value of approximately 1 m/s. Beyond this point it decreases with increasing crack velocity. The maximum value of the strain energy release rate is approximately 46%

higher than the quasi-static value. A logarithmic polynomial expression was fitted to the experimental results to correlate fracture toughness and crack velocity over seven decades of the latter. The characteristic variation of fracture toughness with crack velocity was related to similar positive and negative rate sensitivity exhibited by polymeric materials.

Results for the T300/F-185 graphite/epoxy were obtained over five decades of crack velocity. The strain energy release rate decreases monotonically by up to 36% over this range and it can be expressed as a power law of the crack velocity. The value of $G_{Ic}$ for this material is approximately ten times that for the composite with unmodified epoxy matrix.

## ACKNOWLEDGEMENTS

The work described here was sponsored by the National Aeronautics and Space Administration (NASA)—Langley Research Center, Hampton, Virginia. We are grateful to Mr J. D. Whitcomb of NASA—Langley for his encouragement and cooperation, Mr Scott Cokeing for his assistance with the experimental work and Mrs Stella Greener for typing the manuscript.

## REFERENCES

1. WHITNEY, J. M., BROWNING, C. E. and HOOGSTEDEN, W., A double cantilever beam test for characterizing Mode I delamination of composite materials, *J. Reinf. Plastics and Composites*, **1** (1982), 297–313.
2. BRUSSAT, T. R., CHIN, S. T. and MOSTOVOY, S., Fracture mechanics for structural adhesive bonds, AFML-TR-77-163, Phase II, Wright-Patterson Air Force Base, Ohio, 1978.
3. YANIV, G. and DANIEL, I. M., Height-tapered double cantilever beam specimen for study of rate effects on fracture toughness of composites. To be published in *Composite Materials: Testing and Design*, ASTM STP. (J. D. Whitcomb ed.), Philadelphia, American Society for Testing and Materials, 1987.
4. HUNSTON, D. L. and BASCOM, W. D., Effects of layup, temperature and loading rate in double cantilever beam tests of interlaminar crack growth, *Composites Technology Review*, **5** (4) (1983), 118–19.
5. MILLER, A. G., HERTZBERG, P. E. and RANTALA, V. W., Toughness testing of composite materials, *Proc. 12th National SAMPE Tech. Conf.*, Seattle, WA, Oct. 1980, pp. 279–293.
6. ALIYU, A. A. and DANIEL, I. M., Effects of strain rate on delamination fracture toughness of graphite/epoxy, in: *Delamination and Debonding of Materials*, ASTM STP 876 (W. S. Johnson ed.), Philadelphia, American Society for Testing and Materials, 1985, pp. 336–348.

7. DANIEL, I. M., SHAREEF, I. and ALIYU, A. A., Rate effects on delamination fracture toughness of a toughened graphite/epoxy, in: *Toughened Composites*, ASTM STP 937 (N. J. Johnston ed.), Philadelphia, American Society for Testing and Materials, 1987.
8. AUSER, J. W., Load rate effects on delamination fracture toughness of graphite/epoxy composites, M.S. thesis, Illinois Institute of Technology, Chicago, IL, May 1985.
9. WILLIAMS, J. G., *Fracture Mechanics of Polymers*, New York, John Wiley, 1984, p. 179.

# 20

# An Investigation of Experimental Methods for the Determination of Bearing Strength of CFRP Laminates

YANG BINGZHANG

*Northwestern Polytechnical University,
Xian, People's Republic of China*

## ABSTRACT

*This paper presents a new method for determining the bearing strength of CFRP laminates. There are four different fixtures designed by the author, such as: tension mode bearing loading assembly: (a) single lap, (b) double lap; compression mode bearing loading assembly: (a) single lap joint, (b) double lap joint. And the quantitative measuring method for damage bearing load ( $P_d$ ) and elongation of bearing hole ( $\delta$ ) which may determine the bearing strength of different specimens are presented. Also presented are the effects of variables such as ply orientation, laminate thickness and bolt clamping pressure. The results obtained are summarized as the experimental values of bearing strength in the table and the corresponding P–$\delta$ curves.*

## INTRODUCTION

CFRP (carbon fibre reinforced plastics) laminates have found wide use in aircraft structures. It is necessary to determine the bearing strength of the laminates with bolted joints.

There are many authors who study this topic, such as Collings,[1,8] Johnson and Matthews,[2] Agarwal,[3] Hart-Smith,[4] and Kretsis and Matthews.[5] But there is no standard method of test for determining bearing strength of CFRP laminates. This paper starts from the ASTM standard method of test for bearing strength of plastics (Designation D953-75),[6] and

2.273

taking advantage of the progress made by the above-mentioned authors, the author improves the test method in the following five respects.

1. So design the specimen, which is being considered as the PRC national standard,[7] as to ensure bearing failure without tension failure and shear-out failure.
2. So design new fixtures as to bring about better accuracy and convenience.
3. Improve the measurement accuracy of the hole elongation for a specimen with a good displacement sensor.
4. Selection of the damage load as the failure load is believed to be most appropriate for determining the bearing strength.
5. Determine the best value of the bolt torque.

A series of tests for determining the bearing strength of CFRP laminates with different lay-ups—$(0/\pm45/90)_s$, $(0/\pm45)_s$, $(\pm45)_s$ and $(0/90)_s$—different thicknesses, different diameters of the hole and subjected to different bolt torques is completed.

Tests show that the proposed improved method appears to be suitable for all FRP laminates and the five improvements previously mentioned appear to be actually realisable.

## EXPERIMENTAL DETAILS

### 1. The Standard Specimen

The shape and size of the tension mode bearing specimen are shown in Fig. 1 and Table 1. In Table 1, $D'$ and $d'$ are the diameters of bolts used in the fixed hole and bearing hole, respectively.

The shape and size for compression mode bearing specimens are shown in Fig. 2 and Table 2.

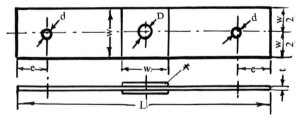

FIG. 1.   Tension mode bearing specimen geometry (* glued with aluminium plates) ($L$, total length; $t$, thickness; $e$, end distance; $W$, width; $d$, bearing hole diameter; $D$, fixed hole diameter).

TABLE 1
*Tension mode bearing specimen size (mm)*

| Type | L | W | D | D' | d | d' | e | t | Pieces |
|------|-----|----------|------------------|------------------|------------------|------------------|----------|---------|--------|
| I | 160 | $30 \pm 0.3$ | $8^{+0.03}_{-0.00}$ | $8^{+0.00}_{-0.03}$ | $5^{+0.03}_{-0.00}$ | $5^{+0.00}_{-0.03}$ | $20 \pm 0.2$ | $2 \pm 0.1$ | 7 |
| II | 180 | $36 \pm 0.4$ | $10^{+0.03}_{-0.00}$ | $10^{+0.00}_{-0.03}$ | $6^{+0.03}_{-0.00}$ | $6^{+0.00}_{-0.03}$ | $24 \pm 0.2$ | $3 \pm 0.1$ | 7 |

TABLE 2
*Compression mode bearing specimen size (mm)*

| Type | L | W | d | d' | e | t | Pieces |
|------|-----|----------|------------------|------------------|----------|---------|--------|
| I | 100 | $30 \pm 0.3$ | $5^{+0.03}_{-0.00}$ | $5^{+0.00}_{-0.03}$ | $20 \pm 0.2$ | $3 \pm 0.1$ | 7 |
| II | 120 | $36 \pm 0.4$ | $6^{+0.03}_{-0.00}$ | $6^{+0.00}_{-0.03}$ | $24 \pm 0.3$ | $4 \pm 0.1$ | 7 |

For each type of laminate and for each hole diameter the values of $e$ and $w$ were also chosen so that failure was in one mode. From Ref. 1, Fig. 3 shows the change of failure mode from one of tension to one of bearing as the ratio $w/d$ is increased, while Fig. 4 shows the change of failure mode from one of shear to one of bearing as the ratio $e/d$ is increased. Hence, we take $w/d = 6$ and $e/d = 4$ for the standard specimen.

## 2. Testing Fixture

For a torsion mode bearing loading assembly, a single lap joint is shown in Fig. 5a. It is suitable for performing thin plate testing. A double lap joint is shown in Fig. 5b. This is suitable for performing thin and thick plate testing.

For a compression mode bearing loading assembly, a single lap joint is shown in Fig. 6a and a double lap joint is shown in Fig. 6b.

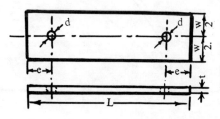

FIG. 2. Compression mode bearing specimen geometry.

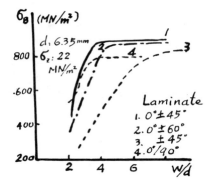

FIG. 3.   Variation of bearing stress at failure with $W/d$ ratio.

## 3. Measurement of the Hole Elongation of CFRP—($\delta$)

A tension mode bearing specimen is shown in Fig. 7a and a compression mode specimen in Fig. 7b. For the tension mode, to measure the displacement of the lowest point B around the hole, it is equivalent to measuring the displacement of B'. Similarly, for the compression mode, to measure the displacement of A, it is equivalent to A'. Of course, all the displacements are relative to the rigid fixture. Then the measured value $\delta$ is the actual deformation of the hole, which is better than measuring the crosshead displacement.[5]

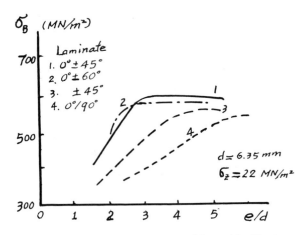

FIG. 4.   Variation of bearing stress at failure with $e/d$ ratio.

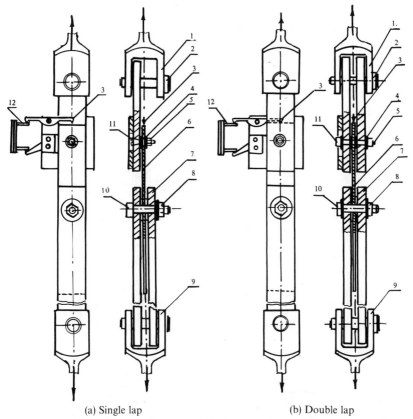

(a) Single lap             (b) Double lap

FIG. 5.   Tension mode bearing loading assembly (1, upper adapter; 2, tie pen; 3, sensor leg; 4, washer; 5, fastening nut; 6, test specimen; 7, fixture plate; 8, fixed nut; 9, lower nut; 10, fixed bolt; 11, bearing bolt; 12, displacement sensor).

The standard specimens are loaded on an Instron 1196 test machine. The bearing load $P$ versus the deformation of the hole $\delta$, i.e. the $p$–$\delta$ curve, is plotted automatically. According to this curve, we can determine the damage load and the maximum load to find the bearing strength.

## 4. Determination of the Bearing Strength

According to the general form of a typical plot shown in Fig. 8, Johnson and Matthews[2] suggest the following various ways of defining failure load:

(a)   the maximum load;
(b)   the first peak in the load/extension plot;

(c) the load corresponding to a specified amount of hole elongation— $0.5\%d$ for CFRP and $1\%d$ for BFRP;

(d) the load at which the load/extension curve first deviates from linearity;

(e) the load at which cracking first becomes audible;

(f) the load at which cracking is initiated;

(g) the load at which cracks become visible outside the washer.

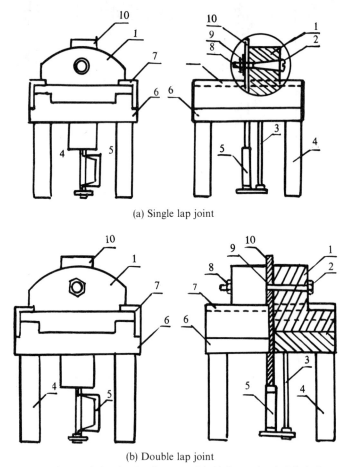

(a) Single lap joint

(b) Double lap joint

FIG. 6. Compression mode bearing loading assembly (1, fixture head; 2, bolt; 3, support of sensor; 4, pedestal; 5, displacement sensor; 6, platform; 7, steel angle; 8, fastening nut; 9, washer; 10, test specimen).

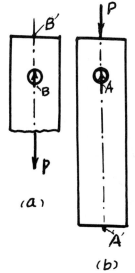

FIG. 7. Loading specimen.

In practice, (a), (b) and (c) are the best ways to define failure load (also called damage load ($P_d$)). All specimens with the same lay-up possess similar load/extension plots, as shown in Fig. 8.

In tests, a damage load $P$ is found by the use of the so-called 'offset method'. This is illustrated in Fig. 9, where a line offset an arbitrary amount of 0·5%$d$ is drawn parallel to the straight line portion of the initial $p$–$\delta$

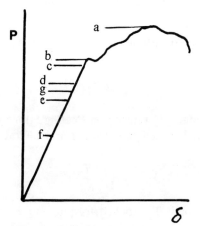

FIG. 8. Typical load/extension plot.

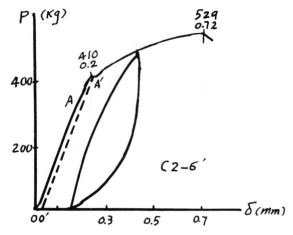

FIG. 9.   Offset method for the damage load of $\pm45°$ CFRP laminate.

diagram. Point A′ is then taken as the damage load of the CFRP at 0·5% offset; we can also determine the $P_{max}$ from the diagram. Then we get an important ratio:

$$\eta = P_d/P_{max} \tag{1}$$

where $\eta = 60–80\%$. Thus, the allowable bearing stress may be written as

$$[\sigma_B] = \frac{\sigma_d}{n} = \frac{P_d}{nA} = \frac{\eta P_{max}}{ntd} \tag{2}$$

where $n$ = safety factor and $A$ = bearing area. Then the bearing strength condition can be stated as follows:

$$\sigma = \frac{P}{td} \le [\sigma_B] \tag{3}$$

## 5.  The Effect of Lateral Constraint on Bearing Strength

The relationship between bolt torque and lateral constraint can be written as

$$P_t = \frac{T}{Kd} \tag{4}$$

where $P_t$ = tensile load; $T$ = applied torque; and $K$ = torque coefficient, $K \doteq 0\cdot2$.

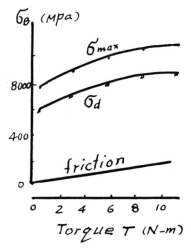

FIG. 10.   Bearing strength and friction at various torques.

For a standard washer, the diameter

$$D = 2·2d \qquad (5)$$

and the lateral constraint can be expressed as a transverse compressive stress, i.e.

$$\sigma_z = \frac{P_t}{(\pi/4)(D^2 - d^2)} \qquad (6)$$

Substituting $D$, $K$ and $P_t$ into (6), we get usually $T = 400\ \text{kgf-mm}$:

$$\sigma_z \doteqdot 1·7\frac{T}{d^3} \qquad (7)$$

for $d = 5\ \text{mm}$; $T = 500\ \text{kgf-mm}$ for $d = 6\ \text{mm}$.

The relation between $\sigma_{max}$, $\sigma_d$ and $T$ is shown in Fig. 10.


## RESULTS AND DISCUSSION

Symbols of lay-up laminates:

$$A = (0/\pm 45/90)_s \qquad B = (0/\pm 45)_s \qquad C = (\pm 45)_s$$
$$D = (0)_s \qquad\qquad E = (0/90)_s$$

Test results are shown in Tables 3–6.

TABLE 3

*Test results I—for double lap joint compression mode bearing loading assembly with CFRP of Liau Yuan fibre*

| Specimen | Torque (kg-cm) | $W$ (mm) | $t$ (mm) | $d$ (mm) | $P_{max}$ (kN) | $\sigma_{max}$ (N/mm²) | $\delta_{max}$ (mm) | $P_d$ (kN) | $\sigma_d$ (N/mm²) | $\delta_d$ (mm) | $p$–$\delta$ curves |
|---|---|---|---|---|---|---|---|---|---|---|---|
| A2-3 | 0 | 36·2 | 4·8 | 6 | 12·1 | 819 | 0·94 | 8·9 | 600 | 0·4 | Fig. 11 |
| B2-4 | 0 | 30·2 | 3·7 | 5 | 12·5 | 680 | 0·82 | 9·6 | 520 | 0·4 | Fig. 12 |
| C2-3 | 0 | 35 | 2·0 | 5 | 4·7 | 470 | 0·64 | 3·8 | 380 | 0·37 | Fig. 13 |
| D2-7 | 0 | 31·6 | 1·7 | 5 | 1·74 | 206 | 0·26 | 1·5 | 175 | 0·17 | Fig. 14 |

TABLE 4

*Test results II—for double lap joint compression ⊖ and tension ⊕ mode bearing loading assembly with CFRP of Shanghai fibre*

| Specimen | Torque (kg-cm) | $W$ (mm) | $t$ (mm) | $d$ (mm) | $P_{max}$ (kN) | $\sigma_{max}$ (N/mm²) | $\delta_{max}$ (mm) | $P_d$ (kN) | $\sigma_d$ (N/mm²) | $\delta_d$ (mm) | $p$–$\delta$ curves |
|---|---|---|---|---|---|---|---|---|---|---|---|
| C2-4' ⊖ | 0 | 30·0 | 2·03 | 5 | 6·0 | 589 | 0·91 | 4·31 | 454 | 0·2 ⎫ | Fig. 15 |
| C2-4 ⊕ | 0 | 36·0 | 2·10 | 6 | 5·96 | 473 | 0·99 | 3·23 | 256 | 0·16 ⎭ | |
| C2-1' ⊖ | 0 | 30·1 | 2·10 | 5 | 5·51 | 525 | 1·00 | 4·02 | 382 | 0·25 ⎫ | Fig. 16 |
| C2-1 ⊕ | 0 | 36·1 | 2·00 | 6 | 5·48 | 456 | 0·64 | 3·04 | 241 | 0·14 ⎭ | |

TABLE 5

*Test results III—for tension mode bearing loading assembly with CFRP of Liau Yuan fibre*

| Specimen | Torque (kg-cm) | $W$ (mm) | $t$ (mm) | $d$ (mm) | $P_{max}$ (kN) | $\sigma_{max}$ (N/mm²) | $\delta_{max}$ (mm) | $P_d$ (kN) | $\sigma_d$ (N/mm²) | $\delta_d$ (mm) | $p$–$\delta$ curves |
|---|---|---|---|---|---|---|---|---|---|---|---|
| Double lap | | | | | | | | | | | |
| E2-2 | 0 | 30·2 | 1·88 | 5·0 | 5·28 | 562 | 0·63 | 2·74 | 292 | 0·20 | Fig. 17 |
| E2-3 | 0 | 30·1 | 1·90 | 5·0 | 5·88 | 625 | 0·64 | 2·94 | 313 | 0·18 | Fig. 18 |
| Single lap | | | | | | | | | | | |
| E2-4 | 0 | 30·3 | 1·90 | 5·0 | 4·37 | 460 | 0·77 | 2·74 | 235 | 0·16 | Fig. 19 |
| E2-6 | 0 | 30·1 | 1·90 | 5·0 | 4·51 | 475 | 1·06 | 2·90 | 303 | 0·16 | Fig. 20 |

TABLE 6

*Test results IV—for double lap joint tension mode bearing loading assembly with CFRP of Liau Yuan fibre in different torque*

| Specimen | Torque (kg-cm) | $W$ (mm) | $t$ (mm) | $d$ (mm) | $P_{max}$ (kN) | $\sigma_{max}$ (N/mm²) | $\delta_{max}$ (mm) | $P_d$ (kN) | $\sigma_d$ (N/mm²) | $\delta_d$ (mm) | $p$–$\delta$ curves |
|---|---|---|---|---|---|---|---|---|---|---|---|
| B2-4-2 | 0 | 30·0 | 2·10 | 5·10 | 4·72 | 487·10 | 0·58 | 3·65 | 376·6 | 0·18 | Fig. 21 |
| B2-5-1 | 20 | 30·2 | 1·95 | 5·10 | 5·36 | 553·42 | 0·52 | 4·93 | 357·8 | 0·15 | (P.4) |
| B2-5-2 | 30 | 30·2 | 2·10 | 5·12 | 7·70 | 718·95 | 0·73 | 5·60 | 522·9 | 0·30 | |
| B2-6-1 | 40 | 30·2 | 2·00 | 5·10 | 8·12 | 796·08 | 0·51 | 6·84 | 670·6 | 0·21 | |
| B2-6-2 | 50 | 30·2 | 2·10 | 5·10 | 8·00 | 746·96 | 0·58 | 6·05 | 564·9 | 0·27 | |

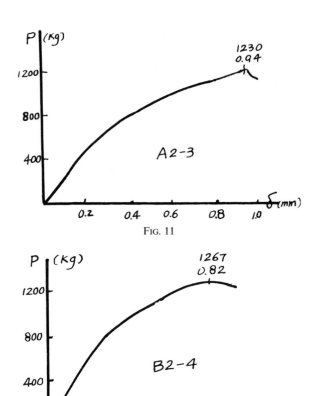

FIG. 11

FIG. 12

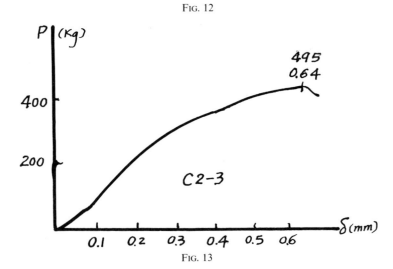

FIG. 13

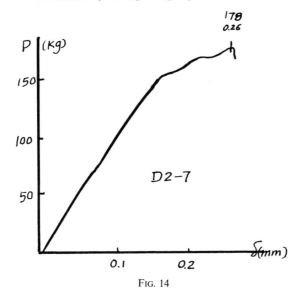

FIG. 14

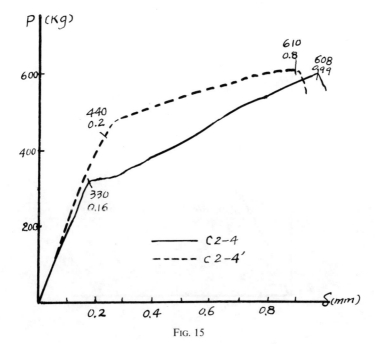

FIG. 15

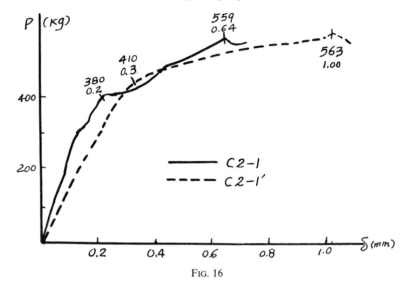

FIG. 16

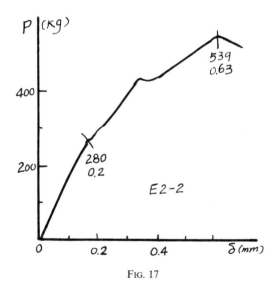

FIG. 17

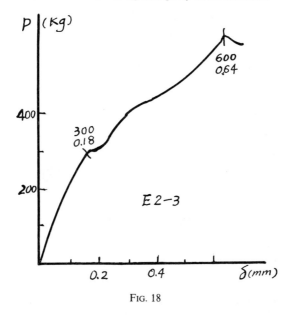

FIG. 18

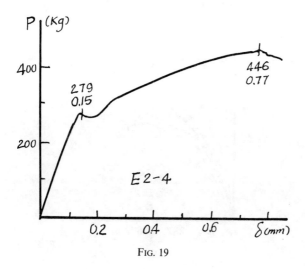

FIG. 19

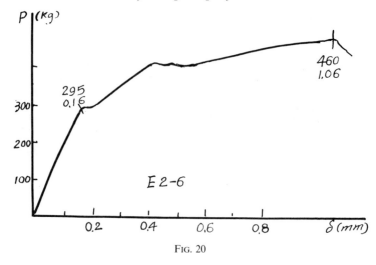

FIG. 20

1. In Table 3 we can see that the different stacking sequence laminate specimens for the same lap joint, the same mode of loading and the same fibre have different bearing strengths, $\sigma_d$ and $\sigma_{max}$, in relation $A > B > C > D$; then laminate $A = (0/\pm 45/90)_s$ is the best.

2. In Table 4 we get $\sigma_{d\,\ominus} > \sigma_{d\,\oplus}$ and $\sigma_{max\,\ominus} > \sigma_{max\,\oplus}$. Therefore, the tension mode fixture is more accurate than the compression mode in the bearing strength.

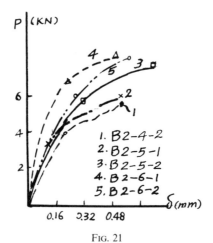

FIG. 21

3. In Table 5, under the other conditions, the bearing strength with double lap loading assembly is greater than that with the single lap.

4. Table 6 and Fig. 19 show that the bearing strength is increased as the bolt torque increases, but when the torque is too large then the strength decreases.

5. The bearing strength of the laminates without clamping pressure decreases linearly with the increase of the proportion of the value $d/t$. For the case of existing clamping pressure, the bearing strength of the laminates decreases with the increasing of the proportion of the value $d/t$, but there does not exist obvious linearity.

6. The laminates with the same lay-up possess the initial slope or $E$ (effective bearing modulus). Therefore, an indicating displacement and $E_B$ can be used to determine the damage bearing load $P_d$, as shown in Fig. 9.


## CONCLUSION

The previous test has shown that the investigation method for determining bearing strength is suitable for all FRP laminates with different lay-up, diameter of the hole and thickness. The designed fixtures according to the load manner can be freely chosen for better accuracy and convenience.


## ACKNOWLEDGEMENTS

The author wishes to express his sincere thanks to Professor M. Z. Yuan, Engineer X. J. An and Mr P. R. Jia for their assistance in experiments and for valuable discussions. He also wishes to acknowledge the support from the Composite Mechanics Research Center of N.P. University.


## REFERENCES

1. COLLINGS, T. A., The strength of bolted joints in multi-directional CFRP laminates, *Composites*, **8**, No. 1 (Jan. 1977), 43–55.
2. JOHNSON, M. and MATTHEWS, F. L., Determination of safety factors for use when designing bolted joints in GRP, *Composites* (April 1979), 73–76.
3. AGARWAL, B. L., *AIAA J.*, **18** (1980), 1371.
4. HART-SMITH, L. J., Bolted joints in graphite/epoxy composite, NASA CR-144899, 1976.

5. KRETSIS, G. and MATTHEWS, F. L., The strength of bolted joints in glass fibre/epoxy laminates, *Composites*, **16**, No. 2 (April 1985), 92–102.
6. ASTM Standard Method of Test for Bearing Strength of Plastics (Designation D953-75).
7. Northwestern Polytechnical University, FRP Laminates—test method for bearing strength by bolted joints (copy for PRC national standard).
8. COLLINGS, T. A., On the bearing strength of CFRP laminates, *Composites*, **13**, No. 3 (July 1982), 241–252.

# 21

# Errors Associated with the Use of Strain Gages on Composite Materials

C. C. Hiel

*Faculty of Engineering, Composites Group, Free University of Brussels (VUB), Pleintaan 2, B-1050 Brussels, Belgium*

ABSTRACT

*This chapter concentrates on the influence of strain gage misalignments when bonded to fiber-reinforced composite materials. The influence of differences in thermal expansion, transverse sensitivity and the non-coincidence of principal stress on principal strain directions are discussed.*

## 1. INTRODUCTION

The use of strain gages on composite materials is an essential element in most mechanical property characterization methods. A recent summary on the state of the art in strain gage technology was written by Perry,[1] one of the leading authorities in the field. Dr Perry concludes that there are fringe areas in strain gage technology which have been only partially explored, such as large strains, reinforcement effects, thermally and hygroscopically anisotropic test materials, etc. They require a much more thorough examination if accuracies of ±10% or better are to be reliably achieved in strain measurements on plastic-based structural materials.

Now that composites are ready for extensive use in strength critical applications, the precise measurement of the materials response is even more important because it is essential that the response which precedes failure be understood so that a larger proportion of the potential material strength can be utilized in design. Indeed, a typical strain value currently

2.291

used in design is 0·4%, which may be compared to the fiber breaking strain of about 1·3%.
Research along these lines has also been initiated by Tuttle and Brinson.[2]

## 2. BASIC STRAIN GAGE RESPONSE

### 2.1. The Uniaxial Strain Gage by Itself

The gage which is schematically shown in Fig. 1 is basically a conductor with resistivity $\rho$, cross-section $S$ and total length $l$. The resistance $R$ between the terminals A and B is given as a function of these three independent variables by eqn. (1):

$$R = \rho \frac{l}{S} \tag{1}$$

This linear relationship is well established in the theory of conductive media.[3]

The strain gage wiring in Fig. 1 will experience elongation when it is pulled. This, of course, can change the conductive properties due to piezoelectric effects.

Alternatively, when the gage is exposed to higher temperatures, the thermoelastic and thermoresistivity effects do contribute to a change in resistance ($R$).

The relative change in resistance is given by eqn. 2:

$$\frac{dR}{R} = (\beta_\rho + 1 + 2v_g)\frac{\Delta l}{l} + (\beta_g - \alpha_g)\,dT \tag{2}$$

where $v_g$ is Poisson's ratio, $\beta_\rho$ is piezoresistivity, $\beta_g$ is thermoresistivity and $\alpha_g$ is thermal expansion.

This equation is obtained by taking the total derivative of eqn. (1) with respect to the elongation $l$ and the temperature $T$.

Equation (2) can also be written as

$$\frac{dR}{R} = K_1\,d\sigma + K_2\,dT \tag{2'}$$

with

$$K_1 = 1 + 2\beta_g + \beta_\rho \qquad K_2 = \beta_g - \alpha_g$$

The change in resistance is due to two contributions, as can be clearly seen in eqns (2) and (2'). It is important to estimate their relative importance. For

FIG. 1.   Strain gage—schematic.

example, for Constantan (45% Ni, 55% Cu)

$$K_1 = 2 \cdot 1 \qquad K_2 = 3 \cdot 3 \times 10^{-6} /°C \qquad (3)$$

Thus, a mechanical deformation of $5 \times 10^{-6}$ (5 microstrain) causes the same change in resistance as a temperature change of $3°C$.

It is important to recognize at this stage that eqn. (2) is a statement about the strain gage wire itself. The situation in engineering practice is one where the wire is bonded on to a deforming structure, which in our case is a composite material. Generally there is a thermal mismatch between the composite and the gage because the thermal expansion of Constantan wire is $6 \times 10^{-6}/°C$, while the thermal expansion for graphite–epoxy (60 volume % of fibers) varies between $0 \cdot 1 \times 10^{-6}/°C$ in the fiber direction and $15 \times 10^{-6}/°C$ transverse to the fiber direction.

## 2.2.  The Uniaxial Gage Bonded on to a Unidirectional Reinforced Material

Suppose a strain gage is bonded on to a fiber-reinforced plastic as schematically shown in Fig. 2. The coefficient of thermal expansion along the fiber direction ($\alpha_L$) is much smaller than $\alpha_T$, which is measured transverse to the fiber direction. Equation (4) then gives the thermal expansion coefficients in the $x$ and $y$ directions:

$$\alpha_x = m^2\alpha_L + n^2\alpha_T \qquad (4a)$$

$$\alpha_y = n^2\alpha_L + m^2\alpha_T \qquad (4b)$$

$$\alpha_{xy} = 2(\alpha_L - \alpha_T)mn \qquad (4c)$$

with $m = \cos\theta_f$ and $n = \sin\theta_f$.

FIG. 2.    Strain gage and fiber orientation.

The coefficient of thermal expansion which is relevant for the strain gage bonded at an angle $\theta_g$ is given by eqn. (5):

$$\alpha_c = \alpha_x \cos^2 \theta_g + \alpha_y \sin^2 \theta_g + 2\alpha_{xy} \sin \theta_g \cos \theta_g \tag{5}$$

For the special case of an isotropic material $\alpha_x = \alpha_y = \alpha$ and $\alpha_{xy} = 0$. Thus

$$\alpha_c = \alpha \tag{6}$$

We subsequently consider the situation where the strain gage is bonded on to the composite material. The characteristics of the composite are its modulus in the direction in which the strain gage is bonded, given by $E_c$, and its coefficient of thermal expansion $\alpha_c$ as given in eqn. (5).

When the hypothesis is made of a perfect adhesion between the strain gage and the composite, we may write

$$\left|\begin{matrix}\text{mechanical} \\ \text{deformation}\end{matrix} + \begin{matrix}\text{thermal} \\ \text{deformation}\end{matrix}\right|_{\substack{\text{strain} \\ \text{gage}}} = \left|\begin{matrix}\text{mechanical} \\ \text{deformation}\end{matrix} + \begin{matrix}\text{thermal} \\ \text{deformation}\end{matrix}\right|_{\text{composite}}$$

or

$$\underbrace{\frac{\Delta \sigma_g}{E_g} + \alpha_g \Delta T}_{\varepsilon_g} = \underbrace{\frac{\Delta \sigma_c}{E_c} + \alpha_c \Delta T}_{\varepsilon_c} \tag{7}$$

The Δs in eqn. (7) represent the variations with respect to the 'stress-free' state which existed at the moment of adhesion between gage and composite. Upon substitution of eqn. (7) into eqn. (2′), we obtain

$$\frac{\Delta R}{R} = K_1 \varepsilon_c + (K_1(\alpha_c - \alpha_g) + K_2)\Delta T \tag{8}$$

The effects of strain and temperature are not fully separated in eqn. (8) because $K_1$ turns out to be a function, however weak, of temperature. It is measured in an experimental set-up.

Substitution of eqn. (5) into eqn. (8) gives

$$\frac{\Delta R}{R} + K_1 \varepsilon_c + [K_1(\alpha_x \cos^2 \theta_g + \alpha_y \sin^2 \theta_g + 2\alpha_{xy} \sin \theta_g \sin \theta_g - \alpha_g + K_2]\Delta T \tag{9}$$

There are thus three ways to assure thermal compensation:

(1) $$\Delta T = 0$$

This is the trivial case, which is more of academic value. One area of practical interest, though, is measurements on shock wave transients or very fast dynamic strains.

(2) $$K_1(\alpha_x \cos^2 \theta_g + \alpha_y \sin^2 \theta_g + 2\alpha_{xy} \sin \theta_g - \alpha_g) + K_2 = 0 \tag{10}$$

It is sufficient to solve eqn. (10) for $\theta_g$. A strain gage mounted along this angle will be completely compensated. This does not bring us any further, though, because the state of strain itself is a function of $\theta_g$ due to the anisotropic nature of the material.

(3) The technique of the 'dummy gage'

The dummy gage is only subject to thermal variations, thus

$$\left(\frac{\Delta R}{R}\right)_{dummy} = [K_1(\alpha_c - \alpha_g) + K_2]\Delta T \tag{11}$$

A wheatstone circuit wired in half bridge enables us to separate out the mechanical contribution

$$\left(\frac{\Delta R}{R}\right)_{active\ gage} - \left(\frac{\Delta R}{R}\right)_{dummy\ gage} = K_1 \varepsilon_c \tag{12}$$

The hypothesis which is implicit in the derivation of eqn. (12) is that the active gage and the dummy gage are both bonded to two perfectly identical composite structures. This hypothesis is violated as soon as there is a small gage misalignment of the dummy with respect to the active.

## 2.3. The Influence of Transverse Sensitivity

The strain gage, which is shown on the composite in Fig. 2, has a certain amount of wire in the direction transverse to the gage ($t$ direction). Consequently the response of the strain gage is not only influenced by the axial strain $\varepsilon_a$ but also by $\varepsilon_t$. (Note that a more consistent notation would be to use $\varepsilon_{tc}$. We will implicitly assume, however, that from now on the material is a composite unless we specify it otherwise. We thus drop the c index.) This statement can be mathematically written as

$$\frac{\Delta R}{R} = F_a \varepsilon_a + F_t \varepsilon_t \tag{13}$$

Equation (13) can also be written as

$$\frac{\Delta R}{R} = F_a(\varepsilon_a + K_{at}\varepsilon_t) \tag{14}$$

with $K_{at} = F_t/F_a$.

This equation can be written more explicitly to model the experiments in the gage calibration phase. We then know precisely what the biaxial strain field is, since the Poisson ratio of the calibration specimen $\nu_0$ is precisely known, thus

$$\frac{\Delta R}{R} = \underbrace{[F_a(1 - K_{at}\nu_0)]}_{K_1}\varepsilon_a \tag{15}$$

For all other situations in which we actually want to characterize the material we cannot measure the true strain immediately but we measure an apparent strain, which is given as

$$(\varepsilon_a)_{apparent} = \frac{\Delta R/R}{K_1} \tag{16}$$

By substituting the results from eqns (14) and (15), we obtain

$$(\varepsilon_a)_{apparent} = \frac{(\varepsilon_a + K_{at}\varepsilon_t)}{(1 - K_{at}\nu_0)} \tag{17}$$

Equation (17) contains two unknowns; consequently we need a second equation, which means that a second measurement has to be taken in the $t$ direction, as schematically indicated in Fig. 3. We thus obtain

$$(\varepsilon_t)_{apparent} = \frac{\varepsilon_t + K_{at}\varepsilon_a}{1 - K_{at}\nu_0} \tag{18}$$

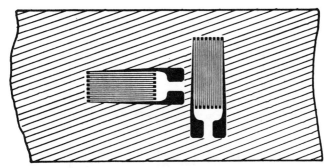

FIG. 3. Gage configuration in the axial and transverse directions.

Equations (17) and (18) can be used to determine the percentage error in the axial strain and in the transverse strain, respectively. The majority of material characterization tests are done on coupons which are loaded in the fiber direction and transverse to the fiber direction, respectively. The percentage error $E$ in the axial and transverse strain indications while loading in the fiber direction is expressed in eqns (19) and (20):

$$E_L(\varepsilon_a) = \frac{K_{at}(v_0 - v_{12})}{1 - K_{at}v_0} \times 100\% \qquad (19)$$

$$E_L(\varepsilon_t) = \frac{K_{at}(v_0 - 1/v_{12})}{1 - K_{at}v_0} \times 100\% \qquad (20)$$

The percentage error $E$ in the axial and transverse strain indications while loading transverse to the fiber direction is expressed in eqns (21) and (22):

$$E_T(\varepsilon_a) = \frac{K_{at}(v_0 - v_{21})}{1 - K_{at}v_0} \times 100\% \qquad (21)$$

$$E_T(\varepsilon_t) = \frac{K_{at}(v_0 - 1/v_{21})}{1 - K_{at}v_0} \times 100\% \qquad (22)$$

Equations (19) and (21) are represented in Fig. 4. The percentage error in the axial strain is plotted as a function of Poisson's ratio.

The values for $K_{at}$ which we choose to plot Fig. 5 are representative for the best strain gages which are available today. The conclusion which can be drawn is that the errors are only of the order of fractions of a per cent. The two shaded areas in the figure are representative for the error on $E_L(\varepsilon_a)$ and $E_T(\varepsilon_a)$, respectively.

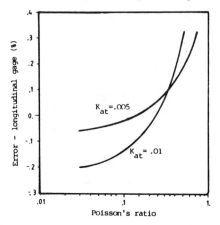

FIG. 4.   The percentage error in the longitudinal strain gage readings.

Figure 5 graphically represents eqns (20) and (22), where we choose the same values for $K_{at}$ as in Fig. 4. Again we marked two shaded areas, A and B. It should be emphasized that both scales in Fig. 5 are plotted logarithmically.

For the minor Poisson's ratio $v_{21}$, which lies in the range 0·015–0·03, we can assume the shaded error region A. The error apparently varies between 16 and 68% depending on the transverse sensitivity of the gage that is actually used and the specific composite material on which it is bonded.

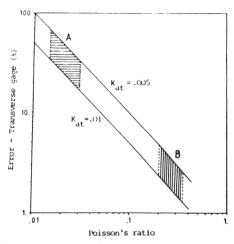

FIG. 5.   The percentage error in the transverse strain gage readings.

The major Poisson's ratio $v_{12}$ usually lies in the range 0·2–0·35. This enables us to define the shaded region B. The transverse error for this type of measurement varies between 1·3 and 4·7%.

In a practical materials testing-type situation, we do not actually know what are the values of the major and minor Poisson's ratios. The correction for transverse sensitivity can be made easily, however, when we solve eqns (17) and (18) for $\varepsilon_a$ and $\varepsilon_t$. We thus obtain

$$\varepsilon_a = \frac{(1 - v_0 K_{at})(\varepsilon_{am} - K_{at}\varepsilon_{tm})}{1 - K_{at}^2} \tag{23}$$

$$\varepsilon_t = \frac{(1 - v_0 K_{at})(\varepsilon_{tm} - K_{at}\varepsilon_{am})}{1 - K_{at}^2} \tag{24}$$

## 3.  THE NON-COINCIDENCE OF PRINCIPAL STRESS AND PRINCIPAL STRAIN DIRECTIONS

We now touch an area in which composites are fundamentally different from isotropic materials.

When a tensile coupon of an isotropic material is loaded uniaxially, we get a strain as schematically shown in Fig. 6.

The type of polar diagram which is shown in Fig. 6 is useful because it enables the experimentalist to get a feel for the influence of strain gage misalignments. We can clearly see from Fig. 6 that a gage misalignment $\beta$ of a few degrees will lead to a slight underestimation of the strain.

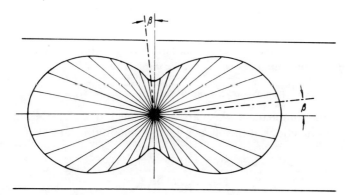

FIG. 6.    Polar diagram for directionally dependent strain induced into an isotropic material.

We can plot similar polar diagrams for unidirectional reinforced composite materials. We have an extra variable now, which is the off-axis angle, as we schematically indicated in Figs 7a, b, c.

Figure 7b indicates that the polar diagram is now rotated over an angle $\theta$ with respect to the direction in which the load is applied. The angle $\theta$ thus measures the deviation between the principal stress and principal strain directions. The gage misalignment ($\beta$) now causes considerable errors in the strain readings.

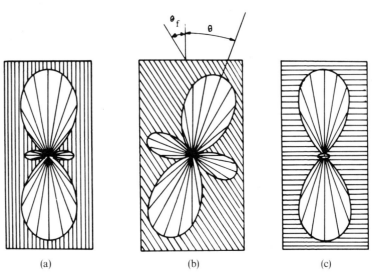

(a)                    (b)                    (c)

FIG. 7.   Polar diagrams for directionally dependent strain induced into a unidirectional composite at various off-axis angles.

We will now derive the dependence of the angle $\theta$ on the off-axis angle $\theta_f$. In the case of a uniaxial stress $\sigma_x$, we can write

$$\varepsilon_x = \overline{S_{11}}\,\sigma_x \qquad \varepsilon_y = \overline{S_{12}}\,\sigma_x \qquad \gamma_{xy} = \overline{S_{16}}\,\sigma_x \qquad (25)$$

with $S_{ij}$ the compliances which are transformed to $x$–$y$ axes. They can be written as functions of the off-axis angle $\theta_f$ and four invariants $U_1, \ldots, U_4$:

$$\overline{S_{11}} = U_1 + U_2 \cos 2\theta_f + U_3 \cos 4\theta_f$$
$$\overline{S_{12}} = U_1 - U_3 \cos 4\theta_f \qquad (26)$$
$$\overline{S_{16}} = U_2 \sin 2\theta_f + 2U_3 \sin 4\theta_f$$

with

$$U_1 = \tfrac{1}{8}(3S_{11} + 3S_{22} + 2S_{12} + S_{66})$$
$$U_2 = \tfrac{1}{2}(S_{11} - S_{22})$$
$$U_3 = \tfrac{1}{8}(S_{11} + S_{22} - 2S_{12} - S_{66}) \tag{27}$$
$$U_4 = \tfrac{1}{8}(S_{11} + S_{22} + 6S_{12} - S_{66})$$

We can find the principal stress directions ($p$–$q$ in Fig. 7b) when we transform the strain vector $(\varepsilon_x, \varepsilon_y, \gamma_{xy})$ into $(\varepsilon_p, \varepsilon_q, \gamma_{pq})$ and impose the requirement that $\varepsilon_{pq}$ has to be equal to zero. Thus

$$\begin{Bmatrix} \varepsilon_p \\ \varepsilon_q \\ \gamma_{pq} \end{Bmatrix} = \begin{vmatrix} m^2 & n^2 & mm \\ n^2 & m^2 & -mn \\ -2mn & 2mn & m^2 - n^2 \end{vmatrix} \begin{Bmatrix} \varepsilon_x \\ \varepsilon_y \\ \gamma_{xy} \end{Bmatrix} \tag{28}$$

with $m = \cos\theta$, $n = \sin\theta$.

The requirement $\gamma_{pq} = 0$ leads to a condition on the angle $\theta$:

$$\mathrm{tg}\, 2\theta = \frac{\gamma_{xy}}{\varepsilon_x - \varepsilon_y} = \frac{S_{16}}{S_{11} - S_{12}} \tag{29}$$

Upon substitution of eqn. (25) into eqn. (29) and taking into account eqn. (26), we obtain

$$\mathrm{tg}\, 2\theta = \frac{U_2 \sin 2\theta_f + 2U_3 \sin 4\theta_f}{(U_1 - U_4) + U_2 \cos 2\theta_f + 2U_3 \cos 4\theta_f} \tag{30}$$

Equation (30) indicates that the principal stress and the principal strain only coincide for $\mathrm{tg}\, 2\theta = 0$. This condition is always satisfied for $\theta_f = 0$ or $\theta_f = \pi/2$.

We can also substitute into eqn. (29) the transformed compliances as a function of the in-plane engineering properties $(E_1, E_2, \nu_{12}, G_{12})$. The result is written as

$$\mathrm{tg}\, 2\theta = \frac{\left[ \left( \dfrac{1 + \nu_{12}}{E_1} - \dfrac{1}{2G_{12}} \right) \cos^2\theta_f + \left( \dfrac{1}{E_2} + \dfrac{12}{E_1} - \dfrac{1}{2G_{12}} \right) \sin^2\theta_f \right] \sin 2\theta_f}{\left( 1 + \dfrac{\nu_{12}}{E_1} \right) \cos^4\theta_g + \left( \dfrac{1}{E_2} + \dfrac{12}{E_1} \right) \sin^4\theta_f} \tag{30'}$$
$$+ \left[ \frac{2}{G_{12}} - \frac{1}{E_2} - \left( \frac{1 + 2\nu_{12}}{E_1} \right) \right] \sin^2\theta \cos^2\theta$$

For some materials there exists an intermediate off-axis angle $\theta_f^*$, in between $\theta$ and $\pi/2$, for which the principal stress and strain directions coincide. This happens when the denominators of eqns (30) and (30′) are equal to zero. Thus

$$U_2 \sin 2\theta_f^* + 2U_3 \sin 4\theta_f = 0 \tag{31}$$

and also

$$\text{tg}^2 \, \theta_f^* = \frac{\left(\dfrac{1+v_{12}}{E_1} - \dfrac{1}{2G_{12}}\right)}{\left(\dfrac{1}{E_2} + \dfrac{v_{12}}{E_1} - \dfrac{1}{2G_{12}}\right)} \tag{31′}$$

Figure 8 indicates that only graphite–epoxy has angle $\theta_f^*$ at about $60°$, while the mechanical properties for the two other materials are such that the right-hand side of eqn. (31′) becomes negative.

Thus $\theta_f^*$ only exists when the numerator and denominator are both positive:

$$\frac{1+v_{12}}{E_1} - \frac{1}{G_{12}} > 0 \quad \text{or} \quad G_{12} > \frac{E_1}{2(1+v_{12})} \tag{32}$$

and

$$\frac{1}{E_2} + \frac{12}{E_1} - \frac{1}{2G_{12}} > 0 \quad \text{or} \quad G_{12} > \frac{E_1}{2\left(v_{12} + \dfrac{E_1}{E_2}\right)} \tag{33}$$

$\theta_f^*$ also exists when the numerator and denominator are both negative. Thus

$$G_{12} < \frac{E_1}{2(1+v_{12})} \tag{34}$$

and

$$G_{12} < \frac{E_1}{2\left(v_{12} + \dfrac{E_1}{E_2}\right)} \tag{35}$$

It is interesting to note that $\theta_f^*$ can only exist when the material is such that $E_x$ is greater than both $E_1$ and $E_2$ (which is implicitly required by eqn. (32)) or $E_x$ is less than both $E_1$ and $E_2$ as required by eqn. (35).

These conditions are graphically represented in Fig. 8.

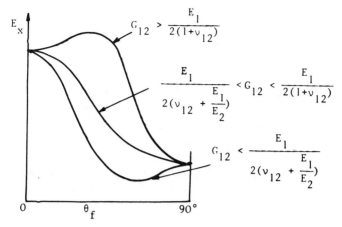

FIG. 8. Graphical representation of the conditions for which $\theta_f^*$ exists.

The angle $\theta$ has been plotted in Fig. 9 (left) as a function of the off-axis angle $\theta_f$ for a graphite–epoxy material system. It is interesting to note that for these composites there is a maximum difference for an off-axis angle of about 10°, while this peak shifts to 20° in the case of glassfiber epoxy.

Another interesting observation is that the graphite–epoxy material when tested at an off-axis angle between 60° and 70° has again principal stress and principal strain directions which coincide.

Figure 9 (right) corresponds to Fig. 9 (left) and gives the percentage error in the strain output of a gage which is misaligned over an angle $\theta$ with respect to the principal stress direction. The figures clearly indicate that the maximum error shows up at the off-axis angle where the deviation between principal stress and principal strain directions is extreme.

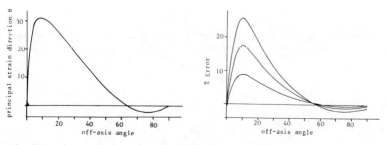

FIG. 9. Dependence of principal strain direction and misalignment error on the off-axis angle.

## 4.  CONCLUSIONS

For every material which has directionally dependent coefficients of thermal expansion there exists a direction for which the strain gage is totally compensated. Corrections for transverse sensitivity should be done routinely. Especially, the errors on the transverse strain measurements should be considered. We get the largest error when we attempt to measure the minor Poisson ratio of a composite. The correction procedures remain important for strain measurements on laminates.

We have to consider the possibility of important errors due to misalignment and the non-coincidence of principal stress and principal strain directions.

### ACKNOWLEDGEMENTS

The author acknowledges the support of the Free University of Brussels and of NASA–Ames Research Center (Dr H. G. Nelson through the Virginia Tech. Summer Faculty Program); Linda Bell and Kevin Hawkins assisted in generating some of the figures.

Special thanks are due to Myriam Bourlau at the Free University of Brussels for her excellent typing.

### REFERENCES

1. PERRY, C. C., The resistance strain gage revisited, *Journal of the Society for Experimental Mechanics*, **24**, no. 4 (1984), 286–299.
2. TUTTLE, M. E. and BRINSON, H. F., Resistance foil strain-gage technology as applied to composite materials, *Journal of the Society of Experimental Mechanics*, **24**, no. 1 (1984), 34–65.
3. THOMPSON, W. (Lord Kelvin), On the electrodynamic qualities of metals, *Proceedings Royal Society*, 1856.

# 22

# Shear Modulus Testing of Composites

JOHN SUMMERSCALES

*Structural Mechanics Section, Royal Naval Engineering College, Manadon, Plymouth PL5 3AQ, UK*

## ABSTRACT

*Fibre reinforced composite materials are generally anisotropic and heterogeneous, and hence possess mechanical properties which are dependent on the structure of the material and on the orientation of the load to the reinforcement. The shear modulus of these materials is usually different in each of the three mutually orthogonal planes and varies within each plane according to the definition of the axes. Reliable and satisfactory methods for the experimental determination of the shear properties of such materials are difficult to develop, because of the requirement for equal and opposite stresses along the four edges of a square planar section. This paper will review the methods used for the determination of the shear moduli of orthotropic materials.*

## 1. INTRODUCTION

The determination of the elastic constants of a composite material is significantly more complicated than the determination of these constants for an isotropic material. The heterogeneous and anisotropic nature of fibre reinforced materials is generally reducible to orthogonal symmetry and hence the material can be defined by three Young's moduli $(E_i, E_j, E_k)$, three Poisson's ratios $(v_{ij}, v_{jk}, v_{ki})$ and three shear moduli $(G_{ij}, G_{jk}, G_{ki})$, where subscripts $i, j, k$ are usually $1, 2, 3$ respectively or occasionally $x, y, z$. Normally the 1- or $x$-direction is the principal fibre direction and the 3- or $z$-direction is the plate thickness. There is thus one in-plane shear modulus $(G_{12})$ and two through-plane shear moduli $(G_{13}$ and $G_{23})$.

A wide variety of experimental methods for the determination of the shear moduli of orthotropic materials have been proposed. In the review which follows the techniques have been divided into seven groups according to the loading method (rails, frames, uniaxial in-plane load, uniaxial flexure, biaxial flexure, torsion, rings). Acoustic, vibration and ultrasonic methods have been specifically excluded.

## 2. TEST METHODS

### 2.1. Rail Tests

The *single-rail* shear test method and specimen configuration were outlined by Floeter and Boller.[1,2] A square or rectangular plate was bolted between two parallel rails with the remaining edges free. Stresses were transmitted to the specimen by a relative displacement of one rail parallel to the other. The existence of free edges causes the stress in the laminate to deviate from pure shear and the narrow gauge section violates St Venant's principle.

Whitney *et al.*[3] presented a detailed theoretical and experimental analysis of the single-rail test. The study validated the use of the test for laminated composites. A uniform shear stress is obtained at a short distance from the free edges if the length-to-width ratio is greater than 10 and if the laminate does not have a high effective Poisson's ratio. Garcia *et al.*[4] used a finite element method to tailor the specimen aspect ratio to a particular test material.

The *symmetric double-rail* shear fixture was proposed by Sims[5] for $0/90_s$ laminates, who found good agreement between double-rail and $\pm 45°$ tension tests shear stress–strain response. The method has an irregular shear zone near the free edges although this does not significantly influence the results when the length-to-width ratio is greater than 10.

Purslow[6] adapted the double-rail technique for testing 10-mm cubes in all three orthogonal planes. The *balanced-rail* shear test was used for both unidirectional CFRP and resin blocks of the same size. Hartshorn *et al.*[7] used the balanced-rail test for GRP samples and reported shear moduli of 1·9 GPa (through-plane) and 3·5 GPa (in-plane).

## 2.2. Frame Tests

The *picture frame* square plate shear test was developed for the determination of the shear modulus of wood and plywood. A square plate with the corners removed is fastened to a clamping frame with pin-jointed corners, and load is applied along one diagonal of the square. Strains parallel and perpendicular to the loading axis are measured to determine the shear modulus.

Farley and Baker[8] used finite element analysis on thin-gauge composite material in the picture frame rig for both square and rectangular tests. Corner stress problems were significantly affected by the location of the corner pins. The distance from the corner fixture pins to the load transfer bolts, and the relative stiffness of the test panel and of the loading tabs, were secondary factors.

ASTM method D805-52 panel shear test[9] was developed by the US Forest Products Laboratory as a standard for the shear strength of veneer, plywood and other glued constructions. The specimen is in the form of a broad cross with the grain running parallel to the panel edges. An almost uniform strain distribution can be obtained within the elastic range. A similar rig for the determination of the in-plane shear stress of plywood is described in British Standard BS 4512 Clause 7.[10]

Bryan[11] used photoelastic procedures to evaluate the shear stress state in the panel shear test for homogeneous isotropic material. The isostatic lines were approximately aligned vertical or horizontal with slight irregularities suggesting pure shear. The isochromatic fringe lines suggested that the shear stress at the edge of the test area was double that at the centre of the test area. The shear stress distribution at the boundary of the actual test area was quite uniform.

The *linked pantograph* technique[12] is also commonly known as the North American in-plane shear test after the aviation company which developed it. The test is essentially the picture frame method with diagonal slots cut in the specimen to isolate the test area. The load is applied along the outer edges and delivered through linkages that are centred on the edges of the test area. The complexity of the test fixture has prevented wide application of the method.

## 2.3. Uniaxial In-plane Loading

Greszczuk[13] proposed that the five principal elastic properties $(2E, 2v, 1G)$ could be obtained from test data on one (or at most two) appropriately instrumented tensile tests by solution of the orthotropic transformation equations. One (theoretical or experimental) Poisson's ratio must be known if just one test is to be used. The shear modulus is obtained from an *off-axis tension* test on a specimen oriented at any angle $(0 < \alpha < 90)$.

Daniel and Liber[14] proposed that $10°$ was the optimum orientation of the off-axis test for intralaminar shear characterisation. Chamis and Sinclair[15] performed detailed theoretical and experimental studies to assess the suitability of this shear test for composites, and recommended the procedure for consideration as a standard test specimen. Eleven advantages and five disadvantages were listed in their paper.

Cron[16] simulated the off-axis test using a finite element method. An almost uniform state of stress could be produced with ideal tab clamping and rotation. The tab should only be clamped for a portion of its area and the pin should be located at least 13 mm behind the clamp edge.

Gulley and Summerscales[17] conducted *separate* $+45°$ *and* $-45°$ *tests* on both unidirectional and woven $0°/90°$ laminates. The shear modulus was calculated using Huber's equation

$$G_{12} = (0{\cdot}5\sqrt{E_{+45}E_{-45}})/(1 + \sqrt{v_{+45}v_{-45}})$$

This calculated shear modulus was used in the calculation of Poisson's ratio from classical laminate theory. Good agreement between theoretical and experimentally measured Poisson's ratios was found when the reference axes were carefully defined.

Petit[18] suggested the use of a $[\pm 45°]_s$ *laminate* in uniaxial tension to obtain the shear stress–strain response. The load, axial and transverse strains, and incremental lamination theory, were used in the analysis. Rosen[19] simplified the Petit analysis by noting that the shear stress is generally half of the applied tensile stress. Shear modulus can then be derived, albeit with certain inherent approximations. The procedure has been adopted by ANSI/ASTM for unidirectional high modulus composites[20] and more recently by CRAG.[21]

Duggan *et al.*[22] proposed a *slotted tension* coupon, which was additionally loaded in transverse compression by a pin-guided assembly containing a hydraulic ram, load cell and reaction frame. A proportional servo-control system applied a transverse force determined from the relative cross sections in each direction and the applied tensile load. The slots were aligned with the tensile loading axis and ran from the edge of the compression loaded region towards the grips. The technique described had no restraints to prevent the imposition of flexural stresses, and is unlikely to find wide applicability because of the cost of the servo-hydraulic system.

Zapel[23] developed the *tensile double shear* technique in which a notched isotropic plate was loaded in tension through pin-joints. The experimental shear moduli were in reasonable agreement with values from elastic theory, but the test was not specifically evaluated for orthotropic materials. Clapper[24] suggested that the size, shape and centring of the notches, the planarity of the sample and the speed of testing are important variables. The relatively high cost of specimen fabrication is likely to preclude extensive use of the test.

Arcan *et al.*[25,26] developed a *double V-notch* fibre-reinforced specimen in an exterior circular fixture for both shear and other 2D stress state tests. The test has been used for composite shear testing at the University of Delaware and at General Dynamics.

Adams and Walrath developed the *Iosipescu* shear test for use with carbon–carbon composites, SMC materials,[27] graphite/epoxy and most recently polyethylene fibre-reinforced epoxy.[28] The test essentially uses axial forces to load a constrained asymmetrical four-point bend notched beam, using individual fixtures on each half of the specimen. A short history and review of the application of the test, with comparisons to the double V-notch and AFPB tests, has been published.[29,30] Finite element analysis was used to establish guidelines for an optimised specimen geometry to minimise stress distribution irregularities in highly orthotropic materials.[31]

Sridhar *et al.*[32] identified acoustic emission amplitude distributions which were characteristic of interlaminar failures in Iosipescu shear testing of glass fibre epoxy composites. The method also allowed a distinction to be made between interlaminar failure in 0/0° and 0/90° composites.

Slepetz *et al.*[33] introduced the *asymmetrical four-point bending* (AFPB) test, essentially the Iosipescu specimen loaded by pairs of rollers contacting the top and bottom surfaces respectively, instead of grips at each end. Specimen notch geometry and stress uniformity were studied by finite element analysis, moiré fringe interferometry and strain gauges.

Sullivan *et al.*[34] compared the Walrath and Adams (W&A) and AFPB fixtures using strain gauge data, photoelasticity and finite element analysis. Abdallah *et al.*[35] compared the AFPB, W&A, and modified Walrath and Adams loading fixtures. Both studies showed that the AFPB fixture produced the purest shear-stress field in the gauge section. Abdallah also showed that the area of constant shearing strain was only 1·5 mm in the original Iosipescu specimen.

Spigel *et al.*[36] used finite element analysis to show that changes of the load locations and notch geometry could lead to unexpected changes in the failure mode of the AFPB test. Prabhakaran and Sawyer[37] used model orthotropic birefringent materials to investigate the shear stress fringe values of unidirectional fibreglass in the AFPB test. For sharp or radiused 90° or 120° notches, reasonable agreement with off-axis tension specimens was found.

### 2.4. Uniaxial Flexure

The shear modulus of a composite material can be established by the four-point pure bending method using a $\pm 45°$ *laminate in flexure*.[38]

*Three-point bend, built-in beam* or *cantilever beam* methods are suitable for the simultaneous determination of both Young's and shear moduli.[39] The methods are characterised by a loading configuration which simultaneously induces flexure and shear deflections. Two different loading-span:specimen-depth ratios, or a single specimen in two loading modes, were tested followed by the solution of two equations each with $E$ and $G$ unknown.

### 2.5. Biaxial Flexure

Nadai[40] suggested that the elastic constants could be determined by the application of a pure bending and twisting moment to a plate. Bergstrasser[41] used the *twisting of a rectangular plate* to determine the shear modulus. Xing Hua[38] presented two equations for the calculation of this shear modulus of composites. A similar method for the determination of the shear modulus of laminated woods is described in ASTM D805-52.[9]

Tsai[42] described a method of using one beam flexure (0° or 90°) and two plate twist (0° and 45°) specimens to obtain all the elastic constants of an orthotropic plate when the principal axes are known. An independent check can be obtained by using both bending tests. Beckett *et al.*[43,44] highlighted the need to ensure that the plate was in *anticlastic bending*. The deflection at the loading corner of the plate then allows a direct measurement of the shear modulus, which may be used independently of the full elastic characterisation.

Whitney[45] used classical laminated plate theory to show that the method is only valid for truly orthotropic materials and 0/90° equivalent laminates. Angle-ply $\pm 45°$ plates were shown to be in serious error. Foye[46] noted that the equations were derived from linear small deflection theory and reported that a point of instability was reached when the corner deflection exceeded several plate thicknesses. A major disadvantage of the method is the requirement for very accurate measurements of the plate thickness, which is cubed in the equation.

Cargen *et al.*[47] used anticlastic bending to monitor the degradation of marine GRP after sea water exposure. A 9% decrease in shear modulus was reported after 100 days immersion, no detectable change of opacity on the defect map and only 0·34% panel weight gain.

Shockey and Waddoups[48] originally described the $\pm 45°$ *cross sandwich beam*. Composite faces were laminated on to a core, with a core elastic modulus two orders of magnitude lower than that of the composite. A cross-shaped configuration was subjected to anticlastic bending, to produce pure shear stress in the membrane at 45° to the cross axis at the very centre of the laminate. Waddoups[49] found that the cross-sandwich beam had a membrane shear strength far in excess of that indicated by the short beam shear specimen. The test is no longer regarded as adequate for shear modulus determination.

## 2.6. Torsion

*Torsion of a thin-walled composite tube*[50-52] is a popular method of shear modulus determination. If the ratio of the tube radius:wall thickness is large and one tube end is free to rotate and to move axially, then a reasonably pure state of stress may exist. Rizzo and Vicario[53] have shown that end attachments may lead to serious stress concentrations. The method is appropriate to applications such as driveshafts, but the expense of fabrication dictates against more general use.

*Torsion of solid circular cross section* rods or discs[54] with unidirectional fibres running parallel to the cylinder axis is an acceptable method for the determination of the shear modulus. The analysis is identical to that for an isotropic cylinder in torsion.

Pagano[55] used an elasticity solution for pure torsion of an orthotropic cylinder to define an experimental technique to determine the shear moduli. Transversely oriented unidirectional reinforcement is used to determine both $G_{12}$ and $G_{23}$ (longitudinal and transverse moduli respectively). Knight[56] proposed that six compliance values could be obtained experimentally from three tests (two orthogonal tension tests and torsion of a solid rod).

## 2.7. Rings

Greszczuk[57,58] proposed that shear modulus of composites could be determined by *out-of-plane four-point twist* of circular rings loaded alternately at 90° intervals. The deflection of the ring is a function of both the flexural and shear moduli. The flexural modulus of the ring was obtained by diametrically opposite uniaxial loading in-plane.

The *Douglas ring*, or split-ring, test involves subjecting the ring to two out-of-plane forces (equal in magnitude but opposite in direction) adjacent to the split.[59] The deflection is predominantly due to shear deformation and hence allows shear modulus determination. As the ratio of the moduli $(E/G)$ increases, the torsional component assumes greater importance, and hence the test is most suitable for high modulus composites. Young's modulus can be obtained by tensile loading in-plane at $\pm 90°$ from the split.

Cheng[60] derived new formulae for the determination of the elastic moduli from the normal deflections measured in both the four-point ring twist and the out-of-plane split ring test. It was noted that flexural moduli in the split ring (radial ring axis) and continuous ring (axis normal to ring-plane) may differ.

Wagner *et al.*[39] used *in-plane tension of a circular thin ring* to induce simultaneous flexure and shear deformation. Using the assumptions in Section 2.4 allows determination of both Young's modulus and shear modulus.

## 3.  CONCLUSIONS

The requirements of an ideal method for the determination of the shear modulus of a composite material are that:

— the specimen be in pure shear, free from excessive edge restraints and stress concentrations;
— reproducible results are obtained;
— a shear stress–strain curve can be plotted;
— the test can be run on a universal testing machine;
— no special equipment be required for specimen preparation;
— the data reduction procedure be simple;
— the test be inexpensive;
— no out-of-plane distortion occurs unless required by the method;
— the test be easily adapted for fatigue and environmental testing.

No single technique would appear to be ideal in all of these respects. It is unlikely that any test will have universal applicability. Double-rail shear and off-axis tension are probably the most popular methods for the testing of flat-plate unidirectional composites. The Arcan/Iosipescu/AFPB variants have received considerable attention in recent years. The $\pm 45°$ laminate in tension has been almost universally accepted for flat-plate bidirectional laminates, although the applicability for other laminate orientations must be resolved. Torsion of a thin-walled tube will be most appropriate when the application uses the composite in tubular form. Lee and Munro[61] used decision analysis to evaluate nine in-plane shear tests:

45° tension and Iosipescu methods gained the highest score, followed by 10° off-axis testing.

In view of the diversity of matrices and reinforcing fibres, of ply-stacking sequences and cure schedules, it would aid comparison of the relative merits of these materials if a limited number of precisely defined test methods could be internationally accepted.

## REFERENCES

1. FLOETER, L. H. and BOLLER, K. H., Use of experimental rails to evaluate edgewise shear properties of glass reinforced plastic laminates, US Forest Products Laboratory Handbook MIL-HDBK-17, 28 April 1967.
2. BOLLER, K. H., A method to measure interlaminar shear properties of composite laminates, AFML-TR-69-311, 1969.
3. WHITNEY, J. M., STANSBERGER, D. L. and HOWELL, H. B., Analysis of the rail shear test—applications and limitations, *J. comp. Mater.* (January 1971), **5**(1), 24–34.
4. GARCIA, R., WEISSHAAR, T. A. and McWITHEY, R. R., An experimental and analytical investigation of the rail shear-test method as applied to composite materials, *Exp. Mech.* (August 1980), **20**(8), 273–279.
5. SIMS, D. F., In-plane shear stress–strain response of unidirectional composite materials, *J. comp. Mater.* (January 1973), **7**(1), 124–128.
6. PURSLOW, D., The shear properties of unidirectional carbon fibre reinforced plastics and their experimental determination, Aeronautical Research Council Current Paper ARC-CP-1381, July 1976.
7. HARTSHORN, R. T., SMITH, M. D. and SUMMERSCALES, J., Determination of through-plane shear moduli for glass reinforced plastic, *Int. Conf. Testing, Evaluation and Quality Control of Composites*, Guildford, September 1983, poster paper. NTIS PB87-106274.
8. FARLEY, G. L. and BAKER, D. J., In-plane shear test of thin panels, *Exp. Mech.* (March 1983), **23**(1), 81–88.
9. Standard methods of testing veneer, plywood, and other glued constructions, American Society for Testing and Materials Designation D805-52, 1952.
10. Methods of test for clear plywood: panel shear, British Standard BS4512, Clause 7, 1969.
11. BRYAN, E. L., Photoelastic evaluation of the panel shear test for plywood, ASTM-STP-289, 1961, pp. 90–95.
12. DICKERSON, E. O. and DiMARTINO, B., Off-axis strength and testing of filamentary materials for aircraft application, *10th National Symposium, SAMPE*, San Diego, November 1966, pp. H23–H50.
13. GRESZCZUK, L. B., New test technique for shear modulus and other elastic constants of filamentary composites, *Symposium on Test Methods, ASTM*, Dayton, 21–23 September 1966, Paper 3, pp. 95–123, AD 801547.
14. DANIEL, I. M. and LIBER, T., Lamination residual stresses in fibre composites, NASA-CR-134826, March 1975, N75-30264.
15. CHAMIS, C. C. and SINCLAIR, J. H., 10° off-axis tensile test for intralaminar shear

characterisation of fibre composites, NASA-TN-D-8215, April 1976, N76-22314.

16. CRON, S. M., Improvement of end boundary conditions for off-axis tension specimen use, Masters thesis, US Air Force Inst. Tech., December 1985, AD A164 321.

17. GULLEY, T. J. and SUMMERSCALES, J., Poisson's ratios in glass fibre reinforced plastics, *Proc. 15th Reinforced Plastics Congress*, BPF, Nottingham, September 1986, pp. 185–189.

18. PETIT, P. H., A simplified method of obtaining the in-plane shear stress–strain response of unidirectional composites, ASTM-STP-460, 1969, pp. 83–93.

19. ROSEN, B. W., A simple procedure for the experimental determination of the longitudinal shear modulus of unidirectional composites, *J. comp. Mater.* (October 1972), **6**, 552–554.

20. Standard recommended practice for in-plane shear stress–strain response of unidirectional reinforced plastics, American National Standard ANSI/ASTM D3518-76, 1976.

21. CURTIS, P. T., CRAG test methods for the measurement of the engineering properties of fibre reinforced plastics, RAE-TR-85099, November 1985.

22. DUGGAN, M. F., McGRATH, J. T. and MURPHY, M. A., Shear testing of composite materials by a simple combined loading technique, *19th Structures, Structural Dynamics and Materials Conference*, AIAA & ASME, Bethesda Md, April 1978, pp. 311–319.

23. ZAPEL, E. J., Design and development of a tensile-loaded shear specimen, ASTM-STP-289, 1961, pp. 26–34.

24. CLAPPER, R. B., Shear and torsion testing of solid materials—a critical discussion, ASTM-STP-289, 1961, pp. 111–120.

25. ARCAN, M., A new method for the analysis of the mechanical properties of composite materials, *3rd Int. Congress Experimental Mechanics*, Los Angeles, 1973.

26. ARCAN, M., HASHIN, Z. and VOLOSHIN, A., A method to produce uniform plane-stress states with applications to fibre reinforced materials, *Exp. Mech.* (April 1978), **18**(4), 141–146.

27. ADAMS, D. F. and WALRATH, D. E., Iosipescu shear properties of SMC composite materials, ASTM-STP-787, 1982, pp. 19–33.

28. ADAMS, D. F., ZIMMERMAN, R. S. and CHANG, H. W., Properties of a polymer-matrix composite incorporating Allied A-900 polyethylene fibre, *SAMPE J.* (September–October 1985), **21**(5), 44–48.

29. WALRATH, D. E. and ADAMS, D. F., The Iosipescu shear test as applied to composite materials, *Exp. Mech.* (March 1983), **23**(1), 105–110.

30. ARCAN, M., The Iosipescu shear test as applied to composite materials—discussion, *Exp. Mech.* (March 1984), **24**(1), 66–67.

31. WALRATH, D. E. and ADAMS, D. F., Analysis of the stress state in an Iosipescu shear test specimen, NASA-CR-176745, June 1983, N86-24755.

32. SRIDHAR, M. K., SUBRAMANIAM, I., AJAY, C. and SINGH, A. K., Acoustic emission monitoring of Iosipescu shear test on glass-fibre–epoxy composites, *J. Acoustic Emission* (April–September 1985), **4**(2/3), S174–S177.

33. SLEPETZ, J. M., ZAGAESKI, T. F. and NOVELLO, R. F., In-plane shear test for composite materials, AMMRC-TR-78-30, July 1978, AD A062 830.

34. SULLIVAN, J. L., KAO, B. G. and van OENE, H., Shear properties and a stress analysis obtained from vinyl-ester Iosipescu specimens, *Exp. Mech.* (September 1984), **24**(3), 223–232.

35. ABDALLAH, M. G., GARDINER, D. S. and GASCOIGNE, H. E., An evaluation of graphite/epoxy Iosipescu shear specimen testing methods with optical techniques, *Proc. Spring Conf. Experimental Mechanics*, SEM, Las Vegas, June 1985, pp. 833–843.

36. SPIGEL, B. S., SAWYER, J. W. and PRABHAKARAN, R., An investigation of the Iosipescu and asymmetrical four-point bend tests for composite materials, *Proc. Spring Conf. Experimental Mechanics*, SEM, Las Vegas, June 1985, pp. 35–44.

37. PRABHAKARAN, R. and SAWYER, W., Photoelastic investigation of asymmetric four-point bend shear test for composite materials, *Comp. Struct.* (1986), **5**(3), 217–231.

38. PING, XING HUA, Measuring in-plane shear properties of glass-fibre/epoxy laminates, *Composites Technology Review* (Spring 1980), **2**(2), 3–13.

39. WAGNER, H. D., MAROM, G. and ROMAN, I., Analysis of several loading methods for simultaneous determination of Young's and shear moduli in composites, *Fibre Sci. Technol.* (January 1982), **16**(1), 61–65.

40. NADAI, A., *Die elastischen Platten*, Berlin, Verlag J. Springer, 1925.

41. BERGSTRASSER, M., Bestimmung der beiden elastischen Konstanten von Plattenformigen Korpern, *Z. techn. Phys.* (1927), **8**, 355–359. NASA-TT-F-11347, October 1967. N67-40139.

42. TSAI, S. W., Experimental determination of the elastic behaviour of orthotropic plates, *J. Engng Ind.* (August 1965), **87**(3), 315–318.

43. BECKETT, R. E., DOHRMANN, R. J. and IVES, K. D., An experimental method for determining the elastic constants of orthogonally stiffened plates, *Proc. 1st International Conference on Experimental Mechanics*, SESA & ONR, New York, November 1961, Oxford, Pergamon Press, 1963, pp. 129–147.

44. BECKETT, R. E., DOHRMANN, R. J. and IVES, K. D., Discussion, *J. Engng Ind.* (August 1965), **87**(3), 317–318.

45. WHITNEY, J. M., Analytical and experimental methods in composite mechanics, *ASCE J. Structural Division* (January 1973), **99**(ST1), 113–129.

46. FOYE, R. L., Deflection limits on the plate twisting test, *J. comp. Mater.* (1967), **1**(2), 194–198.

47. CARGEN, M. R., HARTSHORN, R. T. and SUMMERSCALES, J., The effect of short-term continuous sea water exposure on the shear properties of a marine laminate, *Trans. I. Mar. E* (November 1985), **C97**, 131–134.

48. SHOCKEY, P. D. and WADDOUPS, M. E., Strength and modulus determination of composite materials with sandwich beam tests, American Ceramic Society Meeting, September 1966. General Dynamics Fort Worth Division Report FZM 4691.

49. WADDOUPS, M. E., Characterisation and design of composite materials, in: *Composite Materials Workshop* (Tsai, S. W., Halpin, J. C. and Pagano, N. J., Eds), Stamford Ct, Technomic Publishing, 1968, pp. 254–308.

50. ADAMS, D. F. and THOMAS, R. L., Test methods for the determination of unidirectional composite shear properties, *12th National Symposium, SAMPE*, Anaheim CA, October 1967, Paper AC-5.

51. WHITNEY, J. M., Experimental determination of shear modulus of laminated fibre reinforced composites, *Exp. Mech.* (October 1967), **7**(10), 447–448.
52. WHITNEY, J. M., PAGANO, N. J. and PIPES, R. B., Design and fabrication of tubular specimens for composite characterisation, ASTM-STP-497, 1972, pp. 52–67.
53. RIZZO, R. R. and VICARIO, A. A., A finite element analysis for stress distribution in gripped tubular specimens, ASTM-STP-497, 1972, pp. 68–88.
54. THOMAS, R. L., DONER, D. and ADAMS, D., Mechanical behaviour of fibre reinforced composite materials, AFML-TR-67-96, May 1967, N67-34189.
55. PAGANO, N. J., Shear moduli of orthotropic composites, AFML-TR-79-4164, March 1980, AD A084 975.
56. KNIGHT, M., Three dimensional elastic moduli of graphite–epoxy composites, *J. comp. Mater.* (March 1982), **16**(2), 153–159.
57. GRESZCZUK, L. B., Four-point ring twist test for determining the shear modulus of composites, *24th National Symposium, SAMPE,* San Francisco, May 1979, **24**(1), pp. 791–798.
58. GRESZCZUK, L. B., Application of four-point ring twist test for determining shear modulus of filamentary composites, ASTM-STP-734, 1981, pp. 21–33.
59. GRESZCZUK, L. B., Shear modulus determination of isotropic and composite materials, ASTM-STP-460, 1969, pp. 140–149.
60. CHENG, S., Test method for evaluation of shear modulus and modulus of elasticity of laminated anisotropic composite materials, *J. Testing and Evaluation* (September 1985), **13**(5), 387–389.
61. LEE, S. and MUNRO, M., Evaluation of in-plane shear test methods for advanced composite materials by the decision analysis technique, *Composites* (January 1986), **17**(1), 13–22.

# 23

# Anisotropic Material Identification Using Measured Resonant Frequencies of Rectangular Composite Plates

W. P. De Wilde and H. Sol

*Department of Structural Analysis, Free University of Brussels (VUB), Brussels, Belgium*

## ABSTRACT

*In this chapter a method is presented which determines the elastic properties of a composite material plate using experimentally measured resonant frequencies.*

*The measured frequencies are compared with computed resonant frequencies of a numerical parameter model of the test plate. The parameters in the model are the unknown elastic properties. Starting from an initial guess, the parameters of the numerical model are tuned until the computed resonant frequencies match the measured resonant frequencies. The technique used for the tuning operation is a Bayesian parameter estimation method based on the sensitivities of the resonant frequencies for parameter changes.*

## 1. INTRODUCTION

The appearance of powerful digital computers allows reliable analysis of structures with numerical methods. Among the numerical methods, the finite element method is undoubtedly the most successful.

A finite element analysis of an arbitrary structure mainly requires two kinds of information blocks: geometrical data (nodal point coordinates, element connectivities and boundary conditions) and constitutive properties of the materials used in the structure. Constitutive properties describe the relations between stresses and strains in the material.

Of these two kinds of information blocks, the constitutive material behaviour is the most difficult to establish. In most practical calculations, linear material behaviour is assumed. The main reasons for this simplification are:

— A lot of materials can be approximated in a satisfactory way by a linear model.
— Nonlinearities are difficult to model mathematically and to measure experimentally. In fact, it is already cumbersome to obtain a good estimate of the linear behaviour for someone not equipped with a million dollar laboratory.
— The computer execution time for a nonlinear analysis is several orders of magnitude higher than for a linear analysis.

In recent years some fibre-reinforced polymeric materials, so-called 'composites', have been used increasingly as structural materials. A reason for this success is that composite materials have high strength to weight and high stiffness to weight ratios which can significantly reduce the weight of a structure. Therefore, composite materials have been applied successfully in dynamically loaded structures (vehicles, sporting goods, fast-moving machinery parts…). Composite materials also offer an improved corrosion resistance and fatigue lifetime as compared with traditional structural materials (steel, aluminium…). Perhaps the most important feature of composite materials is that their mechanical properties can be 'tailored' to meet a specific design criterion.

The finite element analysis of laminated fibre-reinforced polymeric structures requires the knowledge of anisotropic elastic properties for each element. Because these material properties depend on the production methods, it is generally impossible to find these properties in tables or databases. Sometimes reasonable values can be found using micro-mechanical models or using the data given by composite materials suppliers, but generally the safest way to establish the properties is to measure them experimentally on test specimens.

In this chapter a method is described which is based on experimentally measured resonant frequencies from completely free rectangular plates. These values are compared with numerically computed resonant frequencies of a Rayleigh–Ritz model of the plate. The parameters in this model are the so-called plate rigidities $D_{ij}$. Starting from an initial guess, these plate rigidities are tuned in an iterative way until the resonant frequencies computed with the numerical model match the measured resonant frequencies. The parameters are updated in every iteration step using a Bayesian parameter estimation method based on the sensitivities of the resonant frequencies for plate rigidity variations.

In the case of isotropic or orthotropic material behaviour (e.g. unidirectional or bidirectional roofings), the starting values for $D_{ij}$ can be generated from the measured resonant frequencies with a direct identification technique. In the case of anisotropic material behaviour, the starting values must be supplied by the experimentalist.

## 2. EXPERIMENT

The test plates are hung on two thin elastic threads (Fig. 1). This configuration simulates completely free boundary conditions. The completely free boundary conditions have several advantages:

— They are easy to accomplish from an experimental point of view.
— The experiment can be brought into excellent agreement with the analytical 'idealized' boundary conditions.
— The resonant frequencies are sensitive for variations of all the plate rigidities.

Hitting the plate with an arbitrary object (e.g. a fingertip or a pencil) generates an electric signal in an accelerometer which is attached to a corner of the plate. This signal is fed into a spectrum analyser and allows the immediate inspection of the resonant frequencies.

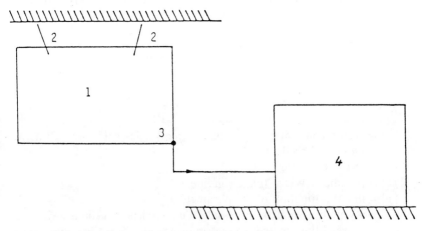

FIG. 1. Experimental set-up (1, test plate; 2, thin threads; 3, accelerometer; 4, spectrum analyser).

### 3.  RAYLEIGH–RITZ MODEL OF THE PLATE

Only thin plates subjected to small lateral deflections are considered (Love–Kirchhoff plate model).

The dynamic equilibrium equations of a freely vibrating plate can be written as an algebraic eigenvalue problem:

$$(K_{ij} - \lambda M_{ij})\{\Phi\} = \{0\} \tag{1}$$

with

$$K_{ij} = \frac{4b}{a^3} D_{11} A_{ij} + \frac{4a}{a^3} D_{22} B_{ij} + \frac{4}{ab} D_{12} C_{ij} + \frac{16}{ab} D_{66} E_{ij}$$

$$+ \frac{8}{a^2} D_{16} F_{ij} + \frac{8}{b^2} D_{26} G_{ij} \tag{2}$$

$$M_{ij} = \rho t \frac{ab}{4} H_{ij} \tag{3}$$

$\rho$ is specific mass; $t$ is plate thickness; and $A, B, \ldots, H$ are constant matrices containing partial derivatives of the shape functions used for the Rayleigh–Ritz approximation of the vibrational behaviour of the plate.

The numerical solution of (1) yields eigenvalues $\lambda^{(k)}$ and corresponding mode-shapes $\{\Phi\}^{(k)}$. The $k$th resonant frequency $f^{(k)}$ can be found from

$$\lambda^{(k)} = (2\pi f^{(k)})^2 \tag{4}$$

### 4.  ITERATIVE DETERMINATION OF THE PLATE RIGIDITIES USING A BAYESIAN PARAMETER ESTIMATION METHOD

In a Bayesian parameter estimation procedure, the discrepancy between the initial model predictions and the test data is resolved by minimizing a weighted error $E$:

$$E = {}^r(\{R\} - \{R^0\})[C_R](\{R\} - \{R^0\}) + {}^r(\{p\} - \{p^0\})[C_p](\{p\} - \{p^0\}) \tag{5}$$

with

$\{R\}$ = vector containing the resonant frequencies calculated with the numerical model;

$\{R^0\}$ = test data vector;

$\{p^0\}$ = parameter vector (plate rigidities);

$\{p\}$ = initial estimation of the parameters;

$[C_R]$ = a weighting matrix expressing the confidence in the test data;

$[C_p]$ = a weighting matrix expressing the confidence in the model parameters.

An updated value for the plate rigidities is found by

$$\{p\} = \{p^0\} + (2[C_p] + {}^t[S][C_R][S])^{-1}\{{}^t[S][C_R](\{R^0\} - \{R\})\} \quad (6)$$

with

$S_{ij}$ = sensitivity of the $i$th resonant frequency for variations of the $j$th plate rigidity;

$i = 1, \ldots, N$ ($N$ = number of measured resonant frequencies);

$j = 1, \ldots, M$ ($M$ = number of unknown plate rigidities)

($N \geq M$ to obtain a numerically well-conditioned sensitivity matrix).

Details about the choice of $[C_p]$ and $[C_R]$ can be found in Ref. 1.

## 5. STARTING VALUES FOR THE PLATE RIGIDITIES

In the case of isotropic or orthotropic material properties, the mode-shapes associated with the resonant frequencies have typical patterns (Fig. 2). Figure 2 shows 3D plots of some typical mode-shape types. For each mode-shape type, the eigenvalue can be expressed as an empirical function of the plate rigidities and the plate sizes:

First torsional mode-shape

$$\lambda_1 = k_1 \frac{D_{66}}{a^2 b^2 \rho t} + f_1\left(\frac{a}{b}, \frac{D_{22}}{D_{11}}, \frac{D_{12}}{D_{11}}, \frac{D_{66}}{D_{11}}\right)$$

First bending mode-shape (1-direction)

$$\lambda_2 = k_2 \frac{D_{11}}{a^2 \rho t} + f_2\left(\frac{a}{b}, \frac{D_{22}}{D_{11}}, \frac{D_{12}}{D_{11}}, \frac{D_{66}}{D_{11}}\right) \quad (7)$$

First bending mode-shape (2-direction)

$$\lambda_3 = k_3 \frac{D_{22}}{b^2 \rho t} + f_3\left(\frac{a}{b}, \frac{D_{22}}{D_{11}}, \frac{D_{12}}{D_{11}}, \frac{D_{66}}{D_{11}}\right)$$

$\vdots \qquad\qquad \vdots$

The correction functions $f_i$ are dependent on the aspect ratio $a/b$ and the ratios between the plate rigidities, $D_{11}$ taken as a reference. The functions $f_i$ are found using curve-fitting techniques.

The starting values for the plate rigidities are the least-squares solutions

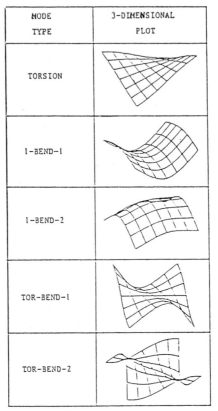

| MODE TYPE | 3-DIMENSIONAL PLOT |
|-----------|--------------------|
| TORSION | |
| 1-BEND-1 | |
| 1-BEND-2 | |
| TOR-BEND-1 | |
| TOR-BEND-2 | |

FIG. 2.   Typical mode-shapes pattern in the case of isotropic or orthotropic material behaviour.

of the set of eqns (7) for all possible mode-shape type sequences. (The number of possible mode-shape type sequences is limited.)

In the case of anisotropic material behaviour, the starting values must be supplied by the experimentalist (e.g. from laminate analysis).

## 6.   EXAMPLE

Ten layers of woven bidirectional glassfibre cloth ($0.230\,\text{kg/m}^2$) were impregnated in a polyvinyl ester resin and hot-pressed into a rectangular plate. The fill-directions of each layer coincide with the 1-direction of the chosen axes system, the warp directions with the 2-direction. The specific mass $\rho$ of the composite was measured as $1776 \, \text{kg/m}^3$. The dimensions and mass of the plate were:

length $= 0.350 \, \text{m}$
width $= 0.300 \, \text{m}$
mass $= 0.4438 \, \text{kg}$

The following resonant frequencies were measured:

$f_1 = 41.5 \, \text{Hz}$
$f_2 = 74.5 \, \text{Hz}$
$f_3 = 100.5 \, \text{Hz}$
$f_4 = 112.5 \, \text{Hz}$
$f_5 = 129.5 \, \text{Hz}$

After four iterations, the following results are obtained with the described procedure:

$D_{11} = 28.9 \, \text{N/m} \, (\pm 2.4\%)$
$D_{22} = 27.8 \, \text{N/m} \, (\pm 1.1\%)$
$D_{12} = 3.6 \, \text{N/m} \, (\pm 0.6\%)$
$D_{66} = 6.15 \, \text{N/m} \, (\pm 1.0\%)$

The relative errors on the final results can be calculated from the experimental errors on the frequency measurements, the sensitivities of the frequencies for plate rigidity variations and the confidence matrices in the Bayesian parameter estimation procedure (details can be found in Ref. 1).

The average thickness of the test plate was calculated from the total mass of the plate and the measured specific mass:

$$t = 0.4488/(0.35 \times 0.30 \times 1776) = 0.002\,41 \, \text{m}$$

The engineering constants $E_1$, $E_2$, $v_{12}$ and $G_{12}$ can be calculated from the expressions of orthotropic plate rigidities:

$$D_{11} = \frac{E_1 t^3}{12(1 - v_{12} v_{12})} \qquad D_{22} = \frac{E_2 t^3}{12(1 - v_{12} v_{12})} \tag{8}$$

$$D_{12} = v_{12} D_{22} \qquad D_{66} = \frac{G_{12} t^3}{12}$$

The results are shown in Table 1 and compared with results from statical tests on the same material.

TABLE 1

*Comparison of results from statical and vibrational tests*

|  | Statical tests | Vibrational tests |
|---|---|---|
| $E_1$ (N/m$^2$) | $(2 \cdot 5 \pm 0 \cdot 1)E + 10$ | $(2 \cdot 44 \pm 0 \cdot 06)E + 10$ |
| $E_2$ (N/m$^2$) | $(2 \cdot 4 \pm 0 \cdot 1)E + 10$ | $(2 \cdot 35 \pm 0 \cdot 03)E + 10$ |
| $v_{12}$ | $0 \cdot 13 \pm 0 \cdot 02$ | $0 \cdot 129 \pm 0 \cdot 002$ |
| $G_{12}$ (N/m$^2$) | $(5 \cdot 2 \pm 0 \cdot 5)E + 09$ | $(5 \cdot 27 \pm 0 \cdot 05)E + 09$ |

## 7. CONCLUSIONS

The presented method shows some disadvantages and some advantages:

*Disadvantages*

— linear material behaviour is assumed;
— no failure information is obtained;
— in the case of anisotropic material behaviour, the experimentalist must provide starting values for the plate rigidities.

*Advantages*

— nondestructive test;
— average results over the domain of the plate;
— no need to machine and prepare tensile specimen;
— simple test set-up;
— fast experiment: several elastic constants are found with one experiment; the measurement and the arrangement of the test plate only requires a few minutes;
— the method is capable of producing very accurate results and a good error estimation (see, for example, Table 1).

The state of the art of the method enables the correct and fast determination of orthotropic (and, of course, isotropic) material constants. Starting values can be generated automatically.

The results for anisotropic plates at this moment are poor. Anisotropic material identification also requires user-supplied starting values. Further investigation will hopefully help to overcome these problems.

## REFERENCE

1. SOL, H., Identification of anisotropic plate rigidities using free vibration data, Ph.D. dissertation, Free University of Brussels (VUB), 1986.

# 24

# Composite Materials for Use in Orthopaedic Applications: Fracture Behavior of Acrylic Bone Cement Reinforced with High Toughness Organic Fibers

H. D. WAGNER

*Department of Materials Research, The Weizmann Institute of Science, Rehovot 76100, Israel*

and

B. POURDEYHIMI

*Department of Textiles and Consumer Economics, University of Maryland, College Park, Maryland 20742, USA*

## ABSTRACT

*A study of the fracture behavior of poly(methyl methacrylate) bone cement reinforced with short ultra-high strength polyethylene fibers is presented. The flexural strength and modulus are apparently not improved by the incorporation of polyethylene fibers in the PMMA cement, probably because of the presence of voids, the poor mixing practice and the weakness of the fiber/matrix interfacial bond. Linear elastic and nonlinear elastic fracture mechanics techniques are used to assess the ultimate behavior of the unreinforced and reinforced cements. Both techniques yield similar trends for the fracture toughness, indicating that a significant reinforcing effect is obtained at fiber content as low as 1% by weight, but beyond that concentration a plateau value is reached and the fracture toughness becomes insensitive to fiber content. The present polyethylene/PMMA composite presents several advantages as compared to other composite cements, but overall the mechanical performance of this system resembles that of Kevlar 29/PMMA cement, with a few differences. Scanning electron microscopy reveals characteristic micromechanisms of energy absorption in PE/PMMA bone cement, including plastic bending and kinking of the PE fibers, pull-out and possibly some splitting. More fundamental modeling treatments are needed to obtain a quantitative estimate of such micromechanisms, within the framework of the fracture behavior of short fiber composites with weakly bonded constituents, as well as to optimize the various mechanical properties with respect to structural parameters and cement preparation techniques.*

## 1. INTRODUCTION

Acrylic bone cement has been used routinely for fixation of metallic prostheses in partial and total joint surgery, and in various other surgical applications, for over two decades.[1-3] The clinical experience with joint replacement surgery, as well as the more recent theoretical (mathematical, analytical and computerized finite element studies) and experimental (*in vitro* and *in vivo* testing) information, have substantially improved our understanding of the behavior of the implant/cement/bone composite structure.[4,5] The main function of the cement is to serve as an interfacial phase between the high modulus metallic implant and the low stiffness bone, and therefore to transfer and distribute body weight loads and their reactions, as well as cyclic loads due to walking movements, from the prosthesis to the bone. Despite a relatively good success rate of implant fixation with acrylic-based bone cement, a number of persistent problems are encountered,[4] mainly related to the relatively poor mechanical behavior of the cement, as compared to bone and implant performances. Consequently, a significant number of surgical repeats are being performed every year.[5]

In recent years, several approaches have been tried to improve the performance of existing surgical cements. Porous poly(methyl methacrylate) (PMMA) bone cement was developed so as to allow bone ingrowth within the cement and thereby improve the strength of the interface.[6] Another approach consists in using advanced composite materials to improve the mechanical properties of the cement, constituting a new development in biomaterials science and technology. Various fiber types have been tried over the past few years: metal wires,[7] carbon and graphite fibers,[8,9] and aramid fibers.[4,10,11] We have recently shown that Kevlar 29, an aramid fiber (manufactured by E. I. du Pont de Nemours & Co.), may be a better candidate than carbon as a reinforcing phase in acrylic cements, in terms of fracture energy dissipated.[4,11] Flexural strength and modulus are, apparently, not improved, though, by the presence of aramid fibers within the cement, which could be explained by a combination of several factors, including poor interfacial bonding, large void content and unsatisfying mixing technique.[4] Biocompatibility tests indicated that exposure of aromatic polyamide fiber to a simulated physiological environment at 37°C for up to 3 years did not significantly degrade the mechanical properties of the materials tested.[12,13] Dog implantation studies have also shown Kevlar poly(paraphenylene terephthalamide) fibers to have a biocompatibility similar to that of Dacron.[14] Despite this, the biocompatibility of aramids has not yet been fully proved to date.

The objective of the present investigation was to examine the fracture properties of bone cement reinforced with ultra-high strength polyethylene (PE) fibers, with the hope of combining the promising mechanical properties of these fibers[15] with the known biocompatibility of polyethylene (assuming that ultra-high polyethylene is not dissimilar in this respect from normal polyethylene). Indeed, as stated before, Kevlar has been reported to possess good biocompatibility behavior in several studies,[12-14] but unlike polyethylene has so far not been used routinely in surgical applications.

## 2. EXPERIMENTAL PROCEDURE

Standard three-point bend fracture toughness (ASTM Standard D-790) specimens were molded using 3·2-mm long Spectra 900 fibers (Allied Corp.), which were mixed with PMMA powder at 1, 4 and 7% by weight concentrations prior to mixing with the monomer. The PMMA powder and liquid monomer (Zimmer, Inc.) were mixed at room temperature using the customary 2:1 ratio of powder to liquid, and the mixing procedure was identical to the procedure used under unreinforced cement under operating room conditions. Dough stage was reached within about 3–5 min after the mixing was started. The cement was then inserted into a mold made from two glass plates separated by 5-mm thick rubber spacers. Polymerization was complete within a few minutes. The mold was placed in a circulating air oven at 37°C and allowed to rest for at least 24 h before any tests were performed. The dimensions of the resulting plate were $15 \times 15 \times 0·5 \, cm^3$. Three-point bending test beams of dimensions $0·5 \times 0·6 \times 5 \, cm^3$ were machined from the plates and sharp notches of various lengths inserted using a sawtooth edge solid carbide circular saw blade (Technology Assoc., Inc.). The span used was 2·54 cm. Eight specimens were prepared at each concentration and tested in three-point bend loading on an Instron 1130 testing machine, at a crosshead displacement rate of 0·5 cm/min.

Flexural strength ($\sigma$) and modulus ($E$) were computed as in our previous studies, using the standard ASTM D-790 formulas. Fracture toughness ($K_{Ic}$) was computed using the following expression, derived from linear elastic theory:[16]

$$K_{Ic} = (3/2)(FLa^{1/2}/bw^2)Y(a/w) \qquad (1)$$

where $F$ is the maximum load, $a$ is the notch depth and $b$ is the beam thickness. $Y$ is defined as follows, for span-to-depth ratios of 4 (which is the case here):

$$Y(a/w) = 1.93 - 3.07(a/w) + 14.53(a/w)^2 - 25.11(a/w)^3 + 25.80(a/w)^4$$

within 0.2% for all values of $a/w$ up to 0.6.[16]

The fracture toughness analysis was performed using another key parameter, the $J$-integral, used in the framework of nonlinear elastic fracture mechanics, and defined as[16]

$$J = -\partial(U/b)/\partial a \qquad (2)$$

evaluated at constant displacement, and where $U$ is the potential energy, $b$ is the thickness and $a$ the crack length. The use of the $J$-integral as a fracture parameter has been described in the literature,[16–19] and its use in the case of randomly oriented, short fiber composites was considered by Agarwal *et al.*[20] In previous work[4,11] we have computed the critical value of the $J$-integral, $J_{Ic}$, using the energy rate interpretation.[17] The same approach is adopted here and full details of the computation technique can be found in our previous studies.[4,11] In the case of plane strain elastic loading, the critical values $K_{Ic}$ and $J_{Ic}$ are related by the following expression:[16]

$$K_{Ic} = (J_{Ic}E/(1 - v^2))^{1/2} \qquad (3)$$

where $E$ is Young's modulus and $v$ is Poisson's ratio.

The determination of $K_{Ic}$ to predict unstable crack extension in a notched body is most useful for the limited case of linear elastic plane strain fracture. By contrast, the $J$-integral is a fracture criterion which also includes elastic-plastic to fully plastic behavior. This criterion is believed to be more appropriate for PMMA composites, whose stress–strain behavior shows some deviation from linearity,[8] the degree of nonlinearity being dependent on such variables as strain rate, temperature, fiber and void contents. The $J$-integral approach is an exact method, whose only drawback is that it sometimes lacks precision due to scatter in load–displacement curves between specimens.[17]

## 3. RESULTS AND DISCUSSION

### 3.1. Flexural Modulus and Strength

In Fig. 1 we report the results for flexural strength and elastic modulus, obtained with unnotched specimens, as a function of fiber content, using mean values ($\pm$ one standard error). In addition, the results for all three-point bend mechanical tests are summarized in Table 1, where a comparison with previously obtained results with Kevlar 29/PMMA bone cement is presented. Very little reinforcing effect is observed for the strength, as observed with Kevlar reinforced cement.[4,11] The trend of the results for the modulus is somewhat different from that of Kevlar reinforced PMMA: a decrease as a function of fiber content in the 0–1% range is first observed, followed by a region where the modulus is insensitive to fiber content (by comparison, results for Kevlar/PMMA show no decrease below the unreinforced cement value). These results are in contrast, though, with those of graphite fiber reinforced PMMA, for which the flexural modulus increased with fiber content.[9] Our present results may be interpreted, as with Kevlar fibers,[4] in terms of poor interfacial bonding, the large amount of voids and unsatisfactory mixing procedures. In particular, the modulus is particularly sensitive to void content, and it can be noticed that when the fibrous component is weakly bonded to the surrounding matrix, the behavior of the resulting structure may resemble

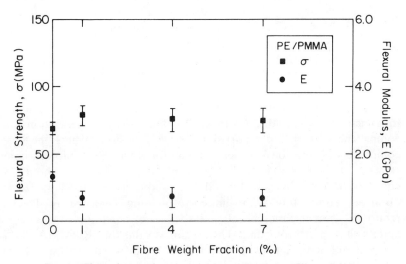

FIG. 1. Flexural strength and modulus as a function of fiber content.

TABLE 1

*Mechanical properties of poly(methyl methacrylate) bone cement reinforced with Kevlar 29 and Spectra 900 fibers*

| $W_f$ (%) | $E$ (GPa)(CV) | | $\sigma$ (MPa)(CV) | | $K_{Ic}$ (MPa m$^{1/2}$)(CV) | | $J_{Ic}$ (kJ/m$^2$) | |
|---|---|---|---|---|---|---|---|---|
| | K/PMMA | PE/PMMA | K/PMMA | PE/PMMA | K/PMMA | PE/PMMA | K/PMMA | PE/PMMA |
| $0^a$ | 1·28 (8·7) | 1·30 (10·9) | 67·4 (7·1) | 68·4 (12·0) | 1·47 (7·5) | 1·47 (7·5) | 1·56 | 1·56 |
| 1 | 1·70 (14·2) | 0·67 (5·3) | 74·2 (11·9) | 77·9 (10·3) | 2·08 (12·1) | 1·83 (5·3) | 2·28 | 5·02 |
| 4 | — | 0·68 (2·5) | — | 74·5 (10·4) | — | 1·86 (5·2) | — | 5·08 |
| 7 | 1·30 (15·6) | 0·66 (7·6) | 82·6 (24·7) | 74·2 (8·7) | 2·61 (17·7) | 1·81 (6·8) | 4·81 | 4·95 |

[a] Only unnotched unreinforced specimens were tested here, giving results for $E$ and $\sigma$ which are very close to our previous results for K/PMMA, as seen. The fracture toughness results of unreinforced cements are taken from our previous study (Ref. 11).

that of a porous matrix, under the condition that the bond is indeed very weak, which is probably the case for PE/PMMA composites. In fact, we conjecture that the interfacial bond in PE/PMMA is even weaker than in Kevlar/PMMA, and thus a decreasing trend for $E_f$ with fiber concentration, which can be assimilated to void concentration, would therefore be expected.[21] This conjecture would as well provide an explanation for the lack of dependence of the strength on PE fiber content, as compared with the weakly increasing dependence for the same property in Kevlar/PMMA.

### 3.2. Fracture Toughness

In Fig. 2 we present data for the mean values ($\pm 1$ SE) of fracture toughness values $K_{Ic}$ and $J_{Ic}$ as a function of fiber weight concentration. For completeness, these are reported in Table 1 as well, where they are compared with previously generated data for Kevlar 29 reinforced bone cement.[11] It is observed that a significant increase occurs at about 1% concentration, but beyond that value no additional reinforcing effect is obtained, contrasting with the corresponding behavior of Kevlar reinforced cement. Fracture toughness values of unreinforced specimens agree well with literature data. The toughness value for PE/PMMA at 7% is comparable with that obtained for Kevlar 29 reinforced bone cement, but intermediate values are much higher for the present PE/PMMA system.

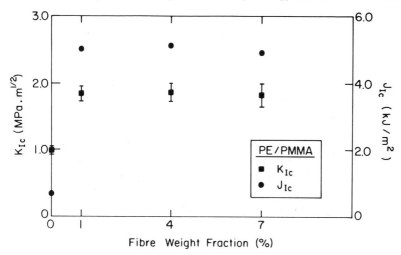

Fig. 2.   Fracture toughness as a function of fiber content.

One difference, thus, between cements reinforced with Kevlar 29 and with ultra-high strength polyethylene concerns the insensitivity of fracture toughness to PE content beyond the initial very strong improvement at 1%. As with Kevlar 29 reinforced bone cement, the present PE/bone cement composites yield high fracture energies as compared with cements where graphite fibers are incorporated. Indeed, 1% by weight reinforcement with the PE fiber (which is about 0·8% by volume) yields a $K_{Ic}$ value of 1·83 MPa m$^{1/2}$, as compared to 1·60–1·88 MPa m$^{1/2}$ for 2% by volume reinforcement with graphite.[22] Although a very limited data base is available at present for such materials systems, these figures may serve as a good indication of a potential trend. Later in this work we present electron photomicrographs so as to attempt to assess the mechanisms of energy absorption which are specific to the PE/PMMA composite system.

Using eqn. (3) with the values presented in Table 1 for Young's modulus and $J_{Ic}$, and $v \approx 0.35$ for all the materials, we obtain $K_{Ic}$ = 1·03, 1·96, 1·98 and 1·93 MPa m$^{1/2}$. We assumed that the incorporation of PE fibers would not affect the value of Poisson's ratio to any great extent, in view of the small fiber concentrations, and the figure of 0·35 employed was that of cast PMMA.[23] Thus, these $K_{Ic}$ values obtained from $J_{Ic}$ agree well with our experimental values of $K_{Ic}$ for 0, 1, 4 and 7% by weight, respectively, although they are slightly higher than these latter, reflecting probably some nonlinearity in material behavior.

Finally, it is worth noting that there is at present no satisfactory micromechanical modeling scheme for the fracture toughness, or fracture energy, of weakly-bonded heterogeneous materials with randomly distributed short fiber reinforcement.

### 3.3. Electron Microscopy

Fracture surfaces of unreinforced and PE reinforced specimens were observed using scanning electron microscopy (Figs 3–5). Fewer and smaller voids were observed as compared to our previous studies with Kevlar 29 as a reinforcing agent, perhaps due to easier mixing, and no clustering of fibers was observed, probably for the same reason. Cracks of thermal origin, observed in previous work with aramid reinforced PMMA,[4] were not observed. Figure 3 shows a gap at the interface between PE and PMMA, which reflects the poor bonding compatibility of these two materials. This is typical of all acrylic-based surgical cements and has been observed for carbon/PMMA and Kevlar 29/PMMA as well. In fact, it has been reported[5] that when attempting to reinforce PMMA with carbon fiber,

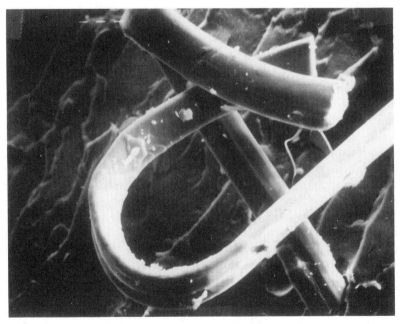

FIG. 3.   SEM photomicrograph of PE/PMMA cement showing poor interfacial bonding between the fibers and the matrix.

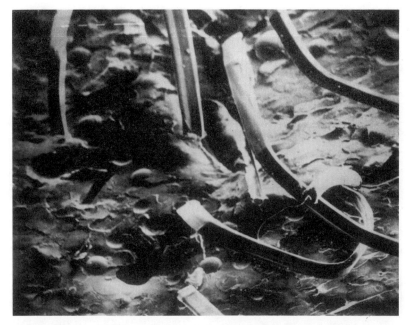

FIG. 4.    SEM photomicrograph showing typical fiber distribution.

stainless steel wire or perforated metal sheet, only the perforated metal sheet provided significant shear strength at its interface with the PMMA matrix. This poor bonding effect is most probably reflected by the low values obtained for the modulus and strength of the present PE/PMMA cement. By contrast, poor interface bonding is perhaps at the origin of increased fracture toughness results for the PE reinforced cement, as compared to pure PMMA. This interesting point was discussed in detail in a previous work[4] involving Kevlar 29/PMMA surgical cement, and this analysis will not be repeated here. Figure 4 shows a typical fracture region from which it is seen that the spatial distribution of the fibers is uniform, reflecting the ease of mixing of the PE fiber with PMMA (which is not the case for Kevlar 29 fibers). Figure 5 shows details of a typical fracture surface of the PE/PMMA composites, from which some of the energy absorption mechanisms can be inferred. These include plastic bending of the PE fiber, and its lateral deformation after pull-out from the matrix. Fiber kinking and perhaps splitting is observed as well (see Figs 3 and 4). Most of these energy absorption mechanisms are not exhibited by PMMA

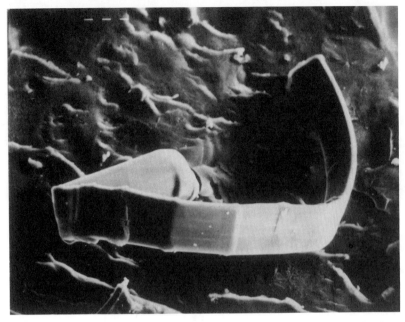

FIG. 5.   SEM photomicrograph showing fiber bending and lateral plastic deformation due to pull-out.

reinforced with brittle fibers such as carbon, which probably explains the higher fracture toughness values obtained for both PE/PMMA and Kevlar 29/PMMA cements.

## 4.  CONCLUSIONS

In this study the elastic and fracture properties of acrylic bone cement reinforced with Spectra 900 polyethylene fibers were analysed and compared with those of both unreinforced and reinforced bone cements. It was found that the flexural strength and modulus are not improved by the incorporation of the fibers, and that these parameters are slightly different for the present system compared to those in Kevlar 29/PMMA composites. It is conjectured that this might be mainly due to the weaker interfacial bond in PE/PMMA, although poor mixing practice might also contribute to the poor results obtained for the strength and the modulus. Both linear elastic and nonlinear elastic fracture mechanics techniques yield similar trends for the fracture toughness. A significant reinforcing effect is measured at a low fiber content of 1% but beyond that concentration a plateau value is reached. SEM observations reveal specific fracture energy absorption mechanisms related to the polyethylene fiber failure modes. Fundamental treatments are needed to quantify these micromechanisms of energy dissipation, within the framework of short fiber composites with weak bonding among the constituents, which, to date, are not well understood. Also, more work is needed to optimize the mechanical properties with respect to structural parameters and cement mixing techniques. Finally, we are presently exploring various chemical ways to improve the bonding properties of PE fibers and will report soon on these developments.

## ACKNOWLEDGEMENTS

This work was partially supported by a grant from the Yeda Fund, the Weizmann Institute of Science. We would like to thank Allied Corp. for supplying short fibers of Spectra 900. Thanks are also extended to Zimmer, Inc., for supplying samples of bone cement.

## REFERENCES

1. CHARNLEY, J., Anchorage of the femoral head prosthesis to the shaft of the femur, *J. Bone Jt Surg.*, **42B** (1960), 28.
2. CHARNLEY, J., Acrylic cement in orthopaedic surgery, *J. Bone Jt Surg.*, **46B** (1964), 518.
3. RAY, A. K., ROMINE, S. J. and PANKOVICH, A. M., Stabilization of pathogenic fractures with acrylic cement, *Clin. Orth.* (1974), 182.
4. POURDEYHIMI, B., ROBINSON, H. H., SCHWARTZ, P. and WAGNER, H. D., Fracture toughness of Kevlar 29/poly(methyl methacrylate) composite materials for surgical implantations, *Ann. Biomed. Engng*, **14** (1986), 277.
5. *Orthopaedic Knowledge Update I*, American Academy of Orthopaedic Surgery, Chicago, Ill., 1984, Chapter 33.
6. RIJKE, A. M. and RIEGER, M. R., Porous acrylic cement, *J. Biomed. Mater. Res.*, **11** (1977), 373.
7. SAHA, S. and KRAAY, M. J., Improved strength characteristics of poly(methyl methacrylate) beam specimens reinforced with metal wires, *J. Biomed. Mater. Res.*, **13** (1979), 443.
8. PILLIAR, R. M., BRATINA, W. J. and BLACKWELL, R. A., Mechanical properties of carbon fiber reinforced poly(methyl methacrylate) for surgical implant applications, ASTM Special Technical Publication STP 636, 1977, p. 206.
9. KNOELL, A., MAXWELL, H. and BECHTOL, C., Graphite fibre reinforced bone cement, *Ann. Biomed. Engng*, **3** (1975), 225.

10. WRIGHT, T. M. and TRENT, P. S., Mechanical properties of aramid fibre-reinforced acrylic bone cement, *J. Mater. Sci. Lett.*, **14** (1979), 503.
11. POURDEYHIMI, B., WAGNER, H. D. and SCHWARTZ, P., A comparison of mechanical properties of discontinuous Kevlar 29 fibre reinforced bone and dental cements, *J. Mater. Sci.*, **21** (1986), 4468.
12. MCKENNA, G. B., BRADLEY, G. W., DUNN, H. K. and STATTON, W. O., Degradation resistance of some candidate composite biomaterials, *J. Biomed. Mater. Res.*, **13** (1979), 783.
13. KING, R. N., MCKENNA, G. B. and STATTON, W. O., Novel uses of fibers as tendons and bones, *J. appl. Polym. Sci.: appl. Polym. Symp.*, **31** (1977), 335.
14. HUXTER, R. H., JAEGER, S. H. and HUNTER, J. M., *Trans. 23rd Ann. Orthop. Res. Soc.*, **2** (1977), 108.
15. SMOOK, J., HAMERSMA, W. and PENNINGS, A. J., The fracture process of ultrahigh strength polyethylene fibres, *J. Mater. Sci.*, **19** (1984), 1359.
16. HERTZBERG, R. W., *Deformation and Fracture Mechanics of Engineering Materials*, New York, John Wiley, 1976.
17. BEGLEY, J. A. and LANDES, J. D., The *J*-integral as a fracture criterion, ASTM Special Technical Publication STP 514, Part II, 1972, p. 1.
18. LANDES, J. D. and BEGLEY, J. A., Recent developments in $J_{Ic}$ testing, ASTM Special Technical Publication STP 632, 1977, p. 57.
19. BUCCI, R. J., PARIS, P. C., LANDES, J. D. and RICE, J. R., *J*-integral estimation procedures, ASTM Special Technical Publication STP 514, Part II, 1972, p. 40.
20. AGARWAL, B. D., PATRO, B. S. and KUMAR, P., *J*-integral as fracture criterion for short fibre composites: an experimental approach, *Engng Fract. Mech.*, **19** (1984), 675.
21. WAGNER, H. D., Elastic response of fibrous composite materials with weak bonding, *C. R. Acad. Sci. Paris, t. 303, Serie II*, no. 14 (1986), 1283.
22. ROBINSON, R. P., WRIGHT, T. M. and BURSTEIN, A. H., Mechanical properties of poly(methyl methacrylate) bone cements, *J. Biomed. Mater. Res.*, **15** (1981), 203.
23. *Modern Plastics Encyclopedia*, 60, 10A, New York, McGraw-Hill, 1983–1984, p. 592.

# 25

# Composite Materials for Bone Fracture Fixation

L. Ambrosio, G. Caprino, L. Nicolais, L. Nicodemo

*University of Naples, Department of Materials and Production Engineering, Piazzale Tecchio, 80125 Naples, Italy*

. S. J. Huang

*University of Connecticut, Institute of Materials Science, Polymer Science Program, Storrs, Connecticut 06268, USA*

G. Guida and D. Ronca

*University of Naples, Institute of Orthopaedics, Naples, Italy*

## ABSTRACT

*The metallic plates, pins and screws currently used in orthopaedics for internal fixation have elastic moduli much higher than that of the bones to which they are connected. This difference in rigidity prevents healing through proliferation of primary callus early in the healing process and in the later process of healing, bone atrophy and osteoporosis occur.*

*Laminate fiber-reinforced composites are the class of materials most recently probed as possible candidates for internal fracture fixation implants. The implementation of correct material choice in combination with micromechanics and lamination theory permits design of a composite laminate of desired mechanical properties for specific applications.*

*Using a unique molding technique, composite materials with complicated geometrical shapes such as screws, pins and plates have been obtained. These prostheses have been implanted on rabbits and some preliminary clinical results are presented.*

## INTRODUCTION

The traditional prostheses currently used in orthopaedics for internal fracture fixation have elastic moduli much higher than that of the bones to which they are connected. The metals most commonly used are 316 stainless steel and cobalt–chromium based alloys. They have elastic moduli up to ten times that of bone.[1] Since the metallic component of an internal fracture fixation implant is more rigid than the bone, it transmits the majority of the stress sustained by the bone–plate systems and stress-shields the bone. This difference in rigidity between the plate and the bone prevents healing through proliferation of primary callus early in the healing process. In the later process of healing, bone atrophy and osteoporosis occur as a result of stress-protection of the bone by the more rigid plate.

Since the recognition of stress-shielding osteoporosis, much effort has been dedicated to finding alternative materials for use in internal fracture fixation. Materials with greater flexibility allowing for more deformation of the implant are sought. In turn, the amount of load carried by the bone will be increased. Titanium alloys, thermoplastic polymers and fiber-reinforced composites have been investigated. Titanium and its alloys still exhibit moduli much higher than the bone and, thus, generate stress-protection osteoporosis. Unreinforced thermoplastics have moduli much lower than the bone but are too weak for optimum bone healing. Fiber-reinforced composites are the class of materials most recently probed as possible candidates for internal fracture fixation implants.[2,3]

However, the use of carbon fiber-reinforced plates is not possible in connection with metallic screws, due to different electrochemical potentials of the two materials producing a corrosion problem.

It has been shown by Gillet *et al.*[1] that the use of PBT and nylon with short carbon fiber composites (CFRP) yields the plates with a low value elastic modulus which facilitated the healing process. The mechanical results reported can be rewritten as given in Table 1.

The histological results show the formation of callus after 6–8 weeks for CFRP/nylon, indicating that the low rigidity of the latter material in tension is responsible for such rapid callus formation. However, the low value of the load to failure of few of the CFRP/PBT composites was responsible for the failure of the implanted prosthesis.

In the present chapter, the design criteria and the technology used to

TABLE 1

| Material | $EA/(EA)b$ | $Pu/(Put)b$ | $EI/(EI)b$ | $Mfu/(Mfu)b$ | $Pu/(Puc)b$ |
|---|---|---|---|---|---|
| CFRP/PBT | 0·21 | 0·34 | 0·11 | 0·075 | 0·24 |
| CFRP/nylon | 0·17 | 0·68 | 0·009 | 0·15 | 0·47 |

produce continuous carbon fiber reinforced pins, screws and plates for internal fracture fixation are discussed.


## DESIGN CRITERIA

The ideal plate, for what concerns the mechanical and medical effectiveness, should have two properties: (i) rigidity in tension smaller than that of the bone and (ii) high stress at break in tension and compression, to avoid breakage in service.

Composite materials are very promising, because proper design gives a wide range of elastic moduli and stresses at break, changing the aspect ratio, the orientation of fibers, the matrix and the relative amount of fibers and matrix. With a good approximation, the stress at break of a composite laminate is given by

$$\sigma_{xu} = E_x \varepsilon_u \tag{1}$$

where $\sigma_{xu}$ is the stress at break of the laminate, $E_x$ its elastic modulus and $\varepsilon_u$ the elongation at break of the fiber. The force at break of the prosthesis, $P_u$, if $A$ is the cross-section, is given by

$$P_u = \sigma_{xu} A = E_x A \varepsilon_u \tag{2}$$

From eqn. (2) it can be seen that the rigidity of the prosthesis is $E_x A$; therefore to obtain at the same time a high force at break and small rigidity for the prosthesis it is necessary that the value of $\varepsilon_u$ is the highest possible. In Table 2 values of $\varepsilon_u$, in tension ($\varepsilon_u^t$) and in compression ($\varepsilon_u^c$), for carbon and aramidic fibers are reported.

Another important problem which has to be taken into account is the bearing resistance of the holes where the screws are inserted to fasten the plate to the bone. The total allowable load, $P_{tb}$, for the resistance at

TABLE 2

*Typical values of elongation at break in tension ($\varepsilon_u^t$) and in compression ($\varepsilon_u^c$) for carbon and aramidic fibers*

| Fiber | $\varepsilon_u^t$ (%) | $\varepsilon_u^c$ (%) |
|---|---|---|
| Carbon | 0·9 | 0·9 |
| Aramidic | 1·6 | 0·4 |

the bearing in the case of one fastening nail or screw for each side of the fracture is given by

$$P_{tb} = dt\sigma_b \qquad (3)$$

where $d$ is the hole diameter, $t$ the thickness of the prosthesis and $\sigma_b$ the stress at bearing break of the material. The maximum value for $\sigma_b$ is obtained when $W/d \geq 4$, where $W$ is the width of the prosthesis,[4,5] and depends upon the bone geometry.

Concerning the screw, the main problem is to obtain the maximum shear resistance perpendicular to the screw. Accordingly, the optimum laminate design for the screw is $0° \pm 90°$. However, the top layer of fibers has to be in the direction of the screw threads so that the fibers will not be excluded from the screw threads during fabrication. The pin design has the same problems of the plate with the exclusion of those relative to holes. In particular, in this case the flexural resistance becomes very important. Therefore, assuming the iso-resistance in flexure, it is possible to design intramedullar pins having a rigidity lower than that of the bone. A detailed account of both micromechanics and lamination theories can be found in Refs 7 and 8.

## MATERIALS AND METHODS

To produce pins and plates a silicon mold technique has been used.[9] This technique consists of filling out a model pin or plate with liquid silicon mixture which will crosslink in 12 h at $T = 25°C$. Subsequently, the model is removed and in this cavity the appropriate prepreg laminate is placed. During the subsequent curing process the rubber mold which is contained in a metallic box tries to expand, developing enough high pressure to reduce the voids content in the laminate. In our study (AS4) carbon fabric or fibers MY720 epoxy resin system were used.

Composite screws, made of high Tg epoxy, and reinforcing continuous carbon fibers were prepared by lamination of prepregs with fibers oriented in directions $0°$, $\pm70°$ and $\pm90°$ with respect to the longitudinal direction of the screws. The final shape is then obtained by using a silicon mold in which the shape is preformed.

The main difficulty faced in making composite screws is to obtain threads reinforced with fibers and to reduce the voids content or microcracks in the formed object. This problem has been overcome by using a silicon rubber mold which expands during the cure cycle, developing enough pressure to reduce the void content and to force the fibers into the screw threads. The preparation of a composite screw is carried out as follows: a small piece of fiber of required length is rolled to make a stick whose diameter is about half the diameter of the screw, fibers or fabric are wound on the fiber stick in the $\pm 90°$ direction, while the last layer of fibers is placed in the direction of the threads. Now the laminate is inserted in the mold and placed in the oven for the curing cycles. After cooling down the composite screw is removed from the mold and postcured in the oven at 180°C for 2 h.

## EXPERIMENTAL RESULTS

The composite pins, prepared with ten laminae of carbon fabric/epoxy resin prepreg, were tested in flexure[10] by using a three-point bending geometry, as shown in Fig. 1. The results compared with the flexural properties of the bone are presented in Table 3. The composite pin presented in Fig. 2 shows a resistance in flexure almost twice that of the bone, while the rigidity is lower.

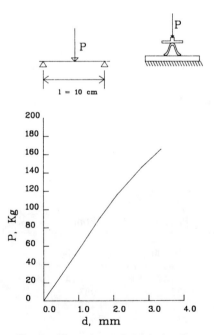

FIG. 1.   Flexural test of composite pin.

TABLE 3
*Flexural properties of the composite pin
compared to the bone*

|  | $M$ (kg mm) | $EI$ (kg mm$^2$) |
|---|---|---|
| Bone | 2 928 | $2 \cdot 93 \times 10^6$ |
| Pin | 4 175 | $1 \cdot 27 \times 10^6$ |

The carbon fiber–fabric/epoxy screw is shown in Fig. 3 along with a metallic screw where the well reproduction of the screw threads is evident. In Fig. 4 a micrograph of a longitudinal section of the composite screw showing the presence of continuous fiber in the thread is presented. To characterize the mechanical properties of the screw, a pure shear test has been performed. The results are shown in Table 4, where three different composite screws are well compared with the metallic one.

Composite plates were prepared with carbon fabric/epoxy resin using laminae at $\pm 90°$ and $\pm 45°$ to achieve the required characteristics.[11] Mechanical results obtained with these plates are compared in Table 5, with

TABLE 4
*Ultimate and specific strengths in pure shear tests of metallic and composite screws*

| Sample | Weight (g) | Breaking strength (kg/cm$^2$) |
|---|---|---|
| Metallic wood screw | 4·14 | 3 536 |
| Carbon fiber/epoxy screw | 1·04 | 2 438 |
| Kevlar fabric/epoxy screw | 0·88 | 2 239 |
| Carbon fabric–carbon fiber/epoxy screw | 0·85 | 2 961 |

TABLE 5
*Mechanical properties of composite plate and bone*

| | Plate | Bone | Property of the plate / Property of the bone |
|---|---|---|---|
| $EA$ (kg) | 210·000 | 131·000 | 1·60 |
| $EI$ (kg mm$^2$) | $0·938 \times 10^6$ | $2·93 \times 10^6$ | 0·32 |
| $GI$ (kg mm$^2$) | $0·47 \times 10^6$ | $1·14 \times 10^6$ | 0·41 |
| $P_{tu}$ (kg) | 1 427 | 983 | 1·45 |
| $P_{cu}$ (kg) | 1 427 | 1 427 | 1·0 |
| $M_{fu}$ (kg mm) | 4 285 | 2 928 | 1·46 |
| $M_{tu}$ (kg mm) | 1 420 | 3 036 | 0·47 |

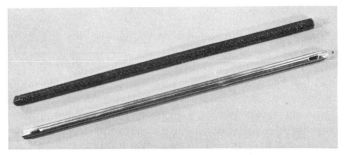

FIG. 2. Carbon fabric/epoxy composite pin compared with the traditional metallic pin.

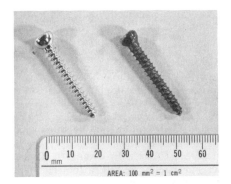

FIG. 3. Composite screw compared with the metallic one.

FIG. 4. Longitudinal section of composite screw.

typical bone data indicating a good matching of rigidity and resistance of these prostheses.

Finally, to improve the biocompatibility of these systems the composite prostheses were coated with a film of polydimethylsiloxane solution. The prepared prostheses have been implanted on white New Zealand rabbits and the results clearly indicate good biocompatibility of these materials, together with a better clinical performance compared with the metallic systems. By using composite plate it has also been possible to follow very clearly the bone healing process due to the slight radio-transparency of the plate.

## REFERENCES

1. GILLET, N., BROWN, S. A., DUMBLETON, J. H. and POOL, R. P., The use of short carbon fibre reinforced thermoplastic plates for fracture fixation, *Biomaterials*, **6** (1985), 113.
2. CLAES, L., HUTTER, W. and WEISS, R., Mechanical properties of carbon fibre reinforced polysulfone plates for internal fracture fixation, in: *Biological and Biomechanical Performance of Biomaterials*, ESP, Amsterdam, 1986.
3. CHRISTAL, P., VERT, M., CHABOT, F., GARREAU, H. and AUDION, M., PGA (polyglycolic acid)-fiber-reinforced-PLA (polylactic acid) as an implant material for bone surgery, *Composites in Biomedical Engineering*, Proceedings, Nov. 1985.
4. GODWIN, E. W. and MATTHEWS, F. L., A review of the strength of joints in fibre-reinforced plastics, *Composites*, July 1980, 155–160.
5. MATTHEWS, F. L., Problems in the joining of GRP, *Developments in GRP Technology—1* (B. Harris ed.), London, Elsevier Applied Science Publishers, 1983.
6. MIGLIARESI, C. and NICOLAIS, L., Tailor made composite materials for biomedical use, in: *Polymeric Biomaterials* (E. Piskin and A. S. Hoffman eds), NATO ASI Series, Martinus Nijhoff Publishers, 1986.
7. NICOLAIS, L., Mechanics of composites, *Polymer Engineering and Science*, **15** (1975), 137.
8. HALPIN, J. C., *Primer on composite materials: analysis*, Lancaster, PA, Technomic Publishing Co., 1984.
9. AMBROSIO, L., HUANG, S. J., LEONE, A. and NICOLAIS, L., Protesi ortopedica per la riduzione di fratture ossee, materiale per la sua fabbricazione e procedimenti per ottenerli, Italian Patent 22451-A86.
10. ASTM D790, Flexural properties of plastics and electrical insulating materials.
11. NICOLAIS, L., GIMIGLIANO, R., GUIDA, G., MIGLIARESI, C., PAGLIUSO, S. and RENTA, V., An isoelastic fiber composite plate for fracture fixation, *2nd World Congress on Biomaterials,* Washington, DC, 1984.

# 26

# The Effects of High Temperature Excursions on Environmentally Exposed CFC

T. A. COLLINGS, D. L. MEAD and D. E. W. STONE

*Royal Aircraft Establishment, Farnborough, Hants GU14 6TD, UK*

## ABSTRACT

*The mechanical properties of carbon fibre composites (CFC) when exposed to constant environmental conditions are well known, but the effects of rapid high temperature excursions (thermal spikes) are less well understood. The effects of such thermal spikes on laminates of CFC subjected to a range of conditions of temperature, humidity and spiking have been investigated. These conditions were intended to represent those that could be met in service by a specific component and formed part of the airworthiness clearance procedure. Details of the experimental programme are reported together with the results of mechanical tests and optical examination. Comparisons are made with laminates which had received no environmental exposure. Both uniform moisture distributions and moisture gradients were investigated.*

*Thermal spiking of laminates containing more than 1·13% moisture was found to increase both the rate of moisture uptake and the equilibrium moisture content. This was considered to be associated with matrix damage but only under the most severe conditions could this be confirmed by optical examination.*

*The combined effect of the various combinations of moisture content, thermal spiking and testing temperature on the mechanical properties is quite complex and the tests reported here were not sufficiently exhaustive to establish the contribution made by each effect. The consequence of thermal spiking is a modified total moisture content and distribution, from one that is uniform to one that has a higher concentration at the surface and is of a higher average content. Thus moisture alone could be responsible for the extra loss in strength, or it could be due to the spike induced damage. Sufficient evidence is available, however, to suggest that, at least for the range of temperatures investigated, the effect of thermal spiking damage is small compared to other environmental effects.*

## 1.  INTRODUCTION

It is known that the mechanical properties of carbon fibre composites (CFC) with epoxy matrices are degraded by the presence of moisture at elevated temperature, and it is necessary to quantify this degradation in order that safe values can be set for design and airworthiness purposes. For conditions of constant humidity and constant temperature the mechanical properties are well known, but the effect of rapid high temperature excursions on these properties is not so well understood. Such excursions, termed thermal spikes, can occur during the service life of a military aircraft. Situations which can give rise to thermal spikes include kinetic heating during supersonic flight, ground-reflected efflux from the engines of VTOL aircraft and hot gases from missile efflux; the short duration of the latter may however mean that its effect is less significant.

For these and other reasons there is a growing interest in understanding the way in which thermal spiking can cause degradation.[1,2] Recently, for example, Collings and Stone[3] showed that interspersing thermal spikes during moisture conditioning of a Fibredux XAS/914 CFC material can be harmful. Microscopic examination revealed the presence of interlaminar cracks, whilst regular monitoring of the moisture uptake showed that the moisture absorption kinetics were altered, and the moisture equilibrium level almost doubled. Subsequent tests at room temperature to measure the interlaminar shear strength (ILSS) showed a reduction of up to 25% compared with the dry unspiked state. Unpublished work elsewhere also indicated that, for the same CFC material, both the flexural strength and the interlaminar shear strength were reduced by thermal spiking, although the original strengths were largely recovered after drying out.

This chapter deals with the effect of thermal spiking due to kinetic heating. The work was stimulated by the airworthiness clearance programme for a major component. The clearance procedure for this component required a full scale test to be carried out at a temperature of 120°C after it had been environmentally conditioned to an average moisture content of 1·0%. However, the results reported in Ref. 3 suggested that the test condition planned might not represent the worst case that would be met in service, because no thermal spiking was called for. For this reason work was undertaken at RAE to establish the extent to which environmental history before and between thermal spiking contributes to the degradation of the composite.

Because the programme was aimed at investigating the probable behaviour of a specific component, and at providing answers within a limited time, the range of specimens and test conditions employed was far from comprehensive. Subsequent analysis of the data has resulted in the identification of a number of areas where further experiments are required. Nonetheless it is considered that the work reported here provides a useful insight into the phenomena involved.


## 2. PROGRAMME DEFINITION


The experimental procedure used for the thermal spiking of laminates made from the Fibredux XAS/914 material, conditioned to various levels and through-the-thickness distributions of moisture content, is reported in the next section. The programme consisted of an examination of the effect of thermal spiking on the notched compression and interlaminar shear strengths of two different laminates at both room temperature and at 120°C. Details of all conditioning and spiking procedures are given in Table 1. Examinations were also carried out to ascertain whether thermal spiking caused permanent damage such as cracking or changes in the final moisture equilibrium levels. Throughout this series of tests the effects of spikes having maximum temperatures of 127°C and 137°C were separately investigated.

The first part of the programme was to examine the effects of from one to ten thermal spikes on material which contained 1·06% moisture (the equilibrium value for 59% RH) essentially uniformly distributed through the thickness. The second was to compare the effects of 10 spikes on laminates that contained levels of moisture content that were progressively higher but still uniformly distributed (corresponding to other RH equilibrium levels). The third was to examine the effect of 10 or 24 spikes on material conditioned at 96% RH and 60°C, the first spike being applied after only 4 days of conditioning. Here there was a moisture gradient, the moisture content varying from a maximum on the outer surfaces to a minimum at the centre of the specimens.

TABLE 1

*Environmental conditioning of specimens for the simulated kinetic heating spike programme*

| | Specimen group | Spike temperature (°C) | Number of spikes | Time between spikes (days) | Environment for pre-conditioning and conditioning between spikes |
|---|---|---|---|---|---|
| Controls | V | — | — | — | Dry |
| | W | — | — | — | 59% RH 60°C 1·06% ($M_\infty$) |
| Part 1 | Y | 127 137 | $\left.\begin{array}{c} 1 \\ 4 \\ 10 \end{array}\right\}$ | 1/2 | 59% RH 60°C 1·06% ($M_\infty$) |
| Part 2 | A | 127 137 | 10 | 2 | 67% RH 60°C 1·2% ($M_\infty$) |
| | B | 127 137 | 10 | 2 | 75% RH 60°C 1·35% ($M_\infty$) |
| | C | 127 137 | 10 | 2 | 86% RH 60°C 1·6% ($M_\infty$) |
| | D | 127 137 | 10[a] | 2 | 96% RH 60°C 1·82% ($M_\infty$) |
| Part 3 | E E+ | 127 | 10 24 | 2 | 96% RH 60°C for 4 days |
| | F F+ | 137 | 10 24 | 2 | 96% RH 60°C for 4 days |

[a] The number of spikes for this group was increased from 10 to 14 to compensate for the spikes carried out during the period when there was a drop in humidity due to a breakdown in the environmental chamber.

3.   EXPERIMENTAL PROCEDURE

**3.1.  Choice of Test Specimens**

The strength performance of composite aircraft structures is, in general, governed by the behaviour of structural features such as fastener holes and cut-outs, and much design data is based on notched compression or notched tensile tests. The former was, however, considered likely to provide the most useful measure of degradation since it was considered unlikely that any matrix degradation would significantly reduce the notched tensile performance. Indeed, as postulated by Potter and Purslow,[4] in the notched tensile test matrix degradation could lead to an increase in strength due to the relaxation of stresses at the notch. Compression strength, on the other hand, is dependent upon the extent to which the matrix is able to support the fibres against buckling, and any degradation in the resin modulus or fibre/resin bond would be manifested in a lowering of compression strength. Interlaminar shear strength testing was also included as it provides a simple test specimen for assessing the strength of the matrix and of the fibre/resin interface bond.

### 3.2. Specimen Preparation

Laminates were fabricated from an XAS carbon fibre (130SC/10 000) preimpregnated with the Ciba-Geigy BSL 914 resin system. The laminates were cured in an autoclave and post-cured in an air circulating oven at 185°C for 12 h.

For the notched compression tests one 24-ply laminate containing 50% 0° plies and 50% ±45° plies was made using the lay-up $(\pm 00/\pm 000/\pm 0)_s$. Notched compression specimens, each measuring 100 × 30 × 3 mm (100 mm dimension measured in the 0° direction) with a central hole of 4.83 mm diameter, were made from this laminate. For the interlaminar shear tests one 16-ply 0° (unidirectional) laminate was made and interlaminar shear specimens measuring 12 × 10 × 2 mm (12 mm dimension measured in the 0° direction) were cut from it. However, to ease handling during conditioning, specimens 55 mm long were initially used. After conditioning three interlaminar shear specimens were cut from each 55 mm length. Pre-travellers and travellers were also cut from each laminate. Pre-travellers were only needed for determining the initial linear part of the moisture versus time diffusion curve, the slope of which is required in order to calculate the moisture diffusion coefficients (Section 3.3). Travellers, however, were first used for monitoring the moisture history of the different groups of specimens throughout their environmental conditioning, and were then used for determining the effect of spiking on moisture equilibrium level (Section 3.7). Both pre-travellers and travellers measured 100 × 30 × 3 mm (but without a central hole) for the notched compression, and 55 × 10 × 2 mm for the interlaminar shear specimens.

Immediately after post-cure all specimens were stored in a desiccator until required for moisture conditioning and testing. Specimens were conditioned as soon as environmental ovens were available but, for reference, the maximum time spent in the desiccator before conditioning was 90 days. Monitoring of specimen weight during storage in the desiccator showed there was no significant loss or gain in weight. On this evidence it was considered that the specimens were dry when they entered the desiccator and that they remained dry during their storage.

### 3.3. Moisture Conditioning

Two aspects of moisture conditioning will be described. The first is the determination of diffusion coefficients for each laminate configuration, and the second is the conditioning of the test specimens and their travellers.

(i) Two pre-travellers from each laminate lay-up and thickness were subjected to an environment of 96% RH and 60°C for 144 days, during which their moisture uptake $M$ was monitored against time $t$. Figure 1 gives the plot of $M$ against $\sqrt{t}$. With this information an average value of the diffusion coefficient $D$ for each laminate was calculated using the following equation:[5]

$$D = \frac{\pi h^2}{16 M_\infty^2} \left( \frac{M_2 - M_1}{\sqrt{t_2} - \sqrt{t_1}} \right)^2 \tag{1}$$

where $M_1$ and $M_2$ are the percentage moisture uptakes at times $t_1$ and $t_2$ respectively, $h$ is the laminate thickness and $M_\infty$ is the moisture equilibrium level for the given relative humidity (taken from previous work[6]). The term

$$\left( \frac{M_2 - M_1}{\sqrt{t_2} - \sqrt{t_1}} \right) \tag{2}$$

is the slope of the linear portion of the plot of $M$ against $\sqrt{t}$ (Fig. 1).

Using these values of $D$ and $M$ the predicted moisture uptake of the pre-travellers in the non-linear region could be calculated. It may be seen from Fig. 1 that there is good agreement between the predicted and measured values of moisture uptake, up to a total time of 144 days. This evidence substantiates the use of the theoretical model for establishing the conditioning programmes used in (ii) below.

The values of $D$ derived from eqn. (1) were obtained by measuring the moisture absorption of specimens of finite size and included therefore the moisture that had been absorbed from all six surfaces. To obtain the one-dimensional diffusion coefficients, $D_\infty$, needed for calculating the true through-the-thickness moisture distribution, a correction factor given by Shen and Springer[5] was used, namely

$$D_\infty = D \left( 1 + \frac{h}{w} + \frac{h}{l} \right)^{-2} \tag{3}$$

where $w$ and $l$ are the width and length of the pre-traveller respectively.

(ii) The aim of the conditioning programmes for all the specimens in groups W, Y, A, B, C and D was to achieve, in minimum time, selected average moisture contents, sensibly uniformly distributed through the

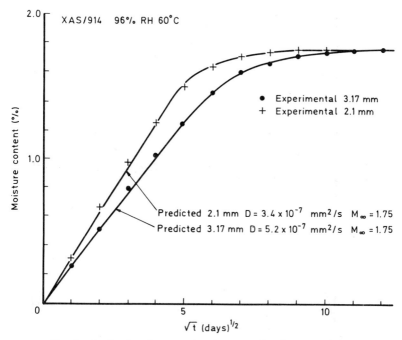

FIG. 1. Predicted and measured moisture uptake of pre-travellers.

thickness of the specimen. This was achieved by employing a three-stage accelerated ageing technique described by Collings and Copley.[7] The three stages for each group were calculated using a computer program by Copley[8] that models the classical Fickian moisture diffusion using a finite difference method. Figure 2 illustrates one of the three-stage accelerated ageing programmes and Table 2 gives the particulars of all of the conditioning programmes for the groups W, Y, A, B, C and D. Six notched compression specimens and a traveller, six interlaminar shear specimens and a traveller, one pre-traveller and one specimen for optical examination were conditioned for each group. After the specimens had been conditioned to the required moisture content they were stored in a controlled RH environment so as to retain the required moisture equilibrium level until ready for thermal spiking.

The travellers were periodically weighed to confirm the moisture uptake predictions. Table 3 shows both the predicted value for each environmental condition and the actual values of mean moisture contents before spiking achieved by the ILSS travellers. It can be seen that the moisture contents

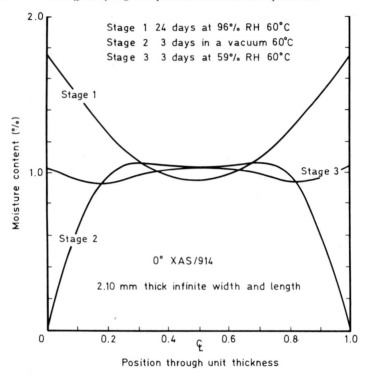

Stage 1  24 days at 96% RH 60°C
Stage 2  3 days in a vacuum 60°C
Stage 3  3 days at 59% RH 60°C

0° XAS/914

2.10 mm thick infinite width and length

Position through unit thickness

FIG. 2.   Typical through-the-thickness moisture profiles obtained using the three-stage accelerated ageing method.

before spiking were very close to those predicted. There was a slight variation between the pre-spike moisture contents of the notched compression specimens and the ILSS specimens (see Tables 4 and 5). It should be noted that the same environmental chambers were used to recondition the specimens between spikes.

Specimens in groups E, E+, F and F+ were simply conditioned at 96% RH and 60°C for 4 days before being thermally spiked. This did not result in a condition of moisture equilibrium being reached; instead a profile of moisture content was produced, as shown in Fig. 3a. It can be seen that although the moisture content at the outer surfaces was 1·8% the mean moisture content was only 0·72%. The distribution shown is that calculated using a one-dimensional model for the interlaminar shear specimen but that in the notched compression specimen is essentially the same.

TABLE 2

*Accelerated ageing of specimens for the simulated kinetic heating spike programme*

| Specimen group | | Moisture equilibrium condition required (%) | Stage 1 | Time at Stage 1 (days) | Stage 2 | Time at Stage 2 (days) | Stage 3 | Time at Stage 3 (days) | Total ageing time (days) |
|---|---|---|---|---|---|---|---|---|---|
| W and Y | ILS | 1·06 | 96% RH 60°C | 24 | Vacuum 60°C | 2 | 59% RH 60°C | 4+ | 30 |
| | Comp'n | | | 31 | | | | | 37 |
| A | ILS | 1·20 | 96% RH 60°C | 25 | Vacuum 60°C | 2 | 67% RH 60°C | 4+ | 31 |
| | Comp'n | | | 40 | | | | | 46 |
| B | ILS | 1·35 | 96% RH 60°C | 31 | Vacuum 60°C | 2 | 75% RH 60°C | 4+ | 37 |
| | Comp'n | | | 50 | | | | | 56 |
| C | ILS | 1·60 | 96% RH 60°C | 43 | Vacuum 60°C | 2 | 86% RH 60°C | 4+ | 49 |
| | Comp'n | | | 67 | | | | | 73 |
| D | ILS | 1·82 | 96% RH 60°C | 100 | — | — | — | — | 100 |
| | Comp'n | | | 144 | | | | | 144 |

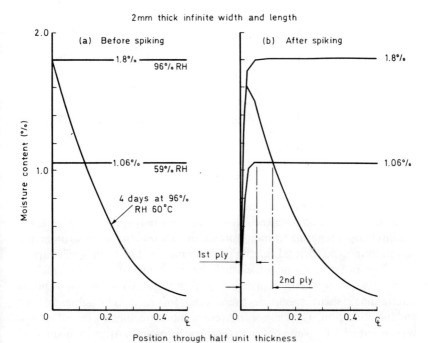

FIG. 3. Calculated moisture profile through-the-thickness of an ILSS specimen after a single 137°C spike.

TABLE 3
*Moisture levels in ILSS specimens*

| Specimen group | Spike temp. (°C) | Pre-conditioning and conditioning between spikes | Moisture content (%) (ILSS traveller) | | Moisture equilibrium reached after post-spike conditioning at 96% RH and 60°C (%) |
|---|---|---|---|---|---|
| | | | Before spiking | After spiking and at test | |
| V | — | Dry | 0 | — | 1·82 |
| W | — | 59% RH 60°C 1·06% $(M_\infty)$ | 1·06 | 1·06 | 1·82 |
| Y | 127 | 59% RH 60°C | 1·06 | 1·0 | 1·82 |
| | 137 | 1·06% $(M_\infty)$ | 1·06 | 0·96 | 1·79 |
| A | 127 | 67% RH 60°C | 1·22 | 1·32 | 1·86 |
| | 137 | 1·2% $(M_\infty)$ | 1·18 | 1·32 | 1·88 |
| B | 127 | 75% RH 60°C | 1·38 | 1·62 | 1·99 |
| | 137 | 1·35% $(M_\infty)$ | 1·35 | 1·63 | 2·02 |
| C | 127 | 86% RH 60°C | 1·62 | 2·08 | 2·20 |
| | 137 | 1·6% $(M_\infty)$ | 1·62 | 2·08 | 2·26 |
| D | 127 | 96% RH 60°C | 1·88 | 2·55 | — |
| | 137 | 1·82% $(M_\infty)$ | 1·88 | 2·51 | — |
| E | 127 | 96% RH 60°C | 0·72 | 1·63 | 1·92 |
| E+ | | for 4 days | 0·72 | 2·32 | 2·18 |
| F | 137 | 96% RH 60°C | 0·72 | 1·76 | 2·08 |
| F+ | | for 4 days | 0·72 | 2·39 | 2·38 |

### 3.4. Thermal Spike Programme

In order to examine the effects of thermal spiking on notched compression strength and interlaminar shear strength, the environmental conditioning and thermal spiking programme listed in Table 1 was carried out. This programme also included the determination of any cracking damage caused by thermal spiking. Sets of six notched compression and six interlaminar shear specimens were used for each environmental and spiking condition. The change in average moisture content for all groups was monitored by weighing the travellers before and after each spike.

With reference to Table 1, part 1 was carried out in order to examine the effects of the number of spikes on strength. Thus, three sets of specimens in

TABLE 4
*Notched compression strength*

| Specimen group | Number of spikes | Temp. of spikes (°C) | Moisture content before spiking (%) | Notched compression strength at RT | | | | Notched compression strength at 120°C | | | | M at test |
| | | | | Mean strength | | Percentage change in net mean strength from RT dry condition | Coefficient of variation (%) | Mean strength | | Percentage change in net mean strength from RT dry condition | Coefficient of variation (%) | |
| | | | | Gross (MN/m²) | Net (MN/m²) | | | Gross (MN/m²) | Net (MN/m²) | | | |
|---|---|---|---|---|---|---|---|---|---|---|---|---|
| V | 0 | — | 0 | 534 | 636 | — | 6·5 | 535 | 637 | +0·1 | 7·7 | 0 |
| W | 0 | — | 1·06 | 551 | 656 | +3·1 | 12·2 | 394 | 469 | −26·3 | 4·2 | 1·06 |
| Y | 10 | 127 | 1·06 | 527 | 627 | −1·4 | 4·8 | 397 | 472 | −25·8 | 5·5 | 1·06 |
| | | 137 | | 569 | 677 | +6·4 | 2·9 | 412 | 491 | −22·8 | 11·0 | 1·06 |
| A | 10 | 127 | 1·20 | 511 | 607 | −4·5 | 7·4 | 341 | 406 | −36·2 | 3·8 | 1·28 |
| | | 137 | | 490 | 582 | −8·4 | 1·7 | 331 | 394 | −38·0 | 5·9 | 1·32 |
| B | 10 | 127 | 1·35 | 489 | 582 | −8·4 | 7·6 | 327 | 390 | −38·7 | 9·8 | 1·62 |
| | | 137 | | 495 | 589 | −7·3 | 2·7 | 363 | 431 | −32·2 | 5·6 | 1·66 |
| C | 10 | 127 | 1·55 | 502 | 597 | −6·1 | 8·0 | 317 | 377 | −40·7 | 2·4 | 2·01 |
| | | 137 | | 502 | 597 | −6·1 | 3·3 | 293 | 347 | −45·4 | 4·4 | 2·14 |
| D | 10 | 127 | 1·80 | 472 | 561 | −11·8 | 10·0 | 275 | 327 | −48·6 | 4·8 | 2·61 |
| | | 137 | | 481 | 572 | −10·0 | 1·3 | 285 | 340 | −46·5 | 8·4 | 2·68 |
| E | 10 | 127 | — | 499 | 595 | −6·4 | 9·8 | 330 | 393 | −38·2 | 1·8 | 1·18 |
| E+ | 24 | 127 | — | 510 | 606 | −4·7 | 13·4 | 332 | 395 | −37·9 | 0·6 | 1·90 |
| F | 10 | 137 | — | 483 | 575 | −9·6 | 8·1 | 315 | 375 | −41·0 | 4·0 | 1·28 |
| F+ | 24 | 137 | — | 479 | 569 | −10·5 | 8·9 | 311 | 370 | −41·8 | 3·9 | 2·06 |

## TABLE 5
### Interlaminar shear strength

| Specimen group | Number of spikes | Temp. of spikes (°C) | Moisture content before spiking (%) | Interlaminar shear strength at RT | | | | Interlaminar shear strength at 120°C | | | | $M$ at test |
|---|---|---|---|---|---|---|---|---|---|---|---|---|
| | | | | Mean strength (MN/m²) | Percentage change in net mean strength from RT dry condition | Failure mode | Coefficient of variation (%) | Mean strength (MN/m²) | Percentage change in net mean strength from RT dry condition | Failure mode | Coefficient of variation (%) | |
| V | 0 | — | 0 | 121 | — | — | 1·0 | 81 | −33·0 | — | 4·3 | 0 |
| W | 0 | — | 1·06 | 109 | −9·9 | — | 2·4 | 64 | −47·1 | — | 1·0 | 1·06 |
| Y | 10 | 127 | 1·06 | 103 | −14·9 | — | 1·0 | 61 | −49·6 | — | 2·4 | 1·0 |
| | | 137 | | 99 | −18·2 | — | 3·8 | 62 | −48·8 | — | 1·6 | 0·96 |
| A | 10 | 127 | 1·20 | 105 | −13·2 | MS | 3·3 | 62 | −48·8 | MS | 2·4 | 1·32 |
| | | 137 | | 106 | −12·4 | MS | 5·2 | 61 | −49·6 | MS | 3·4 | 1·32 |
| B | 10 | 127 | 1·37 | 101 | −16·5 | MS | 2·5 | 54 | −55·4 | MS | 1·1 | 1·62 |
| | | 137 | | 105 | −13·2 | MS | 1·1 | 54 | −55·4 | MS | 3·2 | 1·63 |
| C | 10 | 127 | 1·62 | 103 | −14·9 | MS | 5·6 | 51 | −57·8 | SS | 2·2 | 2·08 |
| | | 137 | | 98 | −19·0 | SS | 0 | 51 | −57·8 | MS | 1·1 | 2·08 |
| D | 10 | 127 | 1·88 | 90 | −25·6 | SS | 1·7 | 46 | −62·0 | Plastic | 2·5 | 2·55 |
| | | 137 | | 92 | −23·9 | SS | 0·6 | 46 | −62·0 | Plastic | 1·3 | 2·51 |
| E | 10 | 127 | — | 107 | −11·6 | SS | 4·8 | 55 | −54·5 | Plastic | 2·1 | 1·63 |
| E+ | 24 | 127 | — | 98 | −19·0 | MS | 2·6 | 51 | −57·9 | Plastic | 1·1 | 2·32 |
| F | 10 | 137 | — | 104 | −14·0 | SS | 4·8 | 54 | −55·4 | Plastic | 1·1 | 1·76 |
| F+ | 24 | 137 | — | 89 | −26·4 | MS | 3·4 | 51 | −57·8 | Plastic | 3·4 | 2·39 |

MS, multi-shear; SS, single-shear.

group Y were given 1, 4 and 10 spikes at 127°C. A further three sets were given similar numbers of spikes at 137°C.

Part 2 was carried out to compare the effects of 10 spikes on laminates containing various evenly distributed levels of moisture content. Spiking was carried out at 127°C and 137°C.

The effect of a moisture gradient on strength performance was examined in part 3. Groups E and F were subjected to 10 spikes and groups E+ and F+ were subjected to 24 spikes.

Two ovens were used to provide the thermal spike environments, one held at 127°C and one at 137°C. Thermal spikes were applied to the specimens by placing them in the appropriate oven for 6 min at a constant temperature. All specimens were mounted in an upright position so that the two major surfaces were heated by the fan-assisted air circulating in the oven. On removal from the ovens all specimens were allowed to cool in air at room temperature. The temperature profiles for the two spikes, measured using a thermocouple mounted halfway through the thickness of a 3 mm thick control specimen, are given in Fig. 4.

It was calculated[8] that, after a single spike, the outermost plies of each specimen would have lost moisture to a depth of about 1·5 × ply thickness (Fig. 3b). The moisture was replaced by returning the specimens to their

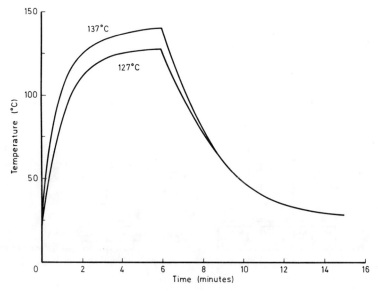

Fig. 4.   Thermal spike profiles.

respective pre-spiking conditioning environment after each spike. A minimum time of approximately 6 h was found to be needed for this moisture to be replaced. The spike repetition frequency for each group of specimens is given in Table 1.

### 3.5. Strength Testing

After the appropriate spiking programme has been completed a series of mechanical tests was performed. Notched compression testing was carried out using the rig[4] shown in Fig. 5. Tests were conducted in a Schenck 63 kN servo-hydraulic machine operating in the displacement mode and controlled using a DEC PDP 11/03 computer. Interlaminar shear testing was carried out in the same testing machine using the conventional short beam three-point bend technique. Control specimens (group V, dry) were tested within 10 min of being removed from the desiccator so their moisture content was taken to be zero. All other groups of specimens were tested shortly (within about 1 h) after removal from their environmentally controlled store. Three specimens were tested from each group except for the unspiked notched compression specimens (groups V and W), where five specimens were tested.

Those specimens tested at 120°C would have undergone some drying of the outer surfaces due to the test environment. The extent to which drying had an effect was calculated on the assumption that each specimen would

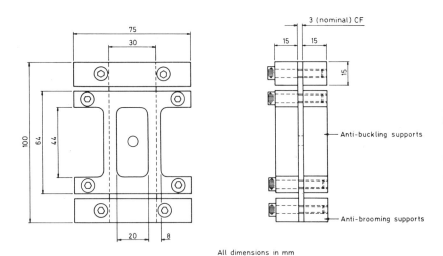

All dimensions in mm

FIG. 5.   Notched compression loading rig.

see, on average, 15 min at 120°C. This would be the most severe condition since some specimens needed less time to reach the test temperature and, moreover, most of the time would be spent reaching that temperature. Figure 6 shows the calculated[8] surface moisture loss for the interlaminar shear specimen for two of the groups.

### 3.6. Optical Examination

One specimen of each type from each group was not mechanically tested to determine its strength, but was examined for evidence of damage due to spiking. Samples were removed from one untested specimen from each group; the edges were polished using a metallographic technique and visually examined under an optical microscope (magnification 160 ×) for evidence of cracking damage.

### 3.7. Determination of the Effect of Spiking on Moisture Equilibrium Level

Another means of assessing the damaging effect of thermal spiking is to measure the change in moisture absorption capacity of the material. This was achieved by exposing the traveller from each group (except D) to a common post-conditioning environment of 96% RH and 60°C until a new

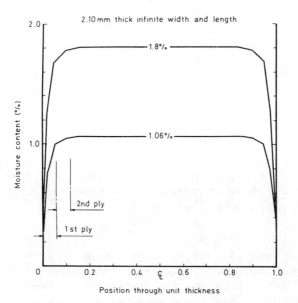

FIG. 6.    Calculated moisture profile through-the-thickness of an ILSS specimen after 15 min at 120°C.

moisture equilibrium had been reached. The moisture level could then be compared with that reached in an unspiked traveller. The results for the ILSS travellers are shown in Table 3.

## 4. EXPERIMENTAL RESULTS

An examination of the results follows and is presented in four parts. The first and second parts deal with the post-spike residual strength measurements, namely the notched compression and interlaminar shear strengths respectively. The third part describes the damage found during microscopic examination, and the fourth part looks at the changes in moisture uptake of the material after thermal spiking.

A full analysis of the results is given in Section 5, but some comments are included in this section in order to clarify the way in which the data is presented and to identify major trends.

### 4.1. Effect of Spiking on Strength

The results of the mechanical testing are given in Tables 4 and 5. For convenience the pre-spike and post-spike (strength test condition) moisture contents are repeated in each table. It is important to be able to distinguish between the loss in strength due to moisture uptake and/or temperature in unspiked specimens, and the additional loss due to damage caused by thermal spiking. Strength losses are therefore expressed throughout as a percentage of the strength of dry specimens tested at room temperature.

### 4.1.1. Notched compression strength

Consider first the unspiked specimens; it is clear from Fig. 7 that neither specimens containing a moisture content of 1·06% and tested at room temperature nor dry specimens tested at 120°C showed any loss in strength. However, a combination of moisture and a test temperature of 120°C reduced the strength by as much as 26%.

For specimens containing 1·06% moisture Table 4 shows that the inclusion of 10 thermal spikes at either 127°C or 137°C in the conditioning programme again did not have any further effect on the notched compression strength at either room temperature or at 120°C. At pre-spike moisture contents above 1·06% there was however a noticeable progressive degradation in strength. For a pre-spike moisture content of 1·8% the losses measured were up to 11% at room temperature and 48% at 120°C.

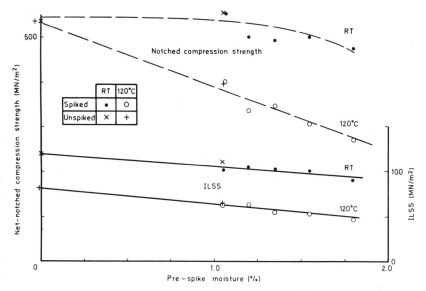

FIG. 7.   Room temperature and 120°C notched compression strength (net) and ILS strength after spiking at 137°C.

No significant difference between the strength after spiking at 127°C or at 137°C was apparent.

For those specimens which contained a moisture gradient (groups E, E+, F and F+) and which were tested at room temperature, there was a decrease in strength of about 5% for those which were spiked at 127°C and approximately 10% for those spiked at 137°C. However, a test temperature of 120°C resulted in a decrease in strength of about 40% with less difference between those spiked at 127°C and those spiked at 137°C.

All specimens failed at the net section, that is in the material either side of the central hole. A typical notched compression failure is shown in Fig. 8a.

### 4.1.2. Interlaminar shear strength

Again consider first the unspiked specimens. The uptake of 1·06% moisture alone reduced the room temperature strength by about 10% (Fig. 7), and testing a dry specimen at 120°C caused a reduction of 33%. A combination of the two conditions caused the strength to fall even further to give a total reduction of 47%.

Table 5 shows that the application of 10 thermal spikes at either 127°C or 137°C to specimens containing 1·06% moisture had only a small additional effect on the interlaminar shear strength tested at room temperature and no effect at 120°C. With a continued increase in the pre-spike moisture content, up to a level of 1·8%, reductions of up to a total of 26% in the room temperature and 62% in the 120°C strength property were measured. Again no significant change in the strength was apparent between spiking at 127°C or at 137°C.

At room temperature the moisture gradient specimens spiked at 127°C exhibited a total strength loss of 12% for 10 spikes, increasing to 19% for 24 spikes. Spiking at 137°C produced somewhat greater losses of 14% and 26% respectively. Testing at 120°C, however, revealed a total strength loss of about 56% with the number of spikes or spiking temperature having little effect.

Three different failure modes were observed (see Fig. 8b): single shear, multiple shear and plastic deformation. Table 5 also gives the failure modes for the different specimen groups.

### 4.2. Effect of Spiking on Laminate Cracking

Microscopic examination revealed visible signs of cracking in specimens taken from groups C and D (see Table 6). In group C, however, cracking was in the form of translaminar cracks in both blocks of three 0° plies in the notched compression specimen. This cracking is typical of cure cracking[3]

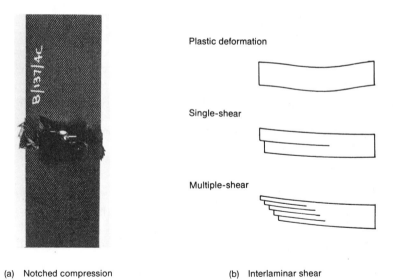

(a) Notched compression          (b) Interlaminar shear

Fig. 8.  Test specimen failures.

TABLE 6
*Damage sustained during simulated kinetic heating spike programme*

| Specimen group | Pre-spike moisture content (%) | Spike temperature (°C) | Damage |
|---|---|---|---|
| V | 0 | — | None |
| W | 1·06 | — | None |
| Y | 1·06 | 127 | None |
|   |   | 137 |   |
| A | 1·20 | 127 | None |
|   |   | 137 |   |
| B | 1·35–1·37 | 127 | None |
|   |   | 137 |   |
| C | 1·55–1·62 | 127 | None |
|   |   | 137 | Translaminar cracking of the block of three 0° plies in the notched compression specimen; none in ILS specimen |
| D | 1·80–1·88 | 127 | Interlaminar cracking |
|   |   | 137 | Interlaminar cracking |
| E | — | 127 | None |
| E+ | — | 127 | None |
| F | — | 137 | None |
| F+ | — | 137 | None |

associated with a blocked ply lay-up and was confirmed to be present in unspiked specimens. It was not therefore associated with thermal spike damage.

It has in fact been shown[9] that the likelihood of transverse cracking increases with the degree of ply blocking and that a block of three plies results in transverse strains close to the critical level. It is likely therefore that a full microscopic examination of all the cross-ply specimens would have revealed transverse cracks in groups other than group C.

Photographs of the type and degree of damage found in the group D specimens are given in Figs 9 and 10, together with schematic drawings summarising the damage found in each laminate configuration. Only group D specimens showed interlaminar cracking which is known to be typical of thermal spike damage.[3]

### 4.3. Effect of Spiking on Moisture Uptake

It is evident from the post-spike moisture contents that there was no significant effect due to spiking at 127°C and at 137°C at a pre-spike

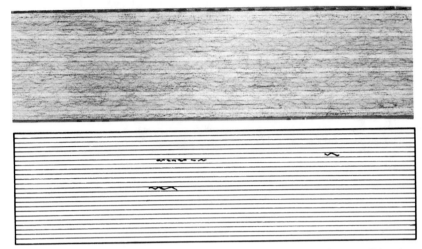

FIG. 9.   Thermal spike damage (notched compression specimen). Specimen conditioned to
1·8% moisture equilibrium level and spiked at 127°C (group D).

moisture content of 1·06% (Fig. 13 and Table 3). At a pre-spike moisture content of 1·2%, however, some change in the moisture kinetics after spiking was evident, as demonstrated by the increase in the post-spike moisture content to 1·32%. Further increases in the pre-spike moisture content showed a corresponding increase in the post-spike moisture content.

FIG. 10.   Thermal spike damage (ILS specimen). Specimen conditioned to 1·88% moisture
equilibrium level and spiked at 127°C (group D).

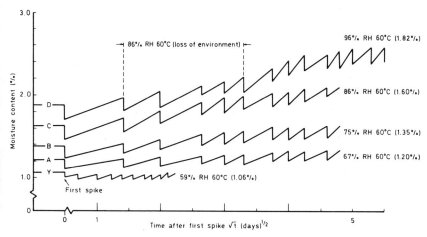

FIG. 11.   Effect of spiking at 127°C on moisture content of XAS/914 ILSS travellers (55 mm × 10 mm × 2 mm thick).

Post-spike conditioning at 96% RH and 60°C also showed that an increase had occurred in the moisture equilibrium levels (Table 3). These increases in moisture absorption are plotted in Fig. 12 as a percentage of the unspiked post-conditioning equilibrium level, Mp, which Table 3 shows to be 1·82%. From this plot it can be inferred that the moisture level at which these CFC laminates were affected by thermal spiking is ≥1·13%.

The effect of spiking on specimens with moisture gradients is shown in Fig. 13. It should be noted that in this figure it has been necessary to locate

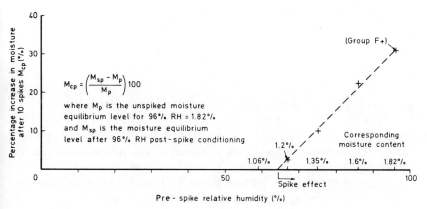

FIG. 12.   Percentage increase in moisture uptake after spiking at 137°C.

the origin of the time axis at the commencement of the conditioning (unlike Fig. 11 where it was located at the commencement of the spiking). It should also be remembered that, since these specimens had not reached moisture equilibrium when spiking commenced, the total moisture content would be expected to increase even if the spiking caused no damage. A comparison has therefore been made with the rate of moisture uptake in the unspiked travellers. Figure 13 shows that after spiking at 127°C the moisture content

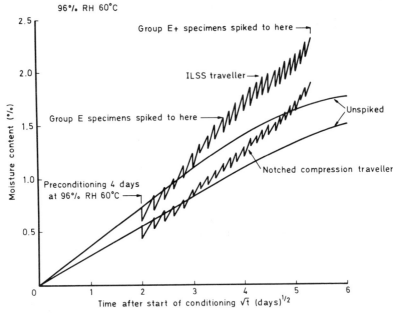

FIG. 13.    Effect of spiking at 127°C on the moisture content of XAS/914 travellers from groups E and E+.

of group E specimens had risen to 1·63% (10 spikes) and those in group E+ to 2·32% (24 spikes). The moisture content of the unspiked travellers had risen to only 1·29% and 1·70% respectively during these periods; similar increases in moisture content were exhibited by the notched compression specimens and their travellers. Similar patterns were observed for specimens spiked at 137°C (groups F and F+), no significant difference in moisture uptake being apparent between these and specimens spiked at 127°C.

## 5.  DISCUSSION

### 5.1.  Introduction

In this rather brief series of tests no attempt was made to dry out any of the specimens after spiking. Thus no assessment can yet be made as to whether the loss in strength is directly attributable to degradation due to the thermal spike excursion or indirectly due to the subsequent increase in the moisture content.

### 5.2.  Moisture Kinetics

It is important to note that for an unspiked specimen, even after prolonged exposure at 96% RH and 60°C, the maximum amount of moisture that can be absorbed (the equilibrium level) is 1·82%, and that the moisture is absorbed by the resin. Now Table 3 shows that in many cases, even at the stage before mechanical testing, moisture contents exceeding 1·82% were observed, and that post-spike conditioning often increased this level still further. It is clear therefore that the equilibrium moisture content of an epoxy matrix CFC can be increased when subjected to thermal spikes of ≥127°C. The pre-spike moisture content of a composite at which thermal spiking begins to affect the kinetics has been shown in Fig. 12 to be ≥1·13%. Since the undamaged resin cannot absorb more than 1·82% the additional moisture uptake must be associated with some type of damage, possibly in the form of micro-cracks. Some observations on the presence of micro-cracks will be given in Section 5.3.

In all cases thermal spiking of specimens containing a moisture gradient modified the moisture kinetics, resulting in an increased rate of moisture uptake. Now, as noted in Section 4.3, the moisture content of the moisture gradient specimens would continue to increase during the spiking period whether or not the spiking had caused any damage. Therefore it is not possible to draw a direct comparison with the behaviour, during the same period, of the specimens containing a uniform distribution of moisture. We can, however, compare the final equilibrium moisture contents after post-spike conditioning. It is, of course, only possible to compare specimens which had the same number of spikes (10). On this basis the moisture gradient specimens given 10 spikes at 127°C (group E) appear equivalent to specimens containing a uniform distribution of about 1·27% by interpolation between groups A and B. Similarly those given 10 spikes at 137°C (group F) appear equivalent to specimens containing a uniform distribution of about 1·4%.

## 5.3. Laminate Cracking

In Section 5.1 it was shown that the moisture uptake figures indicated that some form of micro-cracking damage might have occurred for pre-spike moisture contents of $\geq 1.13\%$. This could not, however, be confirmed by microscopic examination. Only for a uniform pre-spike moisture content of 1·8% was there clear evidence of interlaminar cracking. It is possible therefore that the damage which is induced is of a sub-microscopic nature.

As noted above permanent damage in the form of interlaminar cracks has been shown to start when the pre-spike moisture content (at equilibrium) is somewhere between 1·62% and 1·8%. The reason for the occurrence of this damage can perhaps be postulated using the evidence collected from other published work.[3,10] It has been shown in Ref. 11 that absorbed moisture lowers the glass transition temperature, $T_g$, of a resin matrix. Above $T_g$ there is, of course, a marked drop in the resin-dependent properties so that the effect of moisture uptake is to lower the temperature at which this reduction in properties will take place. In addition, work by Collings and Stone[3] using an identical fibre resin system, conditioned to give a through-the-thickness moisture profile, showed that, as the spike temperature approached $T_g$, damage in the form of interlaminar cracks was initiated. Similar work by McKague[10] supports this hypothesis. Clearly on this reasoning the work presented here suggests that the $T_g$ is sufficiently depressed by the absorbed moisture at an equilibrium level of $>1.62\%$ to permit permanent laminate damage to occur when subjected to thermal spikes at temperatures of $\geq 127°C$.

## 5.4. Notched Compression Strength

Before discussing the detailed results it is first necessary to consider the nature of the failure mechanisms. Potter and Purslow[12] have shown that, in notched compression specimens containing moisture, failure is initiated by out-of-plane buckling of the free surface, and that the onset of buckling is advanced when the modulus of the resin matrix is lowered by the uptake of moisture. Any interlaminar damage close to the free surface would, of course, also encourage the onset of buckling. Thus moisture uptake alone could be responsible for the loss in strength or it could be due to spike-induced damage.

It must, however, be remembered that one effect of thermal spiking is to increase the total moisture content prior to mechanical testing. Thus the group D specimens, which would have contained a uniform distribution of 1·8% moisture had they not been spiked, did in fact contain more than

2·5% at the time they were tested. The way in which this additional moisture was distributed is not known, but it is likely that there was a higher concentration close to the free surfaces. Now it has already been shown in Sections 5.1 and 5.2 that the additional moisture uptake is associated with some form of damage, but there is reason to think that there is little delamination micro-cracking until this damage becomes severe. The exact mechanism may, however, not be important; the distinction which needs to be drawn is between laminates which have not been spiked and which can absorb moisture up to a maximum of 1·82%, and those which have had their properties altered in such a way that they can absorb moisture more rapidly and to a higher equilibrium level. For convenience the latter group will simply be referred to as spike damaged. It is then the behaviour of these two groups of laminates when operating in a given RH environment that needs to be compared.

Unfortunately the series of tests reported here were primarily designed simply to obtain data on the total loss in strength resulting from the spiking of laminates containing various levels of moisture. There is therefore no direct comparison between undamaged and spike-damaged material. Some guidance will, however, be obtained by comparing the data obtained on the moisture gradient specimens with data available elsewhere on unspiked specimens.

Consider first the tests at room temperature (Table 4). No drop in strength occurs when the pre-spike moisture content is ≤1·06%. As the pre-spike moisture content increases so the notched compression strength begins to fall, until at a pre-spike moisture content of 1·8% there is a loss of 11%. It would be interesting to compare this loss with that which would have been exhibited by unspiked specimens exposed to the same environment, but no suitable data appears to be available. Instead we must examine the results obtained on the specimens containing a moisture gradient. Those tested here (groups E to F+) were only subjected to 96% RH at 60°C for 4 days before spiking and at that stage contained a mean moisture content of 0·72%; the decrease in strength after spiking was 5–10%. This may be compared with further results by Potter and Purslow[4] for unspiked specimens of the same material and lay-up which were subjected to the same environment for a longer period so as to produce a mean moisture content of 1%. In this case there was in fact a slight increase in strength. Thus it seems likely that the decrease in strength reported here is indeed due to thermal spiking.

Turning now to the results of the tests at 120°C, Table 4 shows that a uniform moisture content of 1·06% alone is sufficient to cause a drop in strength of 26% and that spiking causes no additional drop. Again, as the pre-spike moisture content increases the notched compression strength falls until at a pre-spike moisture content of 1·8% there is a loss of 48%. Once again we may attempt to quantify the role of spiking damage by examining the results on the moisture gradient specimens. The data reported here shows a drop of 38–42%, and Potter and Purslow[4] also report a drop of 40% on their unspiked specimens. The differences in the various moisture profiles make it impossible to draw any definitive conclusions but it can be postulated that quite a large part of the 48% drop reported above is perhaps due only to the moisture uptake and that the damage caused by the spiking does not degrade the notched compression strength very much further.

It is possible, by examining the results obtained on specimens from groups E and F, to compare the effect of 10 spikes on specimens having a moisture gradient with those containing a uniform through-the-thickness distribution of moisture. Testing at room temperature after spiking at 127°C revealed a drop in strength of about 6%, which is similar to that exhibited by specimens having a uniform distribution of some 1·2–1·3%. The effect of spiking at 137°C, however, was to produce a drop in strength of 10·5%, which is similar to that exhibited by specimens having a uniform distribution approaching 1·8%. Testing at 120°C revealed a pattern which was somewhat similar but less marked, spiking at 127°C producing a drop in strength similar to that exhibited by specimens having a uniform distribution of 1·35%, and spiking at 137°C producing a drop similar to that exhibited by specimens having a uniform distribution between 1·35% and 1·55%.

Thus it would appear that the post-spike notched compression strength of specimens containing a moisture gradient may be governed either by the value of the maximum moisture content or by the moisture content close to the free surface. In the particular specimens employed, of course, these two values were coincident. This is fully consistent with the observation made by Potter and Purslow[12] that failure was initiated by out-of-plane buckling at the free surface, because any delamination close to this surface would advance the onset of buckling. This effect would be less serious in thick material where there would be proportionally more cross-sectional area remaining to support the load. The fact that the moisture gradient effects were less marked in the tests performed at 120°C may perhaps be associated with the fact that, as noted in Section 3.5 and displayed in Fig. 6, the outer plies would have dried out to some extent before the failure load was reached.

## 5.5. Interlaminar Shear Strength

Consider first the room temperature results given in Table 5. A moisture content of 1·06% reduced the strength by about 10% and the addition of spiking reduced this further to 15–18%. For higher pre-spike moisture contents the reduction in strength was actually less than that at 1·06% moisture until a value of 1·62% was reached. The maximum reduction of 25% was, however, still at 1·88% moisture. As for the notched compression tests, there is a limited amount of data available elsewhere which sheds some light on the relative contributions of moisture uptake and spiking damage. Collings and Stone presented results on ILSS specimens of a similar material which had again been conditioned in an environment of 96% RH and 60°C. After 25 days the strength had dropped by 5% and this decrease continued until after 256 days the strength reduction was 17%. At this stage the moisture content of the specimens was sensibly uniform at about 1·8% so, by comparison with the specimens in group D in Table 5, the spiking appears to have had a small but measurable effect. They also thermally spiked some specimens which had been exposed to the environment for 58 days. By interpolation of their data the strength reduction of unspiked material at this stage would be expected to be about 7%. After 24 spikes at 135°C, however, the strength reduction observed was 26%. This is clearly comparable with specimens in group F+ which were subjected to the same spiking programme.

Once again the results are far from definitive but it could be concluded that whilst thermal spiking further reduces the strength of ILSS specimens containing modest amounts of moisture it has only a small additional effect on those that already contain the maximum of about 1·8%.

The results of ILSS tests at 120°C are given in Table 5 but in this case there was no additional evidence with which to separate the two effects. It can only be stated that a temperature of 120°C alone reduces the strength by 33% while combining this with a moisture content of 1·06% reduces it by 47%. At this level thermal spiking has no effect. The worst case is again after spiking at a moisture content of 1·88%, which produced a total reduction of 62%.

Again, by examining the results obtained on specimens from groups E and F, it is possible to compare the effect of 10 spikes on specimens with and without a moisture gradient. Post-spike testing at RT showed a drop in ILSS of only 12% for 127°C spikes and 14% for 137°C spikes, which is much the same as that observed in the specimens containing a uniform distribution of about 1·06% moisture. Testing at 120°C revealed a drop in ILSS of about 55% for both spike temperatures and this may be compared with the drop observed in specimens containing a uniform distribution of about 1·37%. It should, however, be noted that all the moisture gradient specimens exhibited a plastic mode of failure whereas the uniformly distributed specimens exhibiting a similar strength loss failed in a shear mode.

Since the maximum shear stress occurs at the mid-plane, one would expect the ILSS specimens to be less sensitive to the moisture content in the outer plies than were the notched compression specimens.

### 5.6. Failure Modes

It was noted that the failure mode of the notched compression specimens containing moisture was consistent, and was not affected by spiking or elevated temperature. All specimens failed by an out-of-plane buckling of the 0° fibres. The interlaminar shear specimens, on the other hand, failed in one of three modes, single shear, multi-shear or plastic deformation. At low moisture contents ( < 1·37%), at both room temperature and at 120°C, most failures were of the multiple-shear type. At moisture contents of > 1·37% and < 1·55% most failures occurred in single shear. For moisture contents of > 1·62% failure was predominantly single shear with an occasional plastic deformation mode at room temperature. At 120°C the plastic deformation failure was dominant. This change of failure mode from shear to plastic deformation suggests that a combination of spiking and increased moisture uptake had brought about a reduction in the shear modulus of the material and/or a reduction in the strength of the fibre/resin interface bond.

### 5.7. Design Considerations

For thin sections the post-spike notched compression strength depends on the maximum moisture content or on the moisture content close to the free surfaces. Earlier published work[13] suggested that the maximum moisture content likely to be attained in service would be between 1·0% and 1·2%. At this moisture level, although there is a significant reduction in strength at 120°C, the addition of spiking does not result in much further reduction in strength. However, a recent assessment[14] has shown that the worst moisture content worldwide is likely to be nearer 1·6%. At this level the effect of thermal spiking is to increase the moisture content still further, resulting in a reduction in strength additional to that caused by temperature alone. As noted above, this further strength reduction is smaller than that caused by temperature alone; nevertheless, design considerations should take it into account. The presence of a moisture gradient should also be considered because the post-spike notched compression strength of thin sections is considerably reduced by moisture in the outer plies. Because failure is initiated by the out-of-plane buckling of the outer plies the effect is reduced as the thickness of the section increases.

## 6. CONCLUSIONS

It should be noted that the following conclusions do not necessarily hold for any fibre/resin system other than the XAS/914 system employed in these investigations, nor for temperature excursions above 137°C.

### 6.1. Moisture Uptake

(1) Thermal spikes of $\geq 127$°C can increase both the rate of moisture uptake and the equilibrium moisture content.

(2) The pre-spike moisture content at which these thermal spikes affected the equilibrium moisture content was $\geq 1.13\%$.

(3) The implication of (1) and (2) is that spiking must cause matrix damage but no such damage was observed except at a pre-spike moisture content of 1.8%.

(4) Thermal spiking modified the moisture kinetics of specimens containing a moisture gradient, resulting in an increased rate of moisture uptake. The final equilibrium moisture content of these specimens was much the same as that of specimens which had contained a uniform pre-spike moisture content of 1.3%.

### 6.2. Mechanical Strengths

The combined effect of moisture content, thermal spiking and testing temperature on the mechanical strength is quite complex and the experimental programme was not sufficiently comprehensive for definitive conclusions to be drawn. On the available evidence, however, some tentative conclusions have been drawn.

(5) The notched compression strength at room temperature is not greatly affected by either moisture uptake or thermal spiking.

(6) The notched compression strength at 120°C is clearly decreased by moisture uptake. The effect of thermal spiking is less certain but it appears to result in only a modest further decrease in strength.

(7) When moisture gradients are present it seems probable that the notched compression strength is largely governed by the moisture content at the free surface.

(8) The interlaminar shear strength at room temperature of specimens

containing modest amounts of moisture (say 1·06%) is significantly reduced by spiking damage. The effect appears to be less marked, however, on specimens whose strength has already been degraded by high levels of moisture uptake.

(9) The interlaminar shear strength at 120°C is significantly reduced by the uptake of moisture; there is insufficient evidence to state whether spiking damage causes a further decrease.

## REFERENCES

1. McKague, E. L., Halkias, J. E. and Reynolds, J. D., Moisture in composites: the effect of supersonic service on diffusion, *J. Comp. Mater.*, **9** (January 1975).
2. Demuts, E. and Shyprykevich, P., Accelerated environmental testing of composites, *Composites*, **15**, No. 1 (January 1984).
3. Collings, T. A. and Stone, D. E. W., Hygrothermal effects in CFC laminates: damaging effects of temperature, moisture and thermal spiking, *Composite Structures*, **3** (1985), 341–378.
4. Potter, R. T. and Purslow, D., The effect of pre-loading on the environmental degradation of carbon fibre reinforced plastics, RAE Technical Report 83029, 1983.
5. Shen, C. H. and Springer, G. S., Moisture absorption and desorption of composite materials, *J. Comp. Mater.*, **10** (January 1976).
6. Collings, T. A. and Copley, S. M., Moisture diffusion in carbon fibre reinforced plastics, RAE Technical Report (to be published).
7. Collings, T. A. and Copley, S. M., On the accelerated ageing of CFRP, *Composites*, **14**, No. 3 (July 1983).
8. Copley, S. M., A computer program to model diffusion and its application to accelerated ageing of composites, RAE Technical Report 82010, 1982.
9. Collings, T. A. and Stone, D. E. W., Hygrothermal effects in CFC laminates: strains induced by temperature and moisture, *Composites*, **16**, No. 4 (October 1985).
10. McKague, L., Environmental synergism and simulation in resin matrix composites, ASTM STP 658, 1977, pp. 193–204.
11. Bueche, F. and Kelly, N. F., Influence of moisture content on the glass transition temperature of resin, *J. Polym. Sci.*, **45** (1960), 267.
12. Potter, R. T. and Purslow, D., The environmental degradation of notched CFRP in compression, *Composites*, **14**, No. 3 (July 1983), 206–225.
13. Hendrick, I. G. and Whiteside, J. B., Effects of environment on advanced composite structures, AIAA Conference on aircraft composites: the emerging methodology for structural assurance, San Diego, California, March 1977.
14. Collings, T. A., The effect of observed climatic conditions on the moisture equilibrium level of fibre-reinforced plastics, *Composites*, **17**, No. 1 (January 1986), 33–41.

# 27

# Moisture Influence on Edge Delamination in Advanced Composite Structures under Hot-wet Compression Loading

H. CHAOUK and G. P. STEVEN

*Department of Aeronautical Engineering, University of Sydney, New South Wales 2006, Australia*

## ABSTRACT

*The fracture behaviour of graphite/epoxy composites is of current interest, particularly with regard to their durability and damage tolerance. Delamination of such composites will probably constitute one of the most frequent defects and therefore should be extensively studied under all conditions.*

*In this study, an experimental investigation into the influence of moisture on the fracture mechanisms of near surface edge delamination under hot-wet compression was performed. This investigation produced non-linear behaviour between the applied load and the axial shortening, which has required a change in the method of analysis to determine the fracture parameters. An energy release rate approach for this non-linear behaviour on graphite/epoxy AS4/3501-6 composite was used and the results of this are discussed. Delamination was simulated by inserting Teflon film at an appropriate location on one side and close to one surface of the specimen during the lay-up process.*

*In the experimental investigation a special specimen size, a modified Iitri compression testing fixture and a temperature-controlled chamber were implemented. Non-destructive testing was also performed to provide accurate information on delamination initiation. This included acoustic emission, dye penetrant with X-rays and ultrasonic inspection.*

## INTRODUCTION

A commonly observed failure mode in laminated composites is delamination between the composite layers. This unique phenomenon is not found in metals. Delamination usually takes the form of separation of plies and is generally initiated at some form of geometric boundary such as voids, microcracks and foreign inclusions, or at some design detail such as a free edge, ply drop-off, co-cured joints or bolted joints. Such delaminations cause structural degradation, stiffness reduction and lead ultimately to failure at stresses below the design levels for an undamaged laminate. Therefore an understanding of some of the basic delamination behaviour is of great importance in assessing the structural integrity of advanced composite materials and structures.

Earlier studies have been reported on this subject which have dealt extensively with the strain energy release rate ($G$). Some of these studies[1-3] involved tension-loaded specimens for Mode I delamination and linear finite element analysis. Others have different modes of fracture[4-7] and use the compliance method based on a pure linear relationship. Recent work[6] has investigated the effect of moisture and temperature on certain lay-ups under particular loading conditions. In most of these studies, the complexity due to the asymmetry was avoided by creating near-surface delaminations on both surfaces of the specimens.

Compression testing of delaminated specimens encounters several difficulties[8,9] and requires a careful and accurate procedure for data acquisition and result processing. Previous work by the authors[10] has shown that some laminates with certain ply configurations experience closure effects where the delaminated portion closes inwards under compression loading. It was also shown[9] that these particular lay-ups behave non-linearly under this loading condition.

In this chapter, we present our studies on the influence of moisture on the behaviour of near-surface edge delamination in laminates under hot-wet pure compression loading and we apply the concept discussed in Ref. 9 for the analysis of the energy release rate under non-linear response. It is considered that the studies reported herein are very significant in broadening our understanding of delamination initiation behaviour and propagation.

## SPECIMEN PREPARATION AND EXPERIMENTAL PROCEDURE

The specimens used in this investigation were fabricated from the unidirectional AS4/3501-6 graphite/epoxy tapes. Four panels with symmetrical $[+45/-45/0/90/0/+45/-45/0/+45/-45/0]_s$ lay-up were prepared and cured according to the manufacturer's recommendations. Delamination was introduced by inserting Teflon films to have the effect of delamination lengths of 6·35, 9·53 and 12·7 mm between the second and third ply from one surface and on one side only, as shown in Fig. 1. The modified dimensions of the specimens were chosen to allow for a wider width to avoid Euler's buckling in compression.

Twenty specimens were immersed in a temperature-controlled water bath at 60°C for the period of 2 months till the moisture content had exceeded the required value of 1·0% by 0·5% by weight. Figure 2 shows the measured moisture absorption for these specimens in water. Only 1% was required for testing, but the excess was absorbed to compensate for the loss of moisture during the bonding and testing processes. The excess of 0·5% was determined experimentally. These specimens were tested at final 1·0% moisture and 80°C as this has been designated the critical design condition in compression.

Another 20 specimens were tested dry at room temperature. All

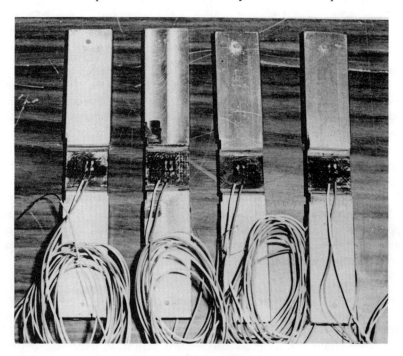

FIG. 1.    Specimen configuration.

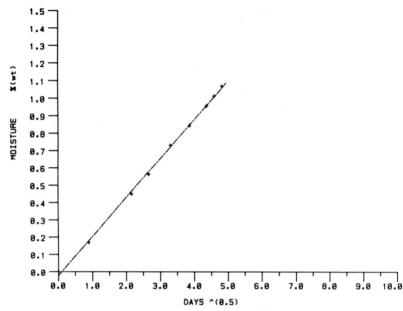

FIG. 2.    Moisture absorption of AS4/3501-6 composite in 70°C distilled water.

specimens having been cut to appropriate dimensions for this test, they were bonded at their ends to aluminium tabs to give a 25·4-mm gauge length and were strain gauged on both sides and clipped gauged along one edge. The bonding material was a hot curing FM300 adhesive. Thicknesses were checked carefully before and after the bonding process to ensure a consistency of thickness distribution over each specimen. Compression was applied in an Instron 1195 series via an Iitri compression jig with modified jaws to accommodate the above-mentioned specimens, as shown in Fig. 3. ·Wet specimens were tested in a controlled heat chamber using the same compression procedure outlined in Ref. 9. Non-destructive testing was also performed to acquire accurate results for delamination initiation. Such non-destructive testing, which is fully described in Ref. 9, included acoustic emission, dye penetrant with X-rays and ultrasonic inspection.

## ANALYSIS AND DISCUSSION

We have observed in this investigation that the above laminates tested behave non-linearly, as previously shown in Ref. 9, under the above-described loading condition. We have also observed that the effect of

Fɪɢ. 3. Compression apparatus.

moisture and temperature on the behaviour of these laminates is not negligible but has a considerable influence on its failure characteristics. Results for the load deflection curve for all the specimens are shown in Fig. 4. These results indicate the same behaviour. Based on this observation, it became evident once again that a linear theory for calculating the energy release rate based on the compliance method could not be used, and that an appropriate term for $G$ should be used to cater for this behaviour. Using the

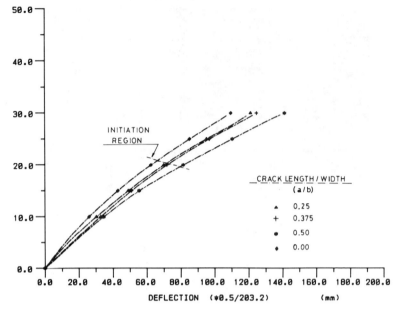

FIG. 4.    Instron load and deflection measurements.

same concepts from Ref. 9 taken from the fundamental theory of fracture mechanics and assuming that the load deflection curve has the following relationship

$$P = C\delta^n \tag{1}$$

where $P$ = load, $\delta$ = deflection and $C$ is a function of crack length ($a$), then the energy release rate is of the form

$$G_c = \left[ \frac{\delta^{n+1}}{n+1} \frac{\partial C}{\partial A} \right]_c \tag{2}$$

where $A$ is the delaminated area.

For linear behaviour this reduces to the usual expression

$$\left[ G = \frac{p^2}{2} \frac{\partial K}{\partial A} \right]_c \quad \text{where } K = \delta/P \tag{3}$$

Using this information and fitting the data for the hot-wet compression loading to eqn. (1), we obtain the following:

$$P = C\delta^{(0.795)}$$

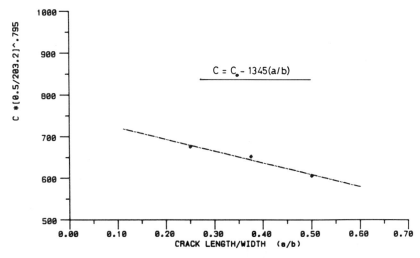

FIG. 5.   *C* versus crack length.

where *C* appears to have a linear relationship with the delamination length/width ratio, as shown in Fig. 5:

$$C = C_0 - m(a/b) \quad \text{where } m = 1345$$

For room temperature compression loading the following results were obtained:

$$P = C\delta^{(0.78)} \quad \text{and} \quad C = C_0 - 3385(a/b)$$

From these calculations, the energy release rates for edge delaminated AS4/3501-6 of this laminate configuration under compression loading were found to be:

For hot-wet    $G_c = 1.58 \, \text{N/m}$
For dry RTD    $G_c = 2.09 \, \text{N/m}$

Our observations have demonstrated clearly the influence of moisture and temperature on these laminates under the above conditions. $G_c$ was reduced by an order of 24% from dry to hot wet conditions. This was shown to be the case for other materials[1] but the effect was of lesser magnitude. Figure 6 shows this comparison for both conditions. The initiation strain had also increased from dry to hot wet, as expected and shown in Fig. 7. It should be emphasized that the specimen preparation and the compression technique can have an influence on the final results and

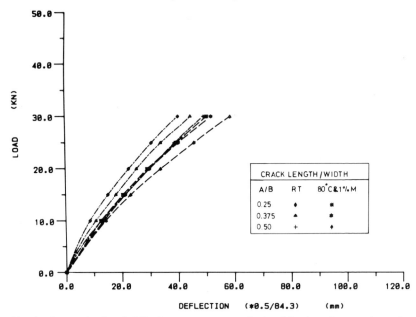

FIG. 6.   Instron load and deflection measurements of edge delaminated composite under
RTD and hot-wet compression loading.

that every effort should be implemented to accurately determine the
initiation region and the moisture absorption and desorption in every test.

Previous work has shown, using 3-D finite element analysis, that Mode II
for this lay-up in compression is of considerable magnitude, which would
indicate that the moisture effects may vary depending on the inter-influence
of the constituents of the mixed mode energy release rate. It was indicated
in Ref. 6 that samples with Mode II behaviour showed a significant
reduction in the fracture energy with increase in temperature and a slight
reduction with moisture content. These reductions were more significant in
these specimens than in other specimens having other modes of failure. This
is directly related to surface ply configuration and material properties.

These values for $G_c$ include the effect of all failure modes and represent
accurately the energy required to initiate edge delamination of this
composite under these conditions. The non-linearity is related to the
asymmetry of the delaminated plies and produces coupling between the
axial and bending stresses which adds more influence to the overall mixed
energy release. This has to be seriously considered. We have also observed
in these experiments that laminates which have predominantly Mode II

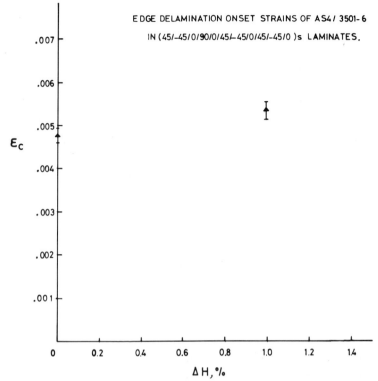

F<small>IG</small>. 7. Effects of temperature and moisture on the onset strain of edge delaminated specimens.

behaviour lack a suitable non-destructive test which accurately locates the front of edge delamination, and we therefore suggest that attention be given to a means of accurately monitoring the size of a delamination during its growth.

## CONCLUSION

Once again we have observed the effect of non-linearity on the behaviour of a composite laminate and the influence of moisture and temperature on its characteristics. Their influences appeared to be of considerable magnitude on laminates which have predominantly Mode II failure behaviour and do not have any Mode I. Therefore, when designing such laminates an appropriate methodology should be applied for the particular ply configuration to achieve the desired weight saving and damage tolerance.

# REFERENCES

1. CROSSMAN, F. W., WARREN, W. T., WANG, A. S. D. and LAW, G. E., Initiation and growth of transverse cracks and delamination in composite laminates, *J. Comp. Mater.*, supplemental volume (1980), 88–106.
2. O'BRIEN, T. K., Characterization of delamination onset and growth in a composite laminate, Damage in Composite Materials, ASTM STP 775, June 1982, p. 140.
3. DE CHARENTENAY, F. X. *et al.*, Characterizing the effects of delamination defects by Mode I delamination test, Effect of Defects in Composite Materials, ASTM STP 836, pp. 84–103.
4. O'BRIEN, T. K., Mixed mode energy release rate effects on edge delamination of composites, Effect of Defects in Composite Materials, ASTM STP 836, 1984, pp. 125–142.
5. JONES, R. and CALLINAN, R. J., Analysis of compression failure in fibre composite laminates, *Progress in Science and Engineering of Composites*, ICCM IV, Vol. 1, 1982, pp. 287–296.
6. RUSSELL, A. J. and STREET, K. N., Moisture and temperature effects on the mixed mode delamination fracture of unidirectional graphite epoxy, Delamination and Debonding of Material, ASTM STP 876, 1985.
7. O'BRIEN, T. K. and RAJU, I. S., Residual thermal and moisture influences on the strain energy release rate analysis of edge delamination, *J. Comp. Technol. Res.*, **8**, No. 2 (1986).
8. CHAMIS, C. C. and SINCLAIR, J. H., Longitudinal compression failure modes in fibre composites: end attachment effects on Iitri-type test specimens, *J. Comp. Technol. Res.*, **7**, No. 4 (1985), 129.
9. CHAOUK, H. and STEVEN, G. P., Non-linearity in edge delamination of advanced composites under compression loading, to be published in the Proceedings of the Sixth International Conference on Composite Materials, 1987.
10. CHAOUK, H. and STEVEN, G. P., Closure effects in delaminated graphite epoxy laminate under compression, to be published in *J. Comp. Struct.* (1986).

# 28

# Sliding Wear and Fretting Fatigue Performance of Polymer Composite Laminates

K. Friedrich

*Polymer and Composites Group, Technical University Hamburg-Harburg, 2100 Hamburg 90, Federal Republic of Germany*

K. Schulte

*Institute for Materials Science, DFVLR, 5000 Cologne 90, Federal Republic of Germany*

and

S. Kutter

*Institute for Materials Science, Ruhr-University Bochum, 4630 Bochum, Federal Republic of Germany*

## ABSTRACT

*In special applications, composite structures can be subjected to wear or fretting fatigue loading conditions. In the present study, two different carbon fiber/epoxy resin laminates were investigated with respect to their sliding wear resistance against steel, and with respect to their fatigue behavior without and with a simultaneous fretting component. Depending on the damage caused in the load bearing 0°-layers of the laminates, an additional fretting wear situation can result in a more or less reduction of the composites fatigue life. A correlation between the sliding wear experiments and the fretting fatigue test results cannot easily be established.*

## INTRODUCTION

As with other materials, composites also can be subjected to different loading conditions when being used as structural components in the design.

Quite often, loading is caused by mechanical stresses (static or fatigue) in more or less aggressive environments and the result can be failure of the components by crack development and final fracture. In some special applications composite structures may experience loading by foreign body contact, leading to wear of the material, followed by dimensional instability of the parts and even finally by failure of total units. Typical examples are bearings, gears or pipelines. If local wear and fatigue load act together on structural parts, often the typical prerequisites for a fretting fatigue situation are fulfilled. The latter exactly occurs when repeated loading of a structural part causes oscillatory sliding movement at material interfaces in the design. This movement, in turn, must induce at or near points of contact between the faces, stresses of sufficient intensity to cause surface damage, for example in form of cracking.[1] It is known for numerous engineering applications that these fretting fatigue requirements are often encountered, and that they have often led to failure of metallic parts in an unpredictable manner. The most common examples are flanges and bolted joints, other design details that can give fretting fatigue problems are multi-layer leaf springs.

As polymer composite laminates are now frequently used for such purposes, it is of interest to know how they perform under these kinds of loading conditions. The present chapter reports about some sliding wear studies and some fretting fatigue experiments carried out with two different carbon fiber–epoxy matrix laminates.

## EXPERIMENTAL

Two different 55 vol % carbon fiber/epoxy laminates were used for testing:

Laminate A: $(\pm 45, 0, \pm 45_3, 90, \pm 45)_{2s}$, thickness $= 6.2$ mm
Laminate B: $(0_2, 90_2, 0_2 90_2)_s$, thickness $= 2.1$ mm

The typical lay-up of laminate B is shown in Fig. 1.

Sliding wear tests of both laminates were performed against rotating 100 Cr6-steel rings in three different directions, at a contact pressure of $p = 2.8$ MPa and a sliding velocity $v = 0.6$ m/s. Sliding in the $x$-direction (reference direction of the laminates) on the $x$-$y$-plane was designated as P, sliding in the $y$-direction on the same plane as AP (cf. Fig. 2). When sliding was performed in the $z$-direction on the $y$-$z$-plane (laminate B) or the $x$-$z$-plane (laminate A), this was designated as N (Fig. 2). The tests were carried out at

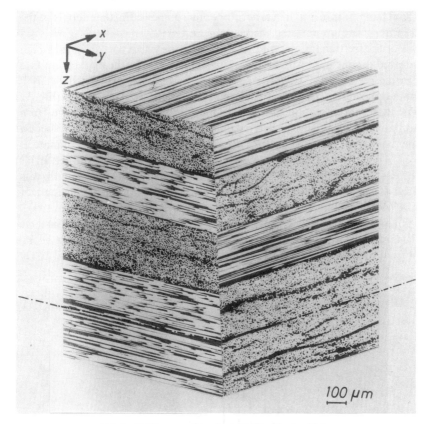

FIG. 1. Half lay-up of laminate B: $(0_2, 90_2, 0_2, 90_2)_s$.

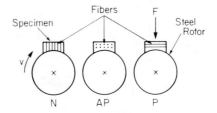

FIG. 2. Definition of fiber orientation in a unidirectional laminate with respect to sliding direction.

room temperature and in a laboratory environment. Further details of the testing procedure are described elsewhere.[2]

For the fretting fatigue studies a special grip system was designed which allowed fatigue of tensile samples and simultaneously the application of a fretting component. This was performed by pressing two fretting pins P from opposite sides onto the surface of the fatigue loaded samples (Fig. 3). The larger cross-section of the 200 mm long tensile bars machined from laminate A ($= 6 \times 6$ mm$^2$) enabled positioning of the fretting pads either on the *x*-*y*-planes or on the *x*-*z*-planes. The thinner samples of laminate B ($= 2 \times 6$ mm$^2$) were only tested with fretting pads on the *x*-*y*-surface. In all the cases, position No. 3 (cf. Fig. 3), i.e. the middle of the gage length, was used for applying the fretting component, which normally consisted of TiAl6V4-pins with a diameter of 5 mm.

In some additional tests, a smaller (2·5 mm) and a larger diameter (10 mm) were used, but in all cases the contact pressure as a result of the normal load $F_N$ was maintained constant as $p_N = 22, 92$ MPa.

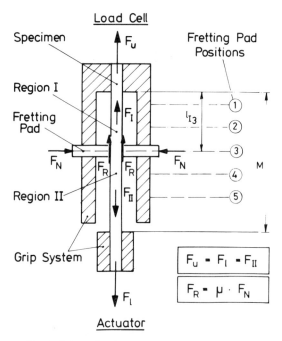

FIG. 3. Schematic of the fretting fatigue test device.

## RESULTS AND DISCUSSION

**Sliding Wear Tests**

The results of the sliding wear experiments are plotted in terms of the specific wear rate $\dot{w}_s$ (in $10^{-6}$ mm$^3$/N m), as derived from measuring the volume loss per sliding distance under a given contact load (Fig. 4). Laminate A does not exhibit any distinct differences in wear rate as a function of sliding direction. With respect to P and AP, this is not surprising, because in both cases fibers are in-plane oriented under 45° to the actual sliding direction. In N-orientation, a wide variety of differently oriented fibers are exposed to the rotating steel counter-surface, and it seems that this mixture of different fiber orientations wears as fast as the 45°-in-plane-oriented fibers in the P- and AP-cases.

A clearly different trend is observed for laminate B. When all the fibers are in plane and perpendicular to the sliding direction (AP), the specific wear rate is highest. On the other hand, the lowest wear is found for the parallel, in-plane fiber arrangement (P). A mixture of fibers oriented normal to the sliding plane (N) with some others in plane and perpendicular to the sliding direction leads to intermediate values of the specific wear rate. These trends are well known from studies of other investigators, i.e. Ref. 3, and will not be further discussed here.

**Fretting Fatigue Studies**

A comparison between laminate A and B on the basis of a sliding wear component in the *x*-direction on the *x-y*-surface, i.e. the P-situation, leads

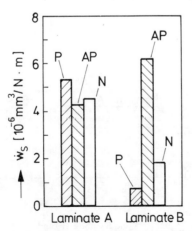

FIG. 4. Specific wear rate $\dot{w}_s$ of laminates A and B, and the effect of sliding direction.

to the conclusion that laminate A wears about seven times faster than laminate B, under the test conditions chosen. This would lead to the conclusion that a fretting wear component applied in the same way to fatigued samples of both laminates can cause much more damage and possibly much earlier fatigue failure of laminate A than of laminate B. Figure 5 illustrates that the opposite result is observed, and this effect is the more pronounced the higher is the contact pressure on the fretting pads.[4] The present figure contains the results obtained under the highest contact

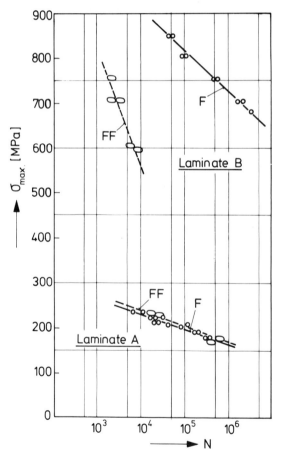

FIG. 5. Fatigue curves of laminates A and B without (F) and with fretting (FF) by two Ti-alloy pads (diameter $D = 5$ mm) under a normal load $F_N = 450$ N.

pressure chosen ($p_N = 22 \cdot 92$ MPa), but almost no effect of the fretting component on the fatigue life is seen for laminate A.

Laminate B having a four-fold higher fatigue strength than A (cf. $A_F$ versus $B_F$) is, however, very sensitive to the fretting load. The latter results in a reduction of the fatigue life of this laminate by about 3 orders of magnitude (cf. $B_F$ versus $B_{FF}$) when fatigue tests are performed at an upper load of $\sigma_{max} = 700$ MPa (sinusoidal tension–tension loading at a ratio of maximum to minimum load of $R = 0 \cdot 1$). Changing the diameter of the fretting pads, thus the contact area $A_0$, but maintaining the contact pressure $p_N$ constant, yields at the same maximum fatigue load drastic changes in the number of cycles to failure of laminate B-samples (Fig. 6). Increasing the diameter reduces the fatigue life and vice versa.

The question why laminate A in spite of its higher wear rate is not affected in its fatigue life by a severe fretting component on the laminate surfaces (neither on $x$-$y$- nor on $x$-$z$-plane) is not yet completely understood. One reason is probably based on the very low or even zero amount of fretting damage of the major tensile load bearing $0°$-layers of this laminate. Otherwise, the well detectable surface damage of $\pm 45°$-layers should

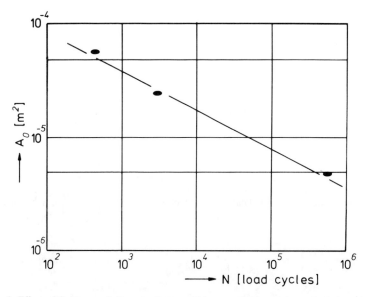

FIG. 6. Effect of fretting pad diameter (expressed in contact area $A_0$) on the fatigue life of laminate B, tested under a normal fretting pressure $p_N = 22 \cdot 92$ MPa and a fatigue stress maximum of $\sigma_{max} = 700$ MPa.

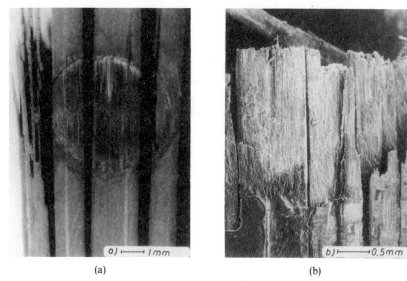

(a)    (b)

FIG. 7. Fretting damage on specimen surfaces. (a) $x$-$z$-plane of laminate A with high amount of wear in the fretted area, but final failure of the specimen away from the fretting position (as always for laminate A; fatigue direction vertical). (b) $x$-$y$-plane of laminate B, showing fretting damage and final rupture of the specimen in the fretted area.

TABLE 1

*Average values of fatigue life under different fretting conditions ($N_{FF}$), reduction factor $\bar{\delta}_-$ (in relation to $N_F = 2289000$ for laminate B at $\sigma_{max} = 700\,MPa$), and the corresponding damage ratio $\alpha/\beta$. For laminate A the highest value of $\alpha/\beta$ achieved in this study was $74 \times 10^{-6}$ and no effect in fatigue life reduction by fretting could be detected under these circumstances*

| $D$ (mm) | $F_N$ (N) | $p$ (MPa) | $\bar{N}_{FF}$ (cycle) | $\bar{\delta}_-$ | $\dfrac{\alpha/\beta}{\left(\dfrac{mm^2/cycle}{mm^2}\right)}$ |
|---|---|---|---|---|---|
| 5 | 150 | 7·64 | 979 945 | 2·3 | $300 \times 10^{-6}$ |
| 5 | 300 | 15·28 | 8 505 | 269 | $600 \times 10^{-6}$ |
| 5 | 450 | 22·92 | 3 185 | 719 | $896 \times 10^{-6}$ |
| 2·5 | 112.5 | 22·92 | 548 420 | 4.2 | $448 \times 10^{-6}$ |
| 10 | 1 288 | 22·92 | 450 | 5 087 | $1 074 \times 10^{-6}$ |

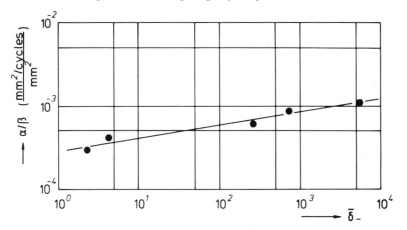

FIG. 8. Plot of damage ratio ($\alpha/\beta$) of load bearing $0°$-layers of laminate B versus the reduction factor $\overline{\delta}_-$ in the fatigue life of this laminate due to fretting ($\sigma_{max} = 700$ MPa).

clearly have an effect (Fig. 7). The idea of the role of load bearing $0°$-layers on the effect of a fretting component on fatigue life reduction can be emphasized by a more careful comparison of the results obtained with laminate B. Table 1 illustrates for this laminate how the average reduction factor $\overline{\delta}_-$, defined as

$$\overline{\delta}_- = N_F/N_{FF} \qquad (1)$$

(where $N_F$ is equal to the average number of cycles to failure under fatigue (F) and $N_{FF}$ under fretting fatigue (FF) conditions) changes with testing parameters ($D$ = fretting pad diameter; $F_N$ normal fretting load). Further included is a ratio $\alpha/\beta$, which characterizes the damaged part of load bearing $0°$-layers per load cycle ($\alpha$ in $10^{-6}$ mm$^2$/load cycle) relative to the share of load bearing $0°$-layers in the total cross-section of laminate B ($\beta$ in mm$^2$).

It becomes obvious that $\overline{\delta}_-$ increases with the ratio $\alpha/\beta$. A double logarithmic graph of this correlation is shown in Fig. 8.

## CONCLUSION

Sliding wear studies against rotating steel rings and tension fatigue experiments without and with an additional fretting component in the form of Ti-alloy fretting pads have been performed with two different carbon fiber/epoxy matrix laminates. Although laminate A with a ($\pm 45, 0, 45_3, 90, \pm 45)_{2s}$ lay-up wears much faster than laminate B with ($0_2, 90_2, 0_2, 90_2)_s$ structure, the latter one is much more sensitive to fretting while fatigued. One reason was found to be based on the amount of damage which occurs in the load bearing $0°$-layers. On the one hand, it is affected by the normal contact pressure and the diameter of the fretting pad, on the other hand, the composite structure in the contact area is of high importance (high degree of damaged $0°$-layers in case of laminate B). Further studies, especially with a more systematic variation of the laminate structure and of the geometry of the test samples are necessary, to strengthen these conclusions.

## ACKNOWLEDGEMENTS

The support of this project by the Bundesministerium für Forschung und Technologie, Bonn (BMFT-YHH-221286) is gratefully acknowledged.

## REFERENCES

1. WATERHOUSE, R. B., *Fretting Fatigue*, London, Elsevier Applied Science Publishers, 1981.
2. FRIEDRICH, K. Wear of reinforced polymers by different abrasive counterparts, in: *Friction and Wear of Polymer Composites* (K. Friedrich ed.), Amsterdam, Elsevier, 1986, pp. 233–287.
3. LANCASTER, J., *Brit. J. appl. Phys. (J. Phys. D)*, 1 (1968), 549.
4. SCHULTE, K., KUTTER, S. and FRIEDRICH, K. Fretting fatigue of carbon fiber reinforced plastic laminates, European Space Agency, Technical Translation ESA-TT-1013, November 1986.

# 29

# Kinematic Equations of Filament Winding on a Curved Tube and Winding Pattern Design

XIAN-LI LI

*Centre for Composite Materials, Wuhan University of Technology, Wuhan, People's Republic of China*

## ABSTRACT

*The moving regularity of a winding curved tube on a computer-controlled filament winding machine is discussed in this chapter. Kinematic equations of the fibre pay-out eye with geodesic and non-geodesic winding on a curved tube are given. Winding pattern design of a curved tube and corresponding calculation formulas are presented.*

## INTRODUCTION

Filament winding is a significant technological method for making high-strength composite materials and it can be widely employed for making rocket generators, tubes and pressure vessels, etc., in various industrial departments (e.g. spaceflight aviation, building materials, chemical, etc.). Reference 1 referred to the winding curved tube, but did not present direct applicable formulas. The moving regularity of a winding curved tube and winding pattern design is discussed in this chapter, while the centre angle of the curved tube is $2\beta_s$ and it rotates around the axis throughout the centroid of the curved tube when the fibre pay-out eye only moves on a plane. A set of calculation formulas is presented and this can be a basis for a winding curved tube on a computer-controlled filament winding machine.

2.395

## MATHEMATIC PATTERN

A curved tube represents a section of a torus. A complex curved surface is formed when the tube rotates around the axis through the centre of mass of the curved tube. We do not discuss the surface, in order to avoid complex calculation when winding. We look upon the motion of the pay-out yarn point M as compounded motion, both because it rotates with the curved tube and translates on the tube.

We choose fixed Cartesian coordinates O-$XYZ$, the origin of which is the centre of mass of the curved tube, as shown in Fig. 1.

The centre plane of symmetry of the tube is coincident with the plane $XOZ$, the plane $YOZ$ is symmetrical about the end planes of the tube, and the angle between the plane $YOZ$ and the end plane is $\beta_s$. The distance from the centre C of curvature of the tube to the origin O is $a$.

Designating radius of curvature of the curved tube as $r_0$ and radius of the curved tube as $R_0$, the relation between the $r_0$ and $R_0$ is $n = r_0/R_0$; then we obtain

$$a = \frac{(3n^2 + 1)R_0 \sin \beta_s}{3n\beta_s} \tag{1}$$

In the sequel we shall examine a dimensionless $R = a/R_0$.

We choose moving Cartesian coordinates $O_1$-$X_1 Y_1 Z_1$, the origin of which is movable on the curved centre line $O_1 O_1$ of the tube. The axis $O_1 Y_1$

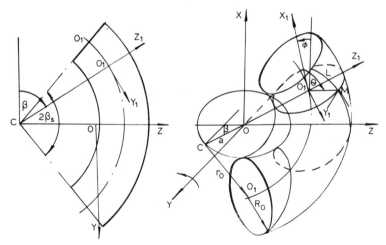

FIG. 1

always is the tangent to the line $O_1O_1$ and rotates around the axis $OY$, together with the product.

Let the axis $O_1X_1$ be parallel to the axis $OX$ in the initial state. When the curved tube rotates angle $\phi$ around the axis $OY$, the position of the point $O_1$, it is the position of the surface $X_1O_1Z_1$, described by angle $\beta$ between the line $CO_1$ and the direction negative to the $Y$ axis. In the Cartesian coordinate system O-$XYZ$, the coordinates of the point $O_1$ are

$$O_1 \{(r_0 \sin \beta - a) \sin \phi, \; -r_0 \cos \beta (r_0 \sin \beta - a) \cos \phi\}$$

The transformation relations between both coordinate systems O–$XYZ$ and $O_1$–$X_1Y_1Z_1$ are

$$\begin{bmatrix} X \\ Y \\ Z \end{bmatrix} = A \begin{bmatrix} X \\ Y \\ Z \end{bmatrix} + B \tag{2}$$

where

$$A = \begin{bmatrix} \cos \phi & 0 & \sin \phi \\ \sin \phi \cos \beta & \sin \beta & -\cos \phi \cos \beta \\ -\sin \phi \sin \beta & \cos \beta & \cos \phi \sin \beta \end{bmatrix}$$

$$B = \begin{bmatrix} (r_0 \sin \beta - a) \sin \phi \\ -r_0 \cos \beta \\ (r_0 \sin \beta - a) \cos \phi \end{bmatrix}$$

In order to calculate the matrix, we have made use of the notation

$$\begin{bmatrix} X \\ Y \\ Z \\ 1 \end{bmatrix} = C \begin{bmatrix} X \\ Y \\ Z \\ 1 \end{bmatrix} \tag{3}$$

where

$$C - \begin{bmatrix} \cos \phi & 0 & \sin \phi & (r_0 \sin \beta - a) \sin \phi \\ \sin \phi \cos \beta & \sin \beta & -\cos \phi \cos \beta & -r_0 \cos \beta \\ -\sin \phi \sin \beta & \cos \beta & \cos \phi \sin \beta & (r_0 \sin \beta - a) \cos \phi \\ 0 & 0 & 0 & 1 \end{bmatrix}$$

## KINEMATIC EQUATIONS OF THE FIBRE PAY-OUT EYE

Let the coordinates of the fibre pay-out eye P be described by $X_p$, $Y_p$, $Z_p$ and the coordinates of the corresponding pay-out yarn point M on the product be described by $X_m$, $Y_m$, $Z_m$. The line MP is tangential to the curve L, which is the fibre path on the product at the point M. Let us denote the unit vectors in the tangent direction by **T**; the line MN is tangential to section of the plane $X_1 O_1 Z_1$ on the curved tube at M, while the section is the circle. Its centre is $O_1$ and radius is $R_0$. The angle between the line MN and the line MP is the winding angle, denoted by $\alpha$.

If the coordinates of the point M are given by the coordinate system $O_1$-$X_1 Y_1 Z_1$: $\{R_0 \cos\theta, 0, R_0 \sin\theta\}$, then in the coordinate system O-$XYZ$ we have

$$\begin{bmatrix} X_m \\ Y_m \\ Z_m \\ 1 \end{bmatrix} = C \begin{bmatrix} R_0 \cos\theta \\ 0 \\ R_0 \sin\theta \\ 1 \end{bmatrix} \tag{4}$$

where $\theta$ is angle between the vector $O_1 M$ and the vector $O_1 X$; it represents the position of the point M in the section of the tube (see Fig. 2).

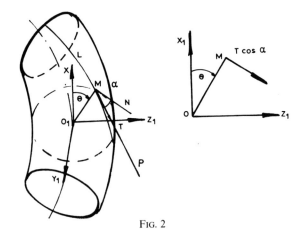

FIG. 2

The vector **T** is described by their projections on the axes of rectangular system $O_1$-$X_1 Y_1 Z_1$:

$$(-\sin\theta \cos\alpha, \sin\alpha, \cos\theta \cos\alpha)$$

Then in the rectangular system O-$XYZ$ we have

$$\begin{bmatrix} T_x \\ T_y \\ T_z \end{bmatrix} = A \begin{bmatrix} -\sin\theta \cos\alpha \\ \sin\alpha \\ \cos\theta \cos\alpha \end{bmatrix} \tag{5}$$

According to the tangent equation of space curve

$$\frac{X_p - X_m}{T_x} = \frac{Y_p - Y_m}{T_y} = \frac{Z_p - Z_m}{T_z}$$

we obtain

$$
\begin{aligned}
X_p &= X_p \\
Y_p &= Y_m + (X_p - X_m)T_y/T_x \\
Z_p &= Z_m + (X_p - X_m)T_z/T_x
\end{aligned}
\tag{6}
$$

The formulas (6) are the kinematic equations of the fibre pay-out eye moved on the plane.

In order to solve the formulas, it is required to establish the relation between the difference $d\Phi$ rotating the curved tube around the $y$ axis and the difference $d\theta$, corresponding to the difference of the pay-out yarn M. We select a differential element, as shown in Fig. 3, in which the plane $O'M_2M$ is

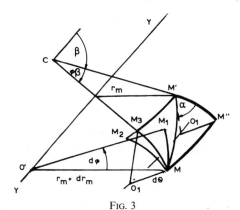

Fig. 3

perpendicular to the $YY$ axis, and the point $M'$ and point M distances from the $YY$ axis are $r_m$ and $r_m + dr_m$, respectively.

According to the geometric relations of the element, we get

$$
\begin{aligned}
MM_3 &= MM_2 \sin \beta = R_0 \, d\theta \\
MM_2 &= (MM_1^2 + \overline{M_1 M_2^2})^{1/2} \\
MM_1 &= (r_m + dr_m) \, d\phi \\
\overline{M_1 M_2} &= dr_m \\
\operatorname{tg} \alpha &= \frac{MM''}{M'M''} = \frac{(n + \sin \theta) \, d\beta}{d\theta}
\end{aligned}
\tag{7}
$$

After translating and neglecting the differences of higher order, we obtain

$$\frac{d\theta}{d\phi} = \frac{-b + (b^2 + ac)^{1/2}}{a} \tag{8}$$

where

$$a = a_1/\sin^2 \beta - [a_2 \operatorname{tg} \alpha/(n + \sin \theta) + a_3]$$
$$a_1 = 4\{(n \sin \beta - R)^2 + 2(n \sin \beta - R)$$
$$\quad \times [\sin \phi \cos(\phi - \theta) - \cos \phi \sin \beta \sin(\phi - \theta)]$$
$$\quad + \cos^2(\phi - \theta) + \sin^2 \beta + \sin^2(\phi - \theta)\}$$
$$a_2 = 2(n \sin \beta - R) \cos \beta [n - \cos \phi \sin(\phi - \theta)]$$
$$\quad + 2n \cos \beta [\sin \phi \cos(\phi - \theta) - \cos \phi \sin \beta \sin(\phi - \theta)]$$
$$\quad + \sin 2\beta \sin^2(\phi - \theta)$$
$$a_3 = 2(n \sin \beta - R)[\cos \phi \sin \beta \cos(\phi - \theta) - \sin \phi \sin(\phi - \theta)]$$
$$\quad + \cos^2 \beta \sin 2(\phi - \theta)$$
$$b = b_1 [a_2 \operatorname{tg} \alpha/(n + \sin \theta) + a_3]$$
$$b_1 = \cos^2 \beta \sin 2(\phi - \theta) - 2(n \sin \beta - R)(1 - \sin \beta) \sin(2\phi - \theta)$$
$$c = b_1^2 + a_1^2/4$$

and

$$\frac{d\beta}{d\phi} = \frac{\operatorname{tg} \alpha}{n + \sin \theta} \frac{d\theta}{d\phi} \tag{9}$$

## 1. Kinematic Equations of the Fibre Pay-out Eye with Geodesic Winding

When the fibre path on the curved tube is geodesic on the tube, according to the Clairant equation $r \sin \alpha = \text{const}$, we obtain $r \sin \alpha = r_0 \sin \alpha_0$, where $\alpha_0$ is the winding angle at $r = r_0$, and this further yields

$$\alpha = \sin^{-1} \left( \frac{n \sin \alpha_0}{n + \sin \theta} \right) \tag{10}$$

The simultaneous equations (6) and (8) to (10) are the kinematic equations of the fibre pay-out eye with geodesic winding. Solving the above equations by the Runge–Kutta method, we have

$$X_p = X(\phi) \qquad Y_p = Y(\phi) \qquad Z_p = Z(\phi)$$

The curves showing the relationship between $\theta, \alpha, \beta, \phi$ are presented in Fig. 4, when $n = 4\cdot34$ and $2\beta_s = 90°$.

The moving curve ($Y_p$–$Z_p$ curve) of the fibre pay-out eye is presented in Fig. 5 with geodesic winding, and $n = 4\cdot34$, $2\beta_s = 90°$, $R_0 = 53$ mm and $X = 100$ mm.

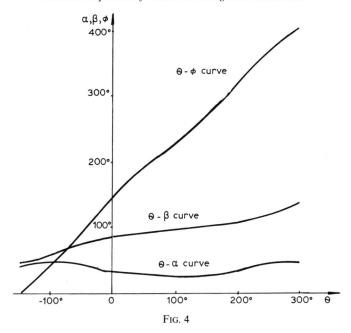

FIG. 4

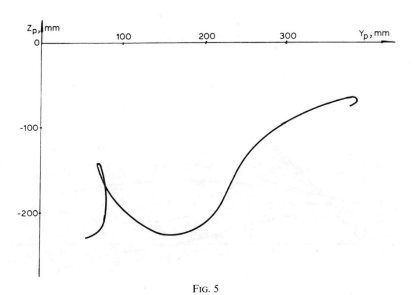

FIG. 5

## 2. Kinematic Equations of the Fibre Pay-out Eye with Non-geodesic Stable Winding

When the curve L is not geodesic on the curved tube, its station is stable because of utilization of the friction forces between the resin-impregnated filament and the bases. It can be shown[2] that the following relation exists:

$$\frac{d\alpha}{d\theta} = -\frac{\mu(\sin\theta + n\cos^2\alpha) + \cos\theta\sin\alpha}{(n + \sin\theta)\cos\alpha} \tag{11}$$

where $\mu$ is the coefficient of friction between the fibre and the core or other fibre.

The simultaneous equations (6) and (8), (9) and (11) are the kinematic

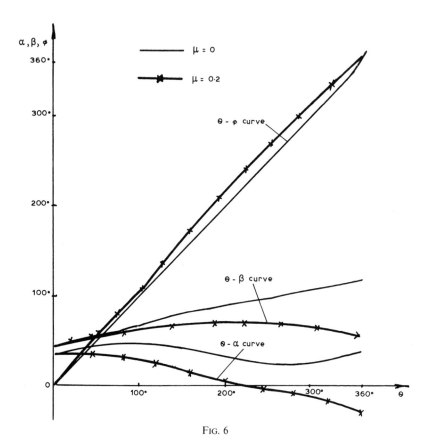

FIG. 6

equations of the fibre pay-out eye with non-geodesic winding. Solving the equations by the Runge–Kutta method, we have

$$X_p = X(\phi) \qquad Y_p = Y(\phi) \qquad Z_p = Z(\phi)$$

The curves showing the relationship between $\theta, \alpha, \beta, \phi$ are presented in Fig. 6, when $\mu$ is equal to 0·2. According to Fig. 6, we know that within a certain range $\beta$ decreases and $\alpha$ changes from positive to negative with increasing $\phi$, and the turn-around of the fibre begins. The operation changes very slowly. In consequence, a non-geodesic stable winding is adapted to winding at the end of a curved tube.

## PATTERN DESIGN

According to the kinematic equations, the normal pattern satisfying the fibre stable condition can be wound on the curved tube. However, the pattern design must be determined with the aim of complete coverage of the mandrel.

### 1. Calculation of Revolving Angle $\Phi$ of the Mandrel when the Fibre Pay-out Eye Traverses Forth and Back Once

The geodesic winding is usually used at the middle of the curved tube and the non-geodesic winding is used at the end of the curved tube (considering turn-around winding). Because the function $\Phi = \Phi(\alpha, \beta, \theta)$ is complex, the revolving angle $\Phi_f$ when the fibre pay-out eye traverses forth is usually not equal to the revolving angle $\Phi_b$ when the fibre pay-out eye traverses back. The revolving angle $\Phi$ of the mandrel when the fibre pay-out eye traverses forth and back once is

$$\Phi = \Phi_f + \Phi_b \tag{12}$$

where $\Phi_f = \Phi_g + \Phi_n$ and $\Phi_b = \Phi'_g + \Phi'_n$; $\Phi_g, \Phi'_g$ are forth and back revolving angles, respectively, with geodesic winding; and $\Phi_n, \Phi'_n$ are forth and back revolving angles, respectively, with non-geodesic winding.

Designating $\beta_g, \beta'_g, \beta_n$ and $\beta'_n$ as the centre angles of the curved tube corresponding to $\Phi_g, \Phi'_g, \Phi_n$ and $\Phi'_n$, respectively, we get the centre angle of the curved tube $2\beta_s$:

$$2\beta_s = \beta_g + \beta_n = \beta'_g + \beta'_n \tag{13}$$

At a junction of geodesic winding with non-geodesic winding we have $\alpha = \alpha_c$, $\theta = \theta_c$ satisfying both eqns (10) and (11). When given these values and a station angle $\alpha = 0$, $\theta = 0_p$ of non-geodesic turn-around winding, according to eqn. (7) in conjunction with eqn. (11), we obtain

$$\frac{d\beta}{d\alpha} = -\frac{\sin \alpha}{\mu(\sin \theta + n \cos^2 \alpha) + \cos \theta \sin \alpha} \tag{14}$$

and yield the centre angle of the curved tube $\beta_n^I$ on the end of the tube with non-geodesic winding by numerical methods. Similarly, we can yield the centre angle of the tube $\beta_n^{II}$ on the other end (see Fig. 7). We have

$$\beta_n = \beta_n^I + \beta_n^{II} \qquad \beta_g = 2\beta_s - \beta_n \tag{15}$$

then yield the corresponding $\Phi_n, \Phi_g$. Similarly, we also get

$$\beta_n' = \beta_n^{I\prime} + \beta_n^{II\prime} \qquad \beta_g' = 2\beta_s - \beta_n' \tag{16}$$

and the corresponding $\Phi_n', \Phi_g'$.

If selected parameters are appropriate, then $\beta_n^I = \beta_n^{II}$ and $\beta_n^{I\prime} = \beta_n^{II\prime}$.

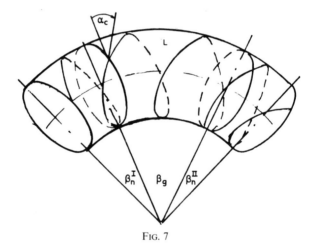

FIG. 7

When winding the curved tube, because the radius of curvature of each latitudinal circle on the tube is different but the fibre quality through the circle is equal, the thickness of the wound curved tube is variable. Usually the value of the selected $n$ is larger, so that the difference of thickness between inner and outer of the curved tube is not too big. Hence, eqn. (14) may be approximated to yield

$$d\beta \approx -\frac{\sin \alpha \, d\alpha}{\mu n \cos^2 \alpha}$$

Integrating both sides of the above equation and considering starting conditions, we get

$$\beta_h \approx \frac{1}{\mu n}\left(\frac{1}{\cos\alpha_c} - \frac{1}{\cos\alpha}\right) \tag{17}$$

With the return trip $\alpha = 0$ and the angle of turn-around $\beta_h$ is

$$\beta_h \approx \frac{1}{\mu n}\left(\frac{1}{\cos\alpha_c} - 1\right) \tag{18}$$

Equation (18) can be used to approximately calculate the turn-around angle so that we preliminarily select $\beta_n^I, \beta_n^{II}, \beta_n^{I'}, \beta_n^{II'}$.

The relationship between $\alpha, \beta, \theta, \phi$ and between $Y_p$ and $Z_p$ are presented, respectively, in Figs 8 and 9, when $n = 4\cdot34$, $2\beta_s = 100°$, $X_p = 100$ mm and the fibre pay-out eye traverses forth and back once.

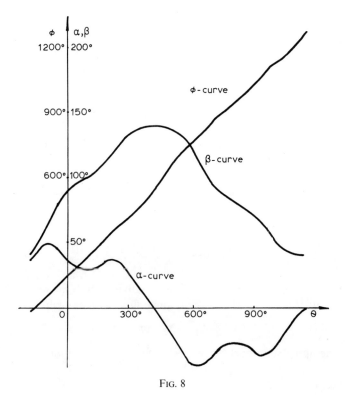

FIG. 8

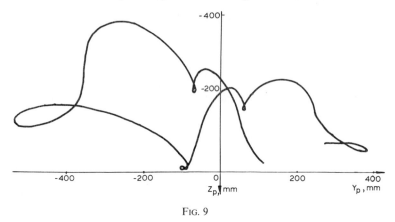

FIG. 9

## 2. Calculation of Increment $\Delta\Phi$ of Revolving Angle

In order to completely cover the surface of the curved tube, after the fibre pay-out eye first traverses forth and back, the pay-out yarn point M must be deflected at the beginning by one fibre bandwidth, as shown in Fig. 10.

Designating fibre bandwidth as $b_m$, then coverage bandwidth in the direction of radius $b$ is $b = b_m/\sin\alpha$. Correspondingly,

$$\Delta\theta = \frac{b}{R_0} = \frac{b_m}{R_0 \sin\alpha} \tag{19}$$

As mentioned above, the radius of curvature of various latitudinal circles are different, but fibre qualities through the circle are equal. We discuss coverage on the basis of the outer side of the curved tube, because its radius of curvature is greatest. (According to coverage of the outer layer, the fibre on other latitudinal circles becomes superimposed.)

Equation (19) becomes

$$\Delta\theta = \frac{(n+1)b_m}{nR_0 \sin\alpha_0} \tag{20}$$

Evaluating the increment $\Delta\Phi$ of the revolving angle corresponding to eqn. (20), we get the total revolving angle $\Phi_1$ under one winding repeat:

$$\Phi_1 = \Phi \pm \Delta\Phi \tag{21}$$

where the upper sign must be taken if the pay-out yarn point is advanced, and the lower sign if the point is hysteretic.

Hence, repeating the above operation, the fibre can cover the mandrel.

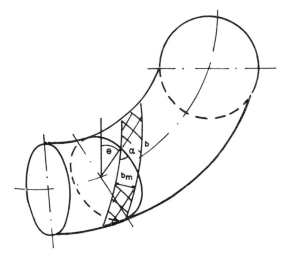

FIG. 10

## CONCLUSION

A set of formulas is presented in this chapter and this can be a basis for a winding curved tube on a computer-controlled filament winding machine.

## REFERENCES

1. MENGES, G. and HILLE, E. A., GRP-pipe fittings produced on a computer controlled filament winding machine, 35th Annual Technical Conference SPI, 1980.
2. XIAN-LI LI, Non-geodesic winding equation on a torus, *ACTA Materiae Compositae Sinica*, **3**, No. 3 (1986), 75–79.

# 30

# Thermo-mechanical Properties of Three-dimensional Fabric Composites

H. HATTA and K. MURAYAMA

*Materials and Electronic Devices Laboratory,*
*Mitsubishi Electric Corporation,*
*1-1-57 Miyashimo, Sagamihara 229, Japan*

## ABSTRACT

*An analytical model to predict the thermo-mechanical properties of 3-D fabric composite was presented based on the Eshelby's equivalent inclusion method. Then fiber orientation distribution by which it becomes possible for us to obtain isotropic properties was explored. Through the numerical results it was shown that the elastic moduli and the thermal expansion coefficients become isotropic when the numbers of fiber orientation axes are over 10 and 3, respectively.*

## 1. INTRODUCTION

One of the most serious disadvantages in conventional laminate-type composites as structural materials is the weakness in the interlaminar strength. For this reason, it is difficult to fabricate structural parts with thick cross-sections. In order to improve the interlaminar properties of composites, several measures were contrived, such as the stitching,[1] the addition of microfibers into the interlamina[2] and the utilization of three-dimensional fabrics (3-D fabrics).[3,4] Among others, the 3-D fabrics are considered to be the most cogent countermeasures because, by the use of the 3-D fabrics, we can not only improve the interlaminar properties but also positively design these properties.

There are many types of 3-D fabrics. The magnaweave invented by Florentine[3,4] is probably the most famous and the composites made of it have been analysed by several authors. Ko and Pastor[5] reported the mechanical properties of the magnaweaved 3-D fabric composite and also proposed the analytical method[6] to predict the tensile strength of the 3-D composite based on the netting analysis. Another group dealing with the magnaweave composite is Chou *et al.* They proposed two kinds of methods to predict the elastic moduli. The first is the extension of the lamination theory to the 3-D fabric composites,[7] and another is the energy method[8] by which the upper and lower bounds of the elastic moduli can be predicted. Aboudi[9] recently proposed different types of model for the tri-orthogonal 3-D fabric composite. He analysed thermo-mechanical properties based on a mosaic model that is essentially the unit cell model. Thus, several models have already been proposed. However, when we apply these models to the actual 3-D composites, we come across several difficulties that stem from the assumptions adopted in the analysis, which are the negligence of matrix contribution[6] and of interaction between fibers,[7] imposition of restriction on the number of fiber axes[9] and the impossibility of dealing with anisotropic fibers.[6-9]

In this chapter an analytical model which can overcome the above-mentioned difficulties will be proposed. The present model for the prediction of the elastic moduli and thermal expansion coefficients of the 3-D fabric composites is based on Eshelby's equivalent inclusion method.[10] We have especially paid attention to the discussion as to how to realize isotropic modulus and isotropic thermal expansion coefficients by use of the 3-D fabrics. Thus, following the description of a general formulation in Section 2, we discuss in detail in Section 3 the number of fiber orientation axes and the fiber orientation angles which are effective in yielding the isotropic properties. To demonstrate the effect of the number of fiber axes on the thermo-mechanical properties, numerical results based on the present formulation will be presented in Section 4. Finally, the conclusion will be given in Section 5.

## 2. FORMULATION

Let us consider the analytical models shown in Fig. 1, where model (a) is the unit cell model that many authors used in the analysis of 3-D fabric composites. If the effect of the fiber location distribution is not so distinctive as is usually assumed, we can replace model (a) by model (b), in which the random fiber location is assumed. Thus, we will analyse model (b) with $N$ fiber orientation angles by the use of Eshelby's equivalent inclusion

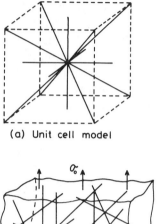

(a) Unit cell model

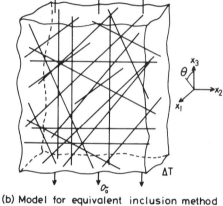

(b) Model for equivalent inclusion method

FIG. 1.    Analytical models for three-dimensional fabric composites.

method[10] on behalf of model (a). In the model, the composite is assumed to be subjected to uniform stress $\sigma_0$ at the far field and uniform temperature change $\Delta T$. Takao *et al.*[11,12] and Takahashi *et al.*[13,14] gave formulae for predicting the effect of fiber orientation distribution on thermo-mechanical properties of a short fiber composite. We adopt the same kind of method regarding the 3-D fabric composites as the limiting case of general fiber orientation distribution, namely we model the 3-D composite as the composite with fiber orientation distribution denoted by the addition of the multiple Dirac delta functions.

Let us focus on a representative fiber $\Omega_i$ in model (b), in which two kinds of coordinate system are used. The global Cartesian coordinates $x_1, x_2$ and $x_3$ are fixed in space. On the other hand, the local coordinates $x'_1, x'_2$ and $x'_3$ are fixed to the fiber, where $x'_3$ is set along the axis of the fiber. Then we apply the equivalency condition to the representative fiber that has the orientation angle $(\theta_i, \phi_i)$:[11,12]

$$\boldsymbol{\sigma}'_f = \mathbf{C}'_f \mathbf{e}'_f = \mathbf{C}'_m(\mathbf{e}'_0 + \bar{\mathbf{e}}' + \mathbf{e}^{Ci'} - \mathbf{e}^{*i'}) = \mathbf{C}'_f(\mathbf{e}'_0 + \bar{\mathbf{e}}' + \mathbf{e}^{Ci'} - \mathbf{e}^{T'}) \tag{1}$$

where the third and fourth equations stand for the stress in the imaginary equivalent inclusion[10] and in the actual fiber, respectively; bold characters denote tensorial quantity; the primes on the tensors denote those that are referred to the local coordinates; and $\mathbf{C}_m$ and $\mathbf{C}_f$ are the elastic moduli of the matrix and fiber, respectively. For the simplicity of computation, we assume that the material properties of the matrix are isotropic and those of the fiber are transversely isotropic. $\mathbf{e}_0$ in eqn. (1) denotes strain induced by the applied stress at the far field $\boldsymbol{\sigma}_0$ and it is defined by

$$\boldsymbol{\sigma}_0 = \mathbf{C}_m \mathbf{e}_0 \tag{2}$$

Note that eqn. (2) is written without a prime to show it is referred to the global coordinates. $\mathbf{e}^{*i}$ in eqn. (1) denotes imaginary strain, so-called eigen strain;[10] $\mathbf{e}^{Ci}$ stands for constrained strain in $\Omega_i$ by the existence of the eigen strain $\mathbf{e}^{*i}$. As is well known, $\mathbf{e}^{Ci}$ is given in terms of the eigen strain as

$$\mathbf{e}^{Ci'} = \mathbf{S}\mathbf{e}^{*i'} \tag{3}$$

where $\mathbf{S}$ is the fourth-order tensor that is a function of the aspect ratio of the fiber and Poisson's ratio of the matrix.[10] $\mathbf{e}^T$ in eqn. (1) denotes misfit strain caused by the difference of the thermal expansion coefficients in the fiber and matrix. Thus

$$\mathbf{e}^T = (\boldsymbol{\alpha}_f - \boldsymbol{\alpha}_m)\Delta T \tag{4}$$

where $\boldsymbol{\alpha}_f$ and $\boldsymbol{\alpha}_m$ denote the thermal expansion coefficients of the fiber and matrix, respectively. Furthermore, $\bar{\mathbf{e}}$ in eqn. (1) denotes the average strain in the matrix induced by the existence of multiple fibers. Following the analysis by Taya and Chou,[15] we can easily obtain

$$\bar{\mathbf{e}} = -\sum_{i=1}^{N} f_i(\mathbf{e}^{Ci} - \mathbf{e}^{*i}) \tag{5}$$

where the subscript $i$ denotes the $i$th orientation angle in the total $N$ fiber orientation angles and $f_i$ denotes the volume fraction of fibers with the $i$th orientation angle. Using the coordinate transfer tensor from the global coordinate to the local coordinate $\mathbf{T}_i^{-1}$, we obtain

$$\bar{\mathbf{e}} = -\sum_{i=1}^{N} f_i \mathbf{T}_i^{-1}(\mathbf{e}^{Ci'} - \mathbf{e}^{*i'}) \tag{6}$$

Then substituting eqns (3) and (6) into eqn. (1)

$$\mathbf{e}^{*i'} = \mathbf{A}^{-1}[(\mathbf{C}'_m - \mathbf{C}'_f)(\mathbf{e}'_0 + \mathbf{e}') + \mathbf{C}'_f \mathbf{e}^{T'}] \tag{7}$$

where $\mathbf{A}$ is the fourth-order tensor defined by

$$\mathbf{A} = (\mathbf{C}'_f - \mathbf{C}'_m)\mathbf{S} + \mathbf{C}'_m \tag{8}$$

Solving $\bar{\mathbf{e}}$ from eqns (6) and (7), we get

$$\bar{\mathbf{e}} = -\sum_{i=1}^{N} \mathbf{D}^{-1}\mathbf{B}_i[(\mathbf{C}'_m - \mathbf{C}'_f)\mathbf{T}_i \mathbf{e}_0 + \mathbf{C}'_f \mathbf{e}^{T'}] \tag{9}$$

where

$$\mathbf{D} = \mathbf{I} + \sum_{i=1}^{N} \mathbf{B}_i(\mathbf{C}'_m - \mathbf{C}'_f)\mathbf{T}_i \qquad \mathbf{B}_i = f_i \mathbf{T}_i^{-1}(\mathbf{S} - \mathbf{I})\mathbf{A}^{-1} \tag{10}$$

and where $\mathbf{I}$ denotes the identity tensor of the fourth order.

## 2.1. Thermal Expansion Coefficient

Effective thermal expansion coefficients of composite materials, $\boldsymbol{\alpha}_c$, can be obtained under the condition of $\mathbf{e}_0 = \mathbf{0}$ in eqn. (1). From the definition

$$\boldsymbol{\alpha}_c = \langle \mathbf{e}_f \rangle = \boldsymbol{\alpha}_m + \langle \mathbf{e}^{*i} \rangle \tag{11}$$

where the $\langle \; \rangle$ denote the averaged value over the entire composite and $\mathbf{e}_f$ denotes actual strain in the fiber. Thus

$$\langle \mathbf{e}^{*i} \rangle = \sum_{i=1}^{N} f_i \mathbf{e}^{*i} = \sum_{i=1}^{N} f_i \mathbf{T}_i^{-1} \mathbf{e}^{*i'} \tag{12}$$

Substitution of eqns (7) and (9) into eqn. (12) yields

$$\langle \mathbf{e}^{*i} \rangle = \sum_{i=1}^{N} f_i \mathbf{T}_i^{-1} \mathbf{A}^{-1} \left[ \mathbf{I} + (\mathbf{C}'_f - \mathbf{C}'_m)\mathbf{T}_i^{-1} \sum_{j=1}^{N} \mathbf{D}^{-1}\mathbf{B}_j \right] \mathbf{C}'_f \mathbf{e}^{T'} \tag{13}$$

Then substituting eqn. (13) into eqn. (11), we obtain the thermal expansion coefficient of the 3-D fabric composite $\boldsymbol{\alpha}_c$.

## 2.2. Elastic Constants

Let us consider the case where $e^T = 0$ in eqns (1)–(10). The effective elastic moduli of the 3-D composite are obtainable from consideration of the strain energy given by[15]

$$W = 0.5\sigma_0 C_c^{-1}\sigma_0 = 0.5\sigma_0 e_0 + 0.5 \sum_{i=1}^{N} f_i \sigma_0 e^{*i} \tag{14}$$

where the stiffness tensor of the composite is denoted by $C_c$. The substitution of eqns (7) and (9) into eqn. (14) yields

$$W = 0.5\sigma_0 C_m^{-1}\sigma_0 + 0.5\sigma_0 \sum_{i=1}^{N} T_i^{-1} A^{-1}(C_m' - C_f')T_i$$

$$\times x\left[ I - \sum_{j=1}^{N} D^{-1}B_j(C_m - C_f)T_j \right] C_m^{-1}\sigma_0 \tag{15}$$

Comparing eqn. (15) with eqn. (14), we obtain

$$C_c^{-1} = C_m^{-1} + \sum_{i=1}^{N} f_i T^{-1} A^{-1}(C_m' - C_f')T_i$$

$$\times x\left[ I - \sum_{j=1}^{N} D^{-1}B_j(C_m - C_f)T_j \right] C_m^{-1} \tag{16}$$

## 3. NUMBER OF FIBER AXES AND ORIENTATION ANGLES

Even restricting to the structures composed of straight fibers there exist a large number of 3-D fabrics. Hence, we should restrict our discussion to 3-D composites which exhibit superior properties. We consider that the possibility of isotropic properties is one of the main advantages of the 3-D fabric composite. Thus, in the following we will discuss the fiber orientation distribution that produces isotropic thermo-mechanical properties.

It may be a necessary condition for the production of the isotropic material properties that the orientation of the fibers is distributed isotropically. In other words, the relative angles between the fiber axes are equivalent to each other. In order to obtain this kind of fiber orientation distribution, one convenient way is to use regular polyhedra. For example, connecting the center of the opposite faces of a cube we get the fiber orientation distribution when the number of fiber axes $N$ is 3 and the connecting opposite apices are that for $N = 4$. By the same|token, we obtain fiber orientation distributions when $N = 6$, 10, 15 and 20 using regular polyhedra, i.e. tetra-, hexa-, octa-, dodeca- and icosahedra. For $N = 3, 4, 6$, 10 and 15, the direction cosines of fiber axes are shown in Table 1. When

TABLE 1
*Direction cosines of fiber axes*

| $N$ | Directional cosine |
|---|---|
| 3 | $(1,0,0)$, $(0,1,0)$, $(0,0,1)$ |
| 4 | $(0,0,1)$, $(-\sqrt{2/3}, -\sqrt{2/3}, -1/3)$, $(\sqrt{2/3}, -\sqrt{2/3}, -1/3)$, $(0, 2\sqrt{2/3}, -1/3)$ |
| 6 | $(0,0,1)$, $(\sqrt{2/3}, 1/2\sqrt{3}, -1/2)$, $(-\sqrt{2/3}, -1/\sqrt{3}, 0)$ |
|  | $(0, \sqrt{3/2}, 1/2)$, $(\sqrt{2/3}, 1/2\sqrt{3}, 1/2)$, $(0, \sqrt{3/2}, -1/2)$ |
| 7 | $(1,0,0)$, $(0,1,0)$, $(0,0,1)$, $(1/\sqrt{3}, 1/\sqrt{3}, 1/\sqrt{3})$ |
|  | $(-1/\sqrt{3}, 1/\sqrt{3}, 1/\sqrt{3})$, $(1/\sqrt{3}, -1/\sqrt{3}, 1/\sqrt{3})$, $(1/\sqrt{3}, 1/\sqrt{3}, -1/\sqrt{3})$ |
| 10 | $(-0.357, 0.491, 0.795)$, $(-0.577, -0.188, 0.794)$, $(0, -0.607, 0.795)$ |
|  | $(0.577, -0.188, 0.795)$, $(0.356, 0.491, 0.795)$, $(-0.577, 0.795, 0.188)$ |
|  | $(-0.934, -0.304, 0.188)$, $(0, -0.982, 0.188)$, $(0.934, -0.304, 0.188)$ |
|  | $(0.577, 0.795, 0.188)$ |
| 15 | $(0, 0.851, -0.526)$, $(-0.809, 0.263, -0.526)$, $(-0.5, -0.688, -0.526)$ |
|  | $(0.5, -0.688, -0.526)$, $(0.809, 0.263, -0.526)$, $(-0.809, -0.588, 0)$ |
|  | $(0.309, -0.951, 0)$, $(1, 0, 0)$, $(0.309, 0.951, 0)$ |
|  | $(-0.809, 0.588, 0)$, $(-0.5, -0.162, -0.850)$, $(0, -0.526, -0.850)$ |
|  | $(0.5, -0.162, -0.851)$, $(0.309, 0.425, -0.851)$, $(-0.309, 0.425, -0.851)$ |

$N = \infty$, the summation in eqns (13) and (16) is replaced by the integration. Thus, we recover the similar formulae that were given by Takao *et al.* for the composites with three-dimensional random fiber orientation distribution.[11]

For the cases of $N = 3, 4$ and 6, the maximum volume fractions of the fibers $V_{f\text{-max}}$ have already been determined.[16] Namely, when the fibers are assumed to be infinite cylinders, $V_{f\text{-max}}$ values are 58·9, 68·0 and 49·4% for $N = 3, 4$ and 6, respectively.

## 4. NUMERICAL RESULTS

For the numerical calculation based on the present formulation, the following values of material properties which simulate a polymer matrix graphite fiber composite[17] will be used:

Fiber:  $E_L = 23\,000\,\text{kg/mm}^2$    $E_T = 2100\,\text{kg/mm}^2$
 $G_{LT} = 4200\,\text{kg/mm}^2$    $v_{LT} = 0.31$
 $v_{TT} = 0.33$    $\alpha_L = -1.5 \times 10^{-6}$
 $\alpha_T = 2.7 \times 10^{-5}$

Matrix:  $E_m = 350\,\text{kg/mm}^2$    $v_m = 0.38$    (17)
 $\alpha_m = 6.6 \times 10^{-5}$

where Young's modulus, shear modulus, Poisson's ratio and thermal expansion coefficient are denoted by $E$, $G$, $v$ and $\alpha$, respectively, and the subscripts L and T denote quantity measured parallel to and perpendicular to the fiber axis, respectively. In the numerical calculation, we set $f_i$ to be constant regardless of $i$. Thus

$$f_i = V_f/N \qquad (18)$$

where $V_f$ is the total volume fraction of the fibers. The Young's modulus of 3-D fabric composite $E_{33}$ is shown in Fig. 2 as a function of the observation angle $\theta$ that is measured from the $x_3$ axis in the $x_1$–$x_3$ plane (see Fig. 1b). The $\theta$ dependency was obtained by the use of the following equation:

$$C_{ijkl}(\theta) = T_{im} T_{jn} C_{mnkl}(0) \qquad (19)$$

where $T_{ij}$ is the coordinate transformation tensor from the reference coordinates to those we are now using. It is obvious from the figure that the 3-D composites with $N = 3$ and 4 (curved solid and dashed lines) have strong anisotropy. However, when $N \geq 6$ the Young's modulus becomes nearly isotropic and above 10 it shows complete isotropy. $N \geq 10$ in the figure, shown by a straight solid line, denotes that $E_{33}$ does not change the value by the number of the fiber axis when $N \geq 10$, i.e. $N = 10, 15, 20$ and $\infty$. In addition to the equivalent fiber angle distributions, the Young's modulus for $N = 7$, the fiber angle of which is the superposition of $N = 3$ and inclined $N = 4$ (see Fig. 1a), is also shown. This composite shows a nearly isotropic modulus. The same kind of angle dependency was calculated for rigidity and Poisson's ratio, and we obtain the same tendency so far as the relation between $N$ and isotropy is concerned.

Next, numerical calculations are conducted for the thermal expansion coefficient of the 3-D fabric composite. Since the thermal expansion

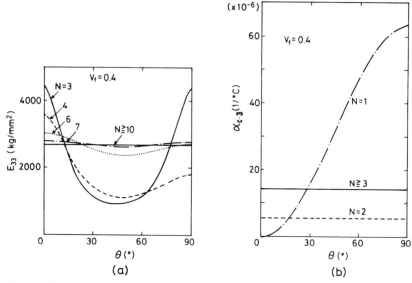

FIG. 2.   The angle dependency of Young's modulus $E_{33}$ (a) and thermal expansion coefficient $\alpha_{c\text{-}3}$ (b) of 3-D fabric composites.

coefficient is a second-order tensor, three components are independent when the principal direction is determined. Accordingly, above $N = 3$ the effective thermal expansion coefficient $\alpha_c$ becomes isotropic, as shown in Fig. 2b, and above $N = 3$ the value of it is constant regardless of the number of fiber axes $N$. In order to compare this value with those of conventional composites, the thermal expansion coefficient of the $x_3$ direction $\alpha_{c\text{-}3}$s for the composite aligned to the $x_3$ axis ($N = 1$) and the cross-ply laminate ($N = 2$, directional cosines $(1, 0, 0)$ and $(0, 0, 1)$) are also shown in Fig. 2b.

## 5.   CONCLUSION

Formulations to predict effective thermal expansion coefficient and elastic modulus of the 3-D fabric composite were derived based on the equivalent inclusion method. The present formulations are advantageous over existing ones at the points where they can deal with arbitrary fiber orientation distribution, arbitrary number of the fiber axis and anisotropic fibers. The following conclusions are obtained through numerical calculations.

   1.   Elastic moduli of 3-D fabric composite become isotropic, provided

the number of the fiber axes is more than 10 and appropriate orientation angles of the fibers are chosen.

2. Elastic moduli of tri-orthogonal 3-D fabric composites show remarkable anisotropy.

3. Thermal expansion coefficients of 3-D fabric composites become isotropic, provided the number of the fiber axes is more than three and equal thickness fiber bundles are placed isotropically.

4. The values of the elastic modulus and thermal expansion coefficient of the 3-D fabric composites do not depend on the number of the fiber axes under the condition that these properties are isotropic.


## ACKNOWLEDGEMENT

This work has been carried out as a part of joint research[18] with the Research Institute of Polymer and Textiles in MITI, Japan. The authors would like to express their appreciation to Mr K. Fukuta, the leader of the project, for encouragement and discussion related to this work.


## REFERENCES

1. TADA, Y. and ISHIKAWA, T., Tentative evaluation of effects of stiching on CFRP laminate specimens, in: *Composites '86: Recent Advances in Japan and United States, Proc. of 3rd Japan–US Conference on Composite Materials* (Kawata, K., Umekawa, S. and Kobayashi, A. eds), Tokyo, Japan Society of Composite Materials, 1986, pp. 351–358.

2. TOWATA, S. and YAMADA, S., Mechanical properties of aluminum alloy composite with hybrid reinforcements of continuous fiber and whisker or particulate, *ibid.*, pp. 497–503.

3. FLORENTINE, R. A., Integrally woven complex shapes for multidimensionally reinforced composites, in: *Proc. 13th National SAMPE Tec. Conf.*, 1981, pp. 625–628.

4. FLORENTINE, R. A., Magnaweave development program: a status report, in: *Proc. 29th National SAMPE Symposium and Exhibition*, 1984, pp. 600–610.

5. KO, F. K. and PASTOR, C. M., Structure and properties of an integrated 3-D fabric composite, ASTM STP 864, 1985, pp. 428–439.

6. KO, F. K., Development of high damage tolerant net shape composites through textile structural design, in: *Proc. of ICCM—5*, 1985, pp. 1609–1616.

7. YAUNG, J. M. and CHOW, T. W., Thermo-elastic analysis of three-dimensional fabric composite, bound volume for ASME WAM, 1983, pp. 61–68.

8. CHOU, T. W., Characterization and modeling of textile structural composite, in: *Proc. ECCM—1*, 1985, pp. 133–138.

9. ABOUDI, J., Minimechanics of tri-orthogonal fiber-reinforced composite: overall elastic and thermal properties, *Fiber Sci. Technol.*, **21** (1984), 277–293.

10. ESHELBY, J. D., The determination of the elastic field of an ellipsoidal inclusion, and related problems, *Proc. Roy. Soc. London, Ser. A*, Vol. 241, 1957, pp. 376–396.

11. TAKAO, Y., CHOU, T. W. and TAYA, M., Effective longitudinal Young's modulus of misoriented short fiber composites, *ASME Trans. J. Appl. Mech.*, **49** (1982), .536–540.

12. TAKAO, Y., Thermal expansion coefficients of misoriented short fiber composites, ASTM STP 864, 1985, pp. 685–699.

13. TAKAHASHI, K., HARAKAWA, K., TANAKA, K. and SAKAI, T., Analysis on the effect of filler orientation distribution in elastic reinforcement theory, *Zairyo* (in Japanese), **26** (1977), 1232–1243.

14. TAKAHASHI, K., SAKAI, T. and HARAKAWA, K., Analysis of the thermal expansion coefficients of particle-filled polymers, *J. Comp. Mater.*, **14**, supplement (1980), 144–159.

15. TAYA, M. and CHOU, T. W., On two kinds of ellipsoidal inhomogeneities in an infinite elastic body: an application to a hybrid composite, *Int. J. Solids Struct.*, **17** (1981), 553–563.

16. MAISTRA, M. A., New multidirectional reinforcement structures made of rigid rod, *Ext. Abs. 14th Biennial Conf. on Carbon*, 1979, pp. 230–231.

17. TAKAHASHI, K., HARAKAWA, K., BAN, K. and SAKAI, T., Analysis of elastic moduli of unidirectional CFRP, *Sen-I Gakkaishi* (in Japanese), **39** (1983), T349–355.

18. FUKUTA, K., TADA, Y., ANAHARA, M. and MURAYAMA, K., The development of automated loom for weaving a three-dimensional fabric, *Proc. 2nd SAMPE Japan*, 1986, pp. 32–35.

# 31

## Stress Concentrations in Orthotropic Laminates Containing Two Elliptical Holes

JIA-KENG LIN

*Nanjing Aeronautical Institute, Nanjing, People's Republic of China*

and

CHARLES E. S. UENG

*School of Civil Engineering, Georgia Institute of Technology, Atlanta, Georgia 30332, USA*

### ABSTRACT

*The titled problem is solved by the method of stress functions of complex variables through the use of single hole results with proper modifications. MacLaurin series expansion is used. Three different cases of orthotropic laminates are investigated here. A parametric study is carried out in order to observe the variation of stress concentration due to a change of the distance between the holes, the size of the hole, the loading direction, and the laminate characteristics.*

### INTRODUCTION

The stress distribution in a laminated plate containing holes and subjected to loads in its plane can be modeled as a plane problem of an anisotropic plate where the degree of anisotropy can be characterized by the complex parameters $\mu_1$ and $\mu_2$. When a given material is orthotropic and the directions of the coordinate axes coincide with the principal directions of elasticity, then the following three possibilities exist for the two complex parameters $\mu_1$ and $\mu_2$:

Case I: $\quad \Delta > 0 \quad \mu_1 = \beta_1 i \quad\quad \mu_2 = \beta_2 i$

Case II: $\quad \Delta < 0 \quad \mu_1 = \alpha + \beta i \quad \mu_2 = -\alpha + \beta i = -\bar{\mu}_1$

Case III: $\quad \Delta = 0 \quad \mu_1 = \mu_2 = \beta i$

where $\Delta = (E_1/G - 2v_1)^2 - 4E_1/E_2$, $E_1$, $E_2$, $v_1$ and $G$ are the four engineering elastic constants of an orthotropic material. The anisotropy of an orthotropic material can be represented by two real parameters $\beta_1$ and $\beta_2$ or $\alpha$ and $\beta$. It is known that many materials, for instance oak, birch and spruce,[1] $[0_4 \pm 45°]_s$ and $[90_4 \pm 45°]_s$ carbon fiber reinforced laminates,[2] belong to Case I. But other orthotropic materials, for example $[\pm 45°]_s$, $[0_3° \pm 45°]_s$ and $[90_3° \pm 45°]_s$ carbon fiber laminates, belong to Case II. Case III is a special case of Case I or Case II. Here if $\beta = 1$, the laminate is said to be quasi-isotropic.

The stress distribution of an infinite isotropic plate containing holes and subjected to in-plane load has been examined by many researchers. Ling[3] used bipolar coordinates and obtained an analytical solution of the stress state of an isotropic plate with two circular holes. Kosmodamiansky[4,5] expanded the stress potential function in complex series expression and solved a number of multiple hole problems. For the orthotropic case, Kosmodamiansky[6] studied the stress state in an orthotropic plate (Case I) with two holes of different sizes and gave the first approximate stress distribution around the contour of the smaller hole. An orthotropic plate which also belongs to Case I with two elliptical holes subject to in-plane normal load has been recently investigated by the authors.[2]

The problem of an infinite laminate consisting of two identical elliptical holes, symmetrically located and subjected to horizontal and vertical in-plane loads at infinity, is studied in this paper. The laminate is treated as an orthotropic plate. Its principal material directions are assumed to coincide with the coordinate axes (Fig. 1). All three cases are included in this investigation.

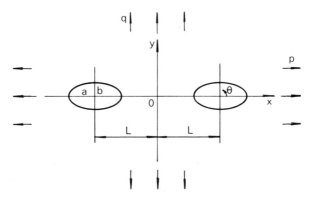

FIG. 1.   Hole and loading arrangement.

## ANALYTICAL METHOD

For an orthotropic plate subjected to a uniform horizontal load $p$ and a vertical load $q$ at infinity, the stress potential functions can be expressed as [1]

$$\phi_0(z_1) = (P_1 + Q_1)z_1 \qquad \psi_0(z_2) = (P_2 + Q_2)z_2 \tag{1}$$

where

$$P_1 = p/D \qquad Q_1 = (\alpha_2^2 + \beta_2^2)q/D$$
$$P_2 = [-1 + i(\alpha_1 - \alpha_2)/\beta_2]p/D$$
$$Q_2 = \{(\alpha_1^2 - \beta_1^2 - 2\alpha_1\alpha_2) + i[\alpha_2(\alpha_1^2 - \beta_1^2) - \alpha_1(\alpha_2^2 - \beta_2^2)]/\beta_2\}q/D \tag{2}$$

and

$$D = 2[(\alpha_1 - \alpha_2)^2 + \beta_2^2 - \beta_1^2]$$

Furthermore, $\alpha_1, \alpha_2, \beta_1$ and $\beta_2$ are the real and imaginary parts of $\mu_1$ and $\mu_2$, respectively; $z_1 = x + \mu_1 y$ and $z_2 = x + \mu_2 y$.

For the case where a single elliptical hole is located at the origin of the coordinate axes, the above functions must be superimposed with $\Delta\phi_0$ and $\Delta\psi_0$, which can be expressed as [7]

$$\Delta\phi_0 = \frac{\mu_2 aq - ibp}{2(\mu_1 - \mu_2)}\frac{1}{\zeta_1} \qquad \Delta\psi = -\frac{\mu_1 aq - ibp}{2(\mu_1 - \mu_2)}\frac{1}{\zeta_2} \tag{3}$$

They satisfy the boundary conditions of the hole and represent the effect due to the presence of such an elliptical hole, in which $a$ and $b$ are the half length of the major and minor axes, $\zeta_1$ and $\zeta_2$ are the complex coordinates of a point in plane $\zeta_1$ and $\zeta_2$, respectively.

If an identical elliptical hole is also present in the plate, it will cause a change of the stress distribution in the plate and around the edge of the first hole. Therefore, the boundary conditions of the first hole will not be satisfied. In order to satisfy the boundary conditions of both holes exactly, an infinite number of modifications of the stress potential functions are necessary. Utilizing the superposition principle, these functions can be expressed as

$$\phi(z_1) = \phi_0(z_1) + \sum_{m=1}^{\infty}\phi_m^R\zeta_1^{-m}(z_1 - L) + \sum_{m=1}^{\infty}\phi_m^L\zeta_1^{-m}(z_1 + L)$$

and

$$\psi(z_2) = \psi_0(z_2) + \sum_{m=1}^{\infty}\psi_m^R\zeta_2^{-m}(z_2 - L) + \sum_{m=1}^{\infty}\psi_m^L\zeta_2^{-m}(z_2 + L)$$

$$\tag{4}$$

where $\zeta_1$ and $\zeta_2$ are chosen as

$$\frac{1}{\zeta_1} = \frac{(z_1 \pm L) - (z_1 \pm L)^2 - 4m_{01}m_{11}}{2m_{11}}$$

and

$$\frac{1}{\zeta_2} = \frac{(z_2 \pm L) - (z_2 \pm L)^2 - 4m_{02}m_{12}}{2m_{12}}$$

(5)

for the purpose that the area outside of a unit circle with center located at the origin of the coordinate axes in plane $\zeta_1$ or $\zeta_2$ corresponds to the area outside of the left or right ellipse in plane $\mu_1$ or $\mu_2$. In eqns (5), the positive and negative signs refer to the left and right hole, respectively. In addition,

$$\begin{aligned} m_{01} &= (a - i\mu_1 b)/2 & m_{11} &= (a + i\mu_1 b)/2 \\ m_{02} &= (a - i\mu_2 b)/2 & m_{12} &= (a + i\mu_2 b)/2 \end{aligned}$$

(6)

and in (4), the superscripts R and L refer to the right hole and left hole, respectively. Since both the geometry and loading are symmetric with respect to the origin, (4) can be simplified as

$$\phi(z_1) = \phi_0(z_1) + \sum_{m=1}^{\infty} \phi_m[\zeta_1^{-m}(z_1 - L) + (-1)^{m+1}\zeta_1^{-m}(z_1 + L)]$$

and

(7)

$$\psi(z_2) = \psi_0(z_2) + \sum_{m=1}^{\infty} \psi_m[\zeta_2^{-m}(z_2 - L) + (-1)^{m+1}\zeta_2^{-m}(z_2 + L)]$$

Once (7) satisfies the boundary conditions of one of the two holes, then the boundary conditions of the other hole will be satisfied automatically. Based on the fact that the holes are free from any load on its contour, the boundary conditions of the right hole can be expressed as follows:

$$\begin{aligned} \text{Re}\,[\phi(z_1 - L) + \psi(z_2 - L)] &= 0 \\ \text{Re}\,[\mu_1\phi(z_1 - L) + \mu_2\psi(z_2 - L)] &= 0 \end{aligned}$$

(8)

Introducing two new variables $z_{r1} = z_1 - L$ and $z_{r2} = z_2 - L$, then by expanding $\zeta_1^{-m}(z_{r1} + 2L)$ and $\zeta_2^{-m}(z_{r2} + 2L)$ into MacLaurin series and finally integrating over a unit circle in $\zeta_1$ and $\zeta_2$ planes, we have a set of polynomial equations in $\zeta$. The coefficients of $\zeta$ with the same power on both sides of the equations must be equal, then we have a system of equations.

Once the stress potential functions are available, the stress distribution in the plate can be determined. The normal stress $\sigma_\theta$ along the opening contour can be calculated by the following formula:[1]

$$\sigma_\theta = \frac{2}{a^2 s^2 + b^2 c^2} \, \text{Re} \left[ (\mu_1 as + bc)^2 \phi'(z_1) + (\mu_2 as + bc)^2 \psi'(z_2) \right] \qquad (9)$$

where $s = \sin\theta$, $c = \cos\theta$, $\phi'(z_1)$ and $\psi'(z_2)$ are the derivatives of $\phi(z_1)$ and $\psi(z_2)$, respectively.

## NUMERICAL RESULTS

The stress around the right elliptical hole in $[\pm 45°]_s$, $[0^\circ_3 \pm 45°]_s$ and $[90^\circ_3 \pm 45°]_s$ laminates (Case II) have been calculated by the above analytical method. The material properties are listed in Table 1. The number of the system of equations is set equal to 4, the highest order of derivatives used and the number of the unknown coefficients is also limited to 4. Different ratios of the major and minor axes as well as different distances between the center of the two holes have been assumed in order to carry out a parametric study. For the purpose of examining the above analytical method, certain results for the case containing circular holes are presented in Table 2. Results for $[0^\circ_4 \pm 45°]_s$, $[90^\circ_4 \pm 45°]_s$ and quasi-isotropic laminates,[2] as well as the corresponding analytical solution for the isotropic case,[3] are also included for comparison purposes.

Certain selected results for the three laminates containing elliptical holes are graphically presented in Figs 2, 3 and 4. The data for the case of a single

TABLE 1
*Material properties of laminates*

| Laminate | $E_1$ (GPa) | $E^2$ (GPa) | $G$ (GPa) | $v_1$ | $\mu_1$ | $\mu_2$ |
|---|---|---|---|---|---|---|
| $[\pm 45°]_s$ | 20·3 | 20·3 | 27 7 | 0·728 | 0·825 0 +0·565 0i | −0·825 0 +0·565 0i |
| $[0° \pm 45°]_s$ | 74·0 | 17·0 | 19·0 | 0·914 | 0·725 6 +1·249 8i | −0·725 6 +1·249 8i |
| $[90^\circ_3 \pm 45°]_s$ | 17·0 | 74·0 | 19·0 | 0·210 | 0·347 8 +0·598 6i | −0·347 8 +0·598 6i |
| $[0^\circ_4 \pm 45°]_s$ | 111·7 | 20·4 | 16·9 | 0·663 | 1·189 8i | 1·966 7i |
| $[90^\circ_4 \pm 45°]_s$ | 20·4 | 111·7 | 16·9 | 0·121 | 0·508 4i | 0·840 6i |

TABLE 2

*Distribution of $\sigma_\theta$ along the circular hole on the right side*

| $L/a$ | Laminate | Bi-direction tension $(p = q = 1 \cdot 0)$ | | Horizontal tension $(p = 1 \cdot 0)$ | Vertical tension $(q = 1 \cdot 0)$ | |
|---|---|---|---|---|---|---|
| | | $\theta = 0$ | $\theta = \pi$ | $\theta = \pm \pi/2$ | $\theta = 0$ | $\theta = \pi$ |
| | $[0_4^\circ \pm 45^\circ]_s$ | 1·977 | 4·878 | 3·118 | 2·416 | 5·243 |
| | $[0_3^\circ \pm 45^\circ]_s$ | 1·672 | 4·985 | 2·983 | 2·127 | 5·604 |
| | $[\pm 45^\circ]_s$ | 0·838 | 5·545 | 1·716 | 1·814 | 6·676 |
| 1·0 | Quasi-isotropic | 2·071 | 5·514 | 2·471 | 3·056 | 6·372 |
| | $[90_3^\circ \pm 45^\circ]_s$ | 1·555 | 5·312 | 1·828 | 3·706 | 6·532 |
| | $[90_4^\circ \pm 45^\circ]_s$ | 2·033 | 6·010 | 1·966 | 4·426 | 7·374 |
| | Isotropic[8] | 2·894 | $\infty$ | 2·569 | 3·869 | $\infty$ |
| | $[0_4^\circ \pm 45^\circ]_s$ | 2·064 | 2·504 | 3·316 | 2·485 | 2·708 |
| | $[0_3^\circ \pm 45^\circ]_s$ | 1·853 | 2·290 | 2·840 | 2·321 | 2·531 |
| | $[\pm 45^\circ]_s$ | 1·286 | 1·641 | 1·862 | 2·255 | 2·272 |
| 1·5 | Quasi-isotropic | 2·190 | 2·755 | 2·526 | 3·163 | 3·299 |
| | $[90_3^\circ \pm 45^\circ]_s$ | 1·655 | 2·357 | 1·957 | 3·671 | 3·760 |
| | $[90_4^\circ \pm 45^\circ]_s$ | 2·098 | 2·968 | 2·075 | 4·360 | 4·492 |
| | Isotropic[8] | 2·255 | 2·887 | 2·623 | 3·151 | 3·264 |
| | $[0_4^\circ \pm 45^\circ]_s$ | 1·974 | 2·018 | 3·705 | 2·373 | 2·336 |
| | $[0_3^\circ \pm 45^\circ]_s$ | 1·767 | 1·798 | 3·203 | 2·217 | 2·182 |
| | $[\pm 45^\circ]_s$ | 1·174 | 1·197 | 2·071 | 2·142 | 2·127 |
| 3.0 | Quasi-isotropic | 2·075 | 2·147 | 2·811 | 3·026 | 3·002 |
| | $[90_3^\circ \pm 45^\circ]_s$ | 1·505 | 1·584 | 2·128 | 3·527 | 3·519 |
| | $[90_4^\circ \pm 45^\circ]_s$ | 1·934 | 2·044 | 2·261 | 4·194 | 4·187 |
| | Isotropic[8] | 2·080 | 2·155 | 2·825 | 3·020 | 2·992 |
| | $[0_4^\circ \pm 45^\circ]_s$ | 1·943 | 1·952 | 3·952 | 2·351 | 2·337 |
| | $[0_3^\circ \pm 45^\circ]_s$ | 1·737 | 1·743 | 3·379 | 2·199 | 2·188 |
| | $[\pm 45^\circ]_s$ | 1·146 | 1·151 | 2·110 | 2·132 | 2·129 |
| 5.0 | Quasi-isotropic | 2·032 | 2·049 | 2·925 | 3·005 | 2·998 |
| | $[90_3^\circ \pm 45^\circ]_s$ | 1·450 | 1·467 | 2·172 | 3·506 | 3·504 |
| | $[90_4^\circ \pm 45^\circ]_s$ | 1·887 | 1·891 | 2·316 | 4·168 | 4·166 |
| | Isotropic[8] | 2·033 | 2·049 | 2·927 | 3·004 | 2·997 |

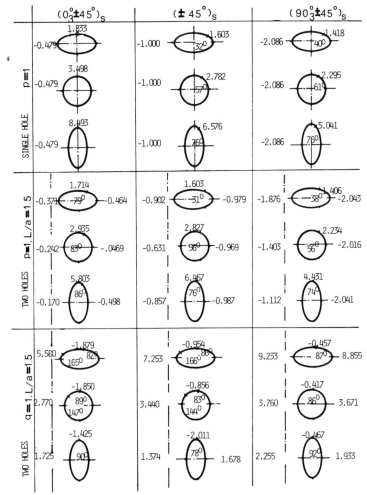

FIG. 2.   $\sigma_{\theta\max}$ and $\sigma_{\theta\min}$ for $a/b = 3$, 1 and 1/3.

hole denoted by $L \to \infty$ are also included for reference. They are calculated by the following formula:[7]

$$\sigma_\theta = p\,\frac{a^2}{a^2 s^2 + b^2 c^2}\,s^2 + p\,\frac{b}{a^2 s^2 + b^2 c^2}$$

$$\times \operatorname{Re}\left\{\frac{e^{i\theta}}{(as - \mu_1 bc)(as - \mu_2 bc)}\,[(\mu_1 + \mu_2)a^3 s^3 + (2 - \mu_1\mu_2)a^2 b s^2 c + b^3 c^3]\right\}$$

$$(10)$$

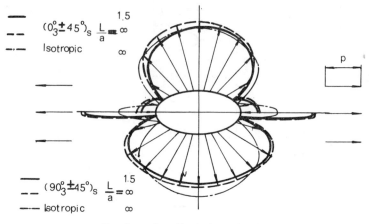

FIG. 3.    $\sigma_\theta$ for $a/b = 3$, due to load $p$.

In order to observe the double hole effect versus a single hole case, a ratio is defined as follows, i.e.

$$K_{p,q} = \left( \frac{\sigma_\theta - \sigma_{\theta,L} \to \infty}{\sigma_{\theta,L} \to \infty} \right)_{p,q} \tag{11}$$

where the subscript $p$ or $q$ refers to the stress on the right-hand side caused by the loading $p$ or $q$, respectively.

Figures 5 and 6 show the variation of $K_p$ (at $\theta = \pi/2$) and $K_q$ (at $\theta = \pi$) versus the ratio $L/a$ for three $a/b$ ratios of two different laminates.

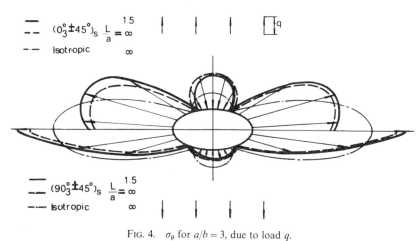

FIG. 4.    $\sigma_\theta$ for $a/b = 3$, due to load $q$.

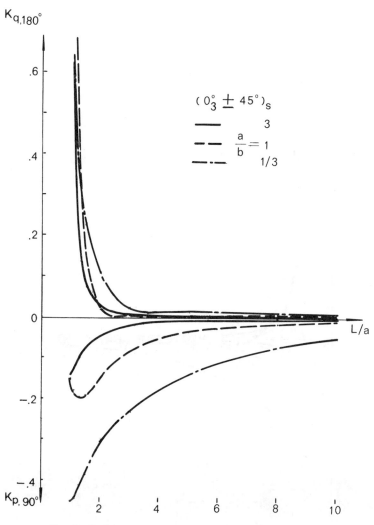

FIG. 5. Variation of $K$ versus $L/a$ for $(0_3^\circ \pm 45^\circ)_s$ laminates.

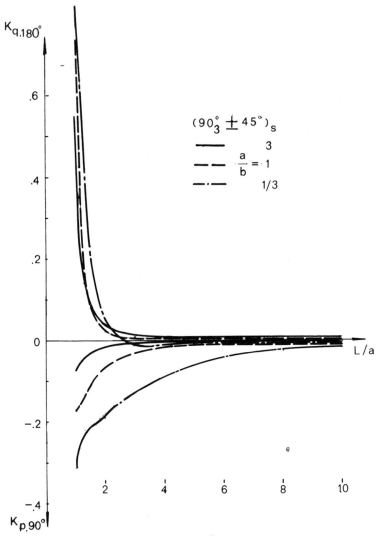

FIG. 6.    Variation of $K$ versus $L/a$ for $(90_3^\circ \pm 45^\circ)_s$.

## DISCUSSION AND CONCLUSIONS

Based on the information from the available literature, no results on the stress distribution of an orthotropic laminate (Case II) containing two holes have been reported. In order to examine the analytical method presented here, the normal stress $\sigma_\theta$ at a number of special points along the circular hole of six laminates which belong to three different cases are included in Table 2. For a relatively small value of $L/a$, the difference between the results of the quasi-isotropic case being reported here and the results obtained by Ling[3] for isotropic case is rather noticeable. But as $L/a$ increases, the deviation gradually diminishes. For $L/a \geq 1.5$, the difference appears to be no more than 5%. Furthermore, from the results of $[0_4^\circ \pm 45^\circ]_s$, $[0_3^\circ \pm 45^\circ]_s$, quasi-isotropic, $[90_3^\circ \pm 45^\circ]_s$ and $[90_4^\circ \pm 45^\circ]_s$ laminates versus different $L/a$ ratios, a similar changing pattern is also observed. Certain maximum and minimum values of $\sigma_\theta$ and their locations along the contour of the hole for the case of a single hole or two holes in three different laminates are shown in Fig. 2.

It is known that for an isotropic plate containing a single hole, the stress distribution is dependent on the loading and the geometry of the hole. This character also exists in an orthotropic plate. But it is more complicated here, as shown in Figs 3 and 4 by the dashed lines. The main reason is the anisotropy of the material.

For the two-hole case (Figs 3 and 4), the stress distribution alone the perimeter of the hole is different from the one-hole case. The amount of change depends on the loading, the degree of orthotropy of the material, the hole geometry, and the distance between the holes. The closer the distance between the two holes, the larger the change will be. When the ratio $a/b$ decreases, this change will increase. For the case under a horizontal loading $p$ and with a ratio of $L/a = 1.5$, the effect of the second hole appears to cause a reduction of $\sigma_\theta$ along most parts of the contour of the first hole, including a reduction of the maximum tensile stress associated with the one-hole case, except a small portion where the original one-hole stresses are rather low. On the other hand, for the case under a vertical loading $q$, the effect of the second hole is to cause an increase of stress along most parts of the first hole.

From Figs 5 and 6, the stress state of a two-hole case converges to the stress state of a single hole when the distance between the two holes increases indefinitely. The convergence is not monotonic but dependent on the orthotropy of the material, the loading, the hole geometry, as well as the location of the point. For example, the value of $\sigma_\theta$ at $\theta = \pi/2$ due to loading

$q$ converges more rapidly than $\sigma_\theta$ at $\theta = \pi/2$ due to loading $p$. On the other hand, $\sigma_\theta$ at $\theta = \pi/2$ due to $p$ converges more slowly for the case where the ratio $a/b$ is fairly small.

From the results shown in this paper for the assumed orthotropy, it can be said that for engineering purposes, in order to provide an adequate structural strength, one may ignore the effect of the second hole in a laminate if the distance $2L$ between the two identical holes is no less than a certain value. This value is dependent on the orthotropy of material, loading and geometry of the hole. For a relatively small value of $L/a$, say 3·0, a detailed calculation of the stress distribution around the edge of the hole will be necessary. The analytic method presented in this paper is useful and relatively simple to follow, and the results obtained here may serve as a useful test case for finite element methods or boundary element methods in situations in which the FEM or BEM might be more useful, such as finite plate geometries, or numerous holes.

## REFERENCES

1. SAVIN, G. N., *Stress Concentration Around Holes* (English translation by W. Johnson), Oxford, Pergamon Press, 1961. Also *Stress Distribution Around Holes*, NASA TTF-607, English translation of *Raspredelemiye Napryazeniy Okolo Otversity*, 1968.
2. LIN, J.-K. and UENG, C. E. S., Stresses in a laminated composite with two elliptical holes, accepted by *J. Comp. Struct.*
3. LING, CHIH-BING, On the stresses in a plate containing two circular holes, *J. appl. Phys.*, **19** (January 1948), 77–81.
4. KOSMODAMIANSKY, A. C., Quasiregularity of infinite systems in a problem on stress concentration beside curvilinear holes, *Prikl. Mekh.*, **1**, No. 1 (1965), 15–21 (in Russian).
5. KOSMODAMIANSKY, A. C. and CHERNIK, V. I., Stress state of a plate weakened by two elliptical holes with parallel axes, *Soviet appl. Mech.*, **17**, No. 6 (June 1981), 576–581.
6. KOSMODAMIANSKY, A. C., The stress state of an anisotropic plate with two non-identical holes, *Akademiia Nauk SSSR. Izvestiia. Otdelenie Teknicheskikh Nauk. Mechaniki i Mashinostroenie*, No. 1 (1961), 175–177 (in Russian).
7. LEKHNITSKII, S. G., *Anisotropic Plates* (English translation by S. W. Tsai and T. Cheron), New York, Gordon and Breach, 1968.
8. MUSKHELISHVILI, N. I., *Some Basic Problems of the Mathematical Theory of Elasticity* (English translation by J. R. M. Radok), P. Noordhoff Ltd, 1953.

# Index of Contributors

# Subject Index

Bifurcational stability. *See* Buckling
Bimodular materials
  analysis theory and formulation,
      *1*.153–4
  definition of, *1*.152
  fibre orientation effects, *1*.155–7
  large-deflection behaviour of thin
      plates, *1*.152–60
  meaning of term, *2*.8
  numerical results for, *1*.154–9
  vibration behaviour of, *2*.8
Bimodulatory ratio, nonlinear
    deflection affected by, *1*.159
BL Rover car component, *1*.343–55
BL Technology, *1*.343
Blades
  types with composite materials,
      *1*.191–2; *2*.2
  wind turbine, *1*.192–210, 218–19
Blast loadings, *2*.2
BMFT research, *2*.394
Bolted joints
  bearing strength determination for,
      *2*.273–89
    torque effects, *2*.283, 288, 289
  creep studies for, *1*.75–85
  glass fibre mat laminates
    comparison of short-term
        strength of, *1*.65–71, 72
    failure modes in, *1*.63
    with inhibition of transverse
        deformation, *1*.61–2, 65,
        66–7
    without inhibition of transverse
        deformation, *1*.60–1, 62, 64–5
  long-term strength behaviour of,
      *1*.71, 73
  prestressing effect on, *1*.65, 68–9
  test specimens for, *1*.61, 76
Bone
  cement
    function of, *2*.326
    polyethylene fiber/fibre reinforced
        acrylic as, *2*.325–35
  fracture fixation implants, *2*.338–44
    biocompatibility of CFRP
        prostheses, *2*.342
    design criteria for, *2*.339–40

Bone—*contd.*
  fracture fixation implants—*contd.*
    experimental results for
        composites, *2*.341–4
    manufacturing method for CFRP
        screws, *2*.340–1
    materials for, *2*.338, 340
Boron–aluminium composites, cost
    comparison for, *2*.83
Boron–epoxy composites
  allowable stresses for, *1*.363
  failure analysis of, *2*.170–3
  plates
    postbuckling behaviour of, *1*.23
    shear buckling behaviour of, *1*.17
  properties of, *1*.363, 365, 441
  vibration behaviour of, *1*.441, 442,
      443, 444
  temperature effects on, *1*.422
Boron fibers/fibres, properties of,
    *2*.169
Box beams, multilevel optimization
    of, *1*.362–6, 398–400, 401–2
BP Research, *1*.343; *2*.44
Braided pipes
  advantages of, *2*.193
  braided angle, effect of, *2*.202–6
  deformation under lateral
      compressive load, *2*.201
  elastic constant of, *2*.199–200
  fabrication of, *2*.193–5
  hybrid pipe, effect of, *2*.206
  lateral compressive stress of,
      *2*.200–1
  numerical results, *2*.202–8
  specimen preparation of, *2*.195–8
  testing of, *2*.198
  theoretical approach, *2*.199–201
  through-the-thickness braided,
      *2*.195, 198, 206–8
Braiding techniques, CFRP
    composites, *1*.347
  vehicle coil springs, *1*.347
Branch systems, *1*.575
  treatment for, *1*.575–6
Branched integration method, *1*.295
Bresse–Timoshenko theory, *2*.7
British Petroleum, *1*.343; *2*.44

Fretting fatigue
  damage to CFRP specimens, 2.392
  examples of failure due to, 2.386
  experimental procedure, 2.387–8
  numerical results for CFRP,
    2.389–93
Frond–wedge–frond crush zone
    morphology, 2.35, 43

Galerkin method
  anisotropic (laminated) plates,
    1.110, 112–15
  dynamic buckling of
    antisymmetrically laminated
    plates, 1.513, 515
  orthotropic plates subjected to
    follower forces, 1.101
  vibrating circular plates, 1.440
  vibrating triangular plates, 1.417
  wind turbine rotor blade vibration
    analysis, 1.198
Gauss quadrature, 1.144
General Dynamics Convair antenna
    design, 1.31, 35
Geodesic winding, effect on filament
    winding kinematic equations,
    2.400–1, 403
Geostationary/geosynchronous orbit
  atmospheric conditions, 1.37
  outgassing parameters in, 1.56
  temperature effects in, 1.54–5
GIFTS (Graphics Interactive Finite-
    element Total System)
    program, 1.172
Glass–epoxy composites
  finite element analysis results for,
    1.581–2
  postbuckling behaviour of plates, 1.23
  properties of, 1.441
  vibration behaviour of, 1.444
    temperature effects on, 1.421, 422
Glass fibre reinforced plastics (GRP)
  braided pipes, 2.195, 197, 205–7
  centrifugally cast pipes made from,
    1.223–34
  cloth laminates, creep behaviour of,
    1.82

Glass fibre reinforced plastics (GRP)—
    contd.
  compared with CFRP, 1.312, 314,
    316, 317, 327
  failure analysis of, 2.170–3
  hybrid (cloth–mat) laminates, 1.75
  creep behaviour of, 1.82
  mechanical properties of, 1.194,
    201, 312
  properties of, 1.322, 375, 507
  pultruded rods, properties of, 1.312,
    314
  stiffened panels, interactive buckling
    effects in, 1.133–6
  strands of pultruded rods,
    properties of, 1.316, 317
  unsaturated polyester mat (GFR-UP-
    M) laminates, bolted joints
    design and testing, 1.60–73, 76–82
  wind turbine rotor blades,
    1.192–205, 208–10, 217,
    218–19
Glass fibres, properties of, 1.311
Glass microspheres, 1.335, 336
  GRP sandwich structures using,
    1.335–8
Graphite–aluminium composites
  column/plate efficiency for, 2.91
  cost comparison for, 2.83
Graphite–epoxy composites
  compression-loaded angle-ply plates
    with holes, 2.176–90
  deflection data for, 1.157, 158
  delamination fracture toughness of,
    2.264–71
  failure analysis of, 2.170–3
  failure modes for compression-
    loaded plates with holes,
    2.179–86
  finite element analysis results for,
    1.580
  impact damage, reduction by
    adhesive layers, 2.20–30
  impact-damaged composite with
    fastener hole, 2.57–67
  off-axis angle plot for, 2.302–3
  plate buckling studies, 1.9–11
  plate postbuckling behaviour, 1.23